Plastic Ocean
Art and Science Responses to Marine Pollution

**Plastic Ocean
Art and Science Responses
to Marine Pollution**

Edition Angewandte
Book Series of the University of Applied Arts Vienna
Edited by Gerald Bast, Rector

edition:'ʌngewʌndtə
Universität für angewandte Kunst Wien
University of Applied Arts Vienna

Plastic Ocean

Art and Science Responses to Marine Pollution

Edited by Ingeborg Reichle

DE GRUYTER

Contents

Acknowledgments

To edit an anthology about art and science responses to marine plastic pollution had been on my mind for some time because there have been important contributions from diverse fields on this pressing issue in recent years, which ought to be known to a wide audience. I am very grateful that a number of excellent authors contributed to the book, and therefore I would like to express my deepest gratitude to everyone involved. Thanks to the strong commitment and great support of Edition Angewandte I was finally able to publish this polyphonic collection of essays from various disciplines. Therefore, I would like to express my sincere gratitude to Anja Seipenbusch-Hufschmied and her committed project management at the University of Applied Arts Vienna and to her colleagues Stefanie Schabhüttl and Barbara Wimmer. I am particularly grateful for the assistance given by Katharina Holas as content and production editor at the publisher De Gruyter, which ensured the successful production of the book. I would also like to express my great appreciation to Gloria Custance, who gave much support during the proofreading process and was crucial in polishing the manuscripts. Many thanks go to Scott Clifford Evans, who oversaw the final corrections of the book, and to Zahra Mirza, who has been of invaluable assistance in corresponding with institutions and individuals concerning copyright issues.

I am particularly grateful to graphic designer Andrea Neuwirth for shaping and designing the book in a most appealing way, fostering unexpected correlations among the contributions due to our most fruitful and collegial exchange. The delightful illustration of the book has created a most coherent style to read across disciplines and very diverse topics. I also thank layout assistant Gabriel Fischer, who carefully adjusted the endnotes.

Finally, I wish to thank my family for understanding my long nights at the computer and for their enthusiastic encouragement and full support while I was editing this anthology during the COVID-19 pandemic. Even though this crisis had a dramatic impact on every aspect of life, it is giving us a rare opportunity to pause and to rethink how we want to live and how we can enable a more sustainable future in which life in our dying oceans will possibly flourish again.

Ingeborg Reichle

»As I gazed from the deck at the surface of
what ought to have been a pristine ocean,
I was confronted, as far as the eye could see,
with the sight of plastic.«

CAPTAIN CHARLES MOORE, DISCOVERER OF THE GREAT PACIFIC GARBAGE PATCH

Who Lives in the *Plastisphere*? Art and Science Responses to Plastic Pollution

Introduction by Ingeborg Reichle

Poison Pills ___ Plastic in the world's oceans has become a global concern as the massive accumulation of micro- and nanoplastic particles has turned into an existential crisis for marine life. These anthropogenic pollutants emerged as a by-product of modernity, but went unnoticed for a long time. However, when they were discovered in the late 1990s as the Great Pacific Garbage Patch, too little was done far too late.[1] Today it is obvious that untold trillions of micro- and nanoplastic particles floating in the world's oceans have created a novel, highly toxic ecological habitat, which threatens marine species and human health.

To describe the massive transformation of marine ecosystems when exposed to micro- and nanoplastics, scientists recently introduced the term *Plastisphere*.[2] Based on samples taken from the northern Atlantic, scientists found that new microbial reefs are emerging in the *Plastisphere*, which differ significantly from any previously known natural conditions of marine environments because toxic micro- and nanoplastic particles provide different conditions for microbes, bacteria, algae, and other micro-organisms than those offered by natural floating marine substrates like wood, feathers, or plants. Also, micro- and nanoplastic particles are highly toxic, serving both as transport medium and »poison pills« due to their ability to absorb toxic chemicals in the marine environment.[3]

Fueling Plastics ___ With Bakelite the first synthetic material made from petro-chemicals (coal) was introduced in 1907; its polymer bonds do not bio-degrade and literally last forever.[4] It is the long polymer strands that make up plastic, which are responsible for the unique characteristics of this material, its durability, versatile malleability, and cost efficiency, charac-teristics that allowed the global plastic industry to increase its annual production from 2 million metric tons in 1950 to approximately 380 million metric tons in 2015. If former trends continue, more than half of this year's plastic production will end up as waste and will be discarded (55 percent), 25 percent will be incinerated, and 20 percent will be recycled. In 2017, the cumulative global plastic production reached estimated 8.3 billion metric tons and it is expected to increase to 34 billion metric tons by 2050.

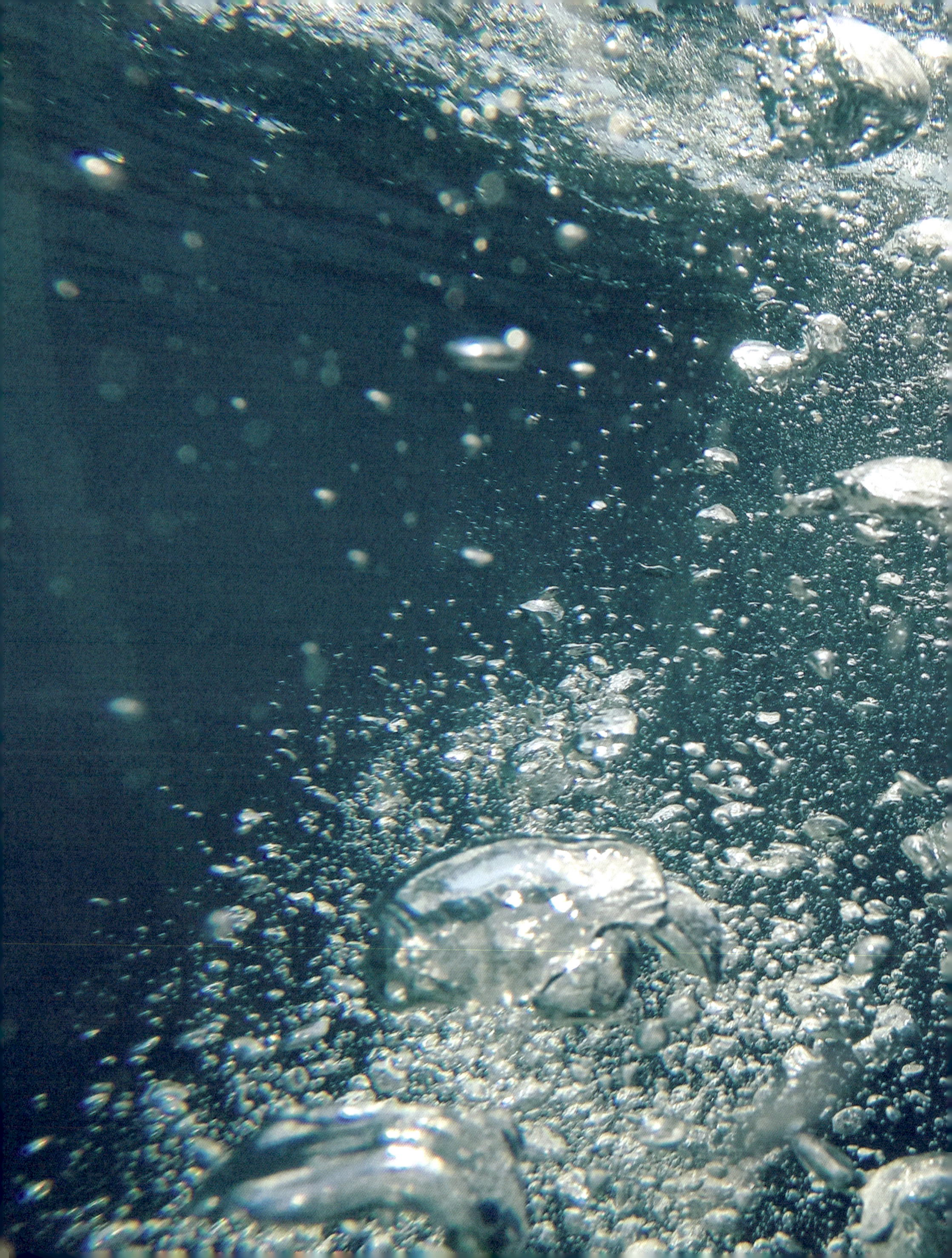

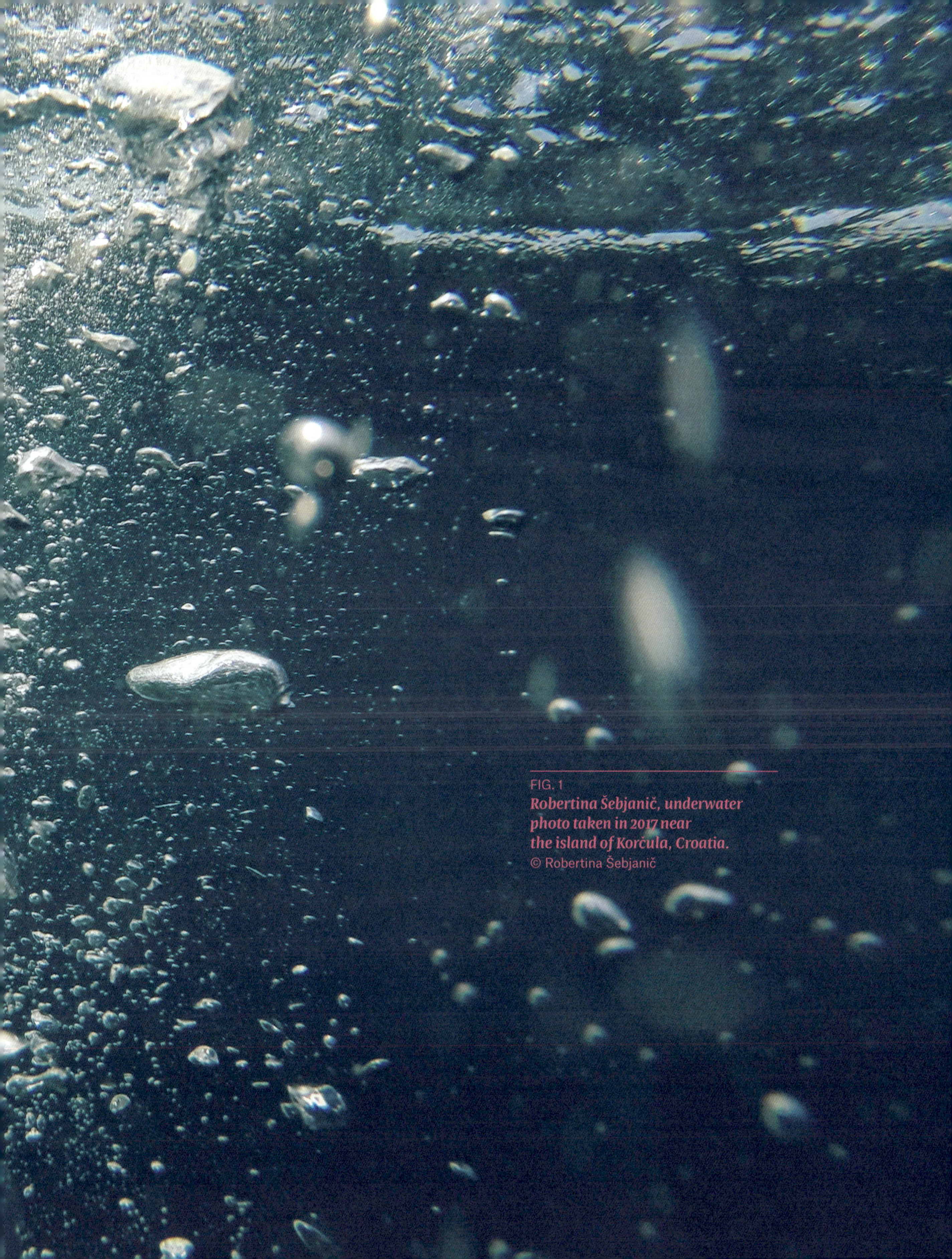

FIG. 1
Robertina Šebjanič, underwater photo taken in 2017 near the island of Korčula, Croatia.
© Robertina Šebjanič

Plastic was first developed around 1800 but it was not until the 1960s and 1970s that plastic, made from chemicals sourced from fossil fuels like coal, natural gas, and crude oil, replaced traditional materials and became the commonest material both in daily use and in the majority of manufacturing sectors. Subsequently, plastic and its toxic chemicals became part of the waste stream. But unlike most types of solid waste, plastic takes centuries to degrade; it never fully decomposes but falls apart into smaller and smaller particles, which still exhibit all the characteristics of plastic. The traditional solid waste disposal infrastructure failed and still fails to control the ever larger amounts of plastic waste, and improperly managed plastic waste disposal has made millions of tons of plastic end up in the world's oceans.[5]

Over the course of the twentieth century, consumption became central to the dynamics of modern capitalist economies. With the rise of mass consumerism most Western countries underwent profound transformations, reorienting almost every activity of daily life toward consumption.[6] After World War II, consumer societies were driven by a system of ever-expanding goods and desires, which became key to the ideal of happiness in the Global North. The »good life« became closely linked to the consumption of material objects and *stuff* on a massive scale, many of which are made of plastic.

Plastic Ocean ___ As a consequence of the vast amount of plastic that has been pouring into marine habitats for decades, the world's oceans are now on the verge of collapse. Because of the urgency to foster positive change in a situation where no escape seems possible, *Plastic Ocean: Art and Science Responses to Marine Pollution* offers an interdisciplinary perspective by including the arts and a wide range of disciplines, sounding the alarm about this developing crisis. In recent decades, artists have joined forces with scientists to advance the necessary mobilization and have teamed up with environmental activists, philosophers, and curators to become ecoactivists raising awareness about this imminent environmental disaster and the future conflicts it will create.

With her artistic strategies, and by cofounding the Plastic Pollution Coalition (PPC), artist Dianna Cohen is committed to influencing the national and international political agenda concerning plastic pollution. She is particularly keen to change policy and legislation about single-use plastic items in everyday use.

Pinar Yoldas explores, in her artistic practice, how marine debris is transforming the world's oceans into a future post-human ecosystem with the goal of driving activism forward in search of environmental justice. She provides us with compelling visual narratives of highly speculative species, which will be able to adapt to the toxicological effects of plastics.

DIY/DIWO (do-it-yourself/do-it-with-others) protocols for hacking hormones are on the mind of artist and biohacker Mary Maggic when

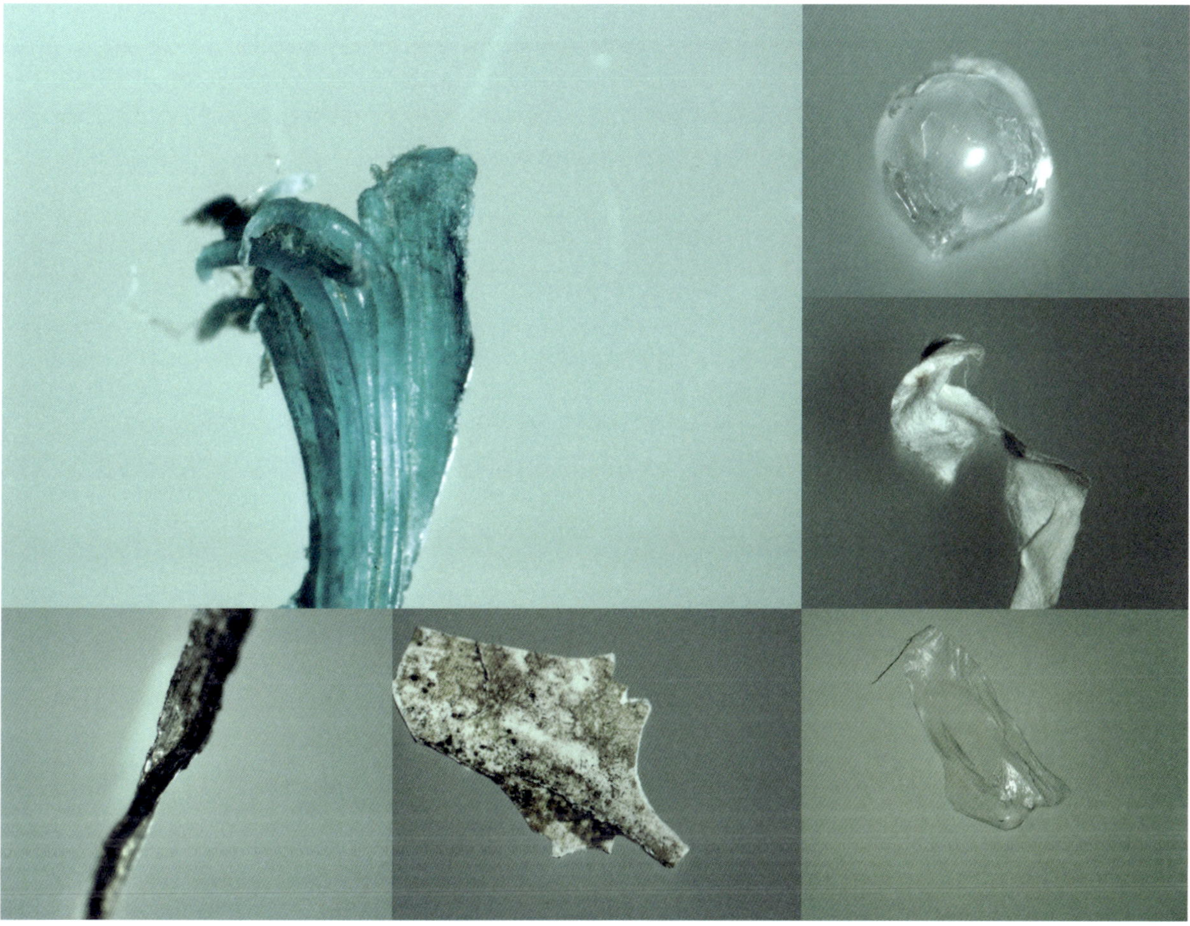

FIG. 2
Max Liboiron, COD OBJECTS (INGESTION STUDY), 2016, digital photographs of items taken from the guts of ATLANTIC COD caught in 2015 on the island of Newfoundland, Canada. COD OBJECTS (SP141, PH66, SP31, PH85, PH67, and PH33) is part of the larger series SEEING LIKE A SCIENTIST.
© Max Liboiron

critically reflecting about how gendered bodies are controlled and managed through corporate and institutional science on the one hand, and on the other hand are permanently exposed to endocrine-disrupting chemicals due to toxicity in the environment. Is there hope for disobedient bodies amongst capitalist and ecological ruins? Endocrine disruption affects not only human health but all living organisms because basic features of biology are shared across all life forms, in particular damaging sexual development and fertility. According to the latest research findings, the failure to regulate endocrine disruptors will lead to reproductive abnormalities: by 2040 most Western males will not be able to father a child in the *old-fashioned way*, which will almost certainly lead to great socio-cultural upheavals.[7]

The loss of biodiversity in the Gulf of Mexico due to the Deepwater Horizon oil spill in 2010, which is the largest environmental disaster in the history of the United States and the largest known petrochemical spill by volume in modern history, is the focus of artist Brandon Ballengée, who seeks to promote systemic change with his collaborative eco-action and outreach events. Key to his art and action responses is the sharing of scientific knowledge about the tremendous negative impact of current oil production methods with the coastal communities affected to increase community resilience in their local ecosystems.

A different important topic is raised by artist Robertina Šebjanič: the rising level of anthropogenic noise in the world's oceans due to increasing gas and oil exploration and extraction, underwater mining, construction work on the seafloor, and shipping. The negative effects of high noise levels on marine wildlife are especially injurious to those marine organisms who communicate through sound. In close collaboration with marine biologists, the artist investigates the phenomenon of noise pollution through collecting invasive sound samples on her extensive research trips.

With her participatory art projects, Victoria Vesna explores how to immerse her audience in an act of deep listening, foregrounding the importance of making the invisible and inaudible aspects of marine life accessible. Using the most compelling science visualizations, which were produced in a laborious process by digital artist Martina R. Fröschl and scientist Alfred Vendl, based on accurate digital 3D models from live images of planktonic organisms and processed by biologist Stephan Handschuh and zoologist Thomas Schwaha, Vesna transforms scientific data into an art-science based experience to render the effects of anthropogenic noise on plankton accessible to the human senses. While the effects of the infiltration of plastic-associated toxins into marine food webs, given the key role played by plankton in ocean ecosystems, is a huge concern of current scientific research, noise pollution is as yet a minor subject that has been largely neglected so far. Both artist and scientists are anxious to draw attention to the fact that planktonic organisms are absolutely vital for ocean food web dynamics, and they also play an integral role in regulating the global climate.

Another perspective on planktonic organisms is offered by artist Reiner Maria Matysik. He uses microscopic images of unicellular diatoms, which live in the oceans, waterways, and soils of the world, to design complex three-dimensional forms while at the same time reflecting on the characteristics of form solutions presented by natural systems. Matysik works in collaboration with biologists who are particularly interested in investigating diatom micro- and nanostructures in the field of functional morphology.

Curators Regine Rapp and Christian de Lutz have, in recent years, mounted a number of exhibitions that focus on artistic responses to the severe consequences of climate change and the huge crisis the world's oceans are facing due to massive plastic pollution. At the same time they also promote new performative artistic strategies like *doing it with others*

(DIWO) actions by artists or participatory hands-on workshops. Performing philosophy through art is evident in the installation discussed by philosophers María Antonia González Valerio and Rosaura Martínez Ruiz, in which art deconstructs the biological category of species to encourage the development of transdisciplinary thinking about problems of ontology, biopolitics, taxonomy, epigenetics, and evolution.

The last chapter of *Plastic Ocean: Art and Science Responses to Marine Pollution*, by industrial microbiologist Michael Sauer, is devoted to laying out alternatives to long-established fossil-based plastic. Bioplastic, for example, which is produced through synthetic biology applications, would not harm marine ecosystems in the same disastrous way. However, his key argument points out that it is not so much the development of new materials which is the stumbling block to creating environmentally benign plastics, but the fact that we still do not know how to define and test, in a comprehensive way, whether a material is environmentally benign or not.

This anthology aims to rethink ocean plastic pollution through the lens of art and science to foster a critical debate about whether this urgent global challenge should be regarded as an epistemological, sociopolitical, or technical problem. If we do not seriously reevaluate the destructive effects of mass consumerism on the environment, the impact of irresponsible use of valuable resources on future generations—if we fail to foster and promote green science and do not raise awareness and show commitment to performing real actions—it will become increasingly difficult to live a more sustainable life to avert the collapse of our marine ecosystems. For it is expected that the amount of plastic pouring into the world's oceans will nearly triple by 2040 to 29 million metric tons every year. Plastic is everywhere, it is in the air, water, soil, and inside us, too; therefore, it seems appropriate to extend the term *Plastisphere* to the entire planet. Who will be able to live in the Plastisphere in years to come—which species will be able to adapt remains an open question.

1　See Charles G. Moore and Cassandra Phillips, *Plastic Ocean. How a Sea Captain's Chance Discovery Launched a Quest to Save the Oceans*, New York: Avery (2011).

2　Erik R. Zettler, Tracy J. Mincer, and Linda A. Amaral-Zettler, »Life in the ›Plastisphere‹: Microbial Communities on Plastic Marine Debris,« *Environmental Science & Technology* 47, no. 13 (2013): 7137–7146 and see Linda A. Amaral-Zettler, Erik R. Zettler, and Tracy J. Mincer, »Ecology of the Plastisphere,« *Nature Reviews Microbiology* 18 (2020): 139–151.

3　Max Liboiron, »The Plastisphere and Other 21st Century Waste Ecosystems,« *Discard Studies Blog*, July 22, 2013.

4　Esther Leslie, *Synthetic Worlds: Nature, Art and the Chemical Industry*, London: Reaktion Books (2005).

5　Max Liboiron, *Redefining Pollution: Plastics in the Wild*, PhD dissertation, New York University (2012).

6　David Banash, *Collage Culture: Readymades, Meaning, and the Age of Consumption*, Amsterdam: Rodopi (2013), 11.

7　Shanna H. Swan, *Count Down: How Our Modern World Is Threatening Sperm Counts, Altering Male and Female Reproductive Development, and Imperiling the Future of the Human Race*, New York: Scribner (2021).

Since 2009, Plastic Pollution Coalition (PPC)
has drawn on the talents and resources
of many living artists, using popular music,
film, books, installations, comic books,
animation, and participatory art installations
to communicate to the broadest possible audience
the complexities of plastic's impact on the
planet. PPC's effective use of this range of
the arts has influenced the national and international political agenda through support of
policy and legislation around single-use plastic
packaging. But just as importantly, the aesthetic
resources of art have helped PPC to address
citizens where they are and to educate them on
an environmental crisis hidden in plain sight.

»Downhill from Everywhere«: Plastic Pollution Coalition and the Force of Art

Dianna Cohen and Jennifer Wagner-Lawlor

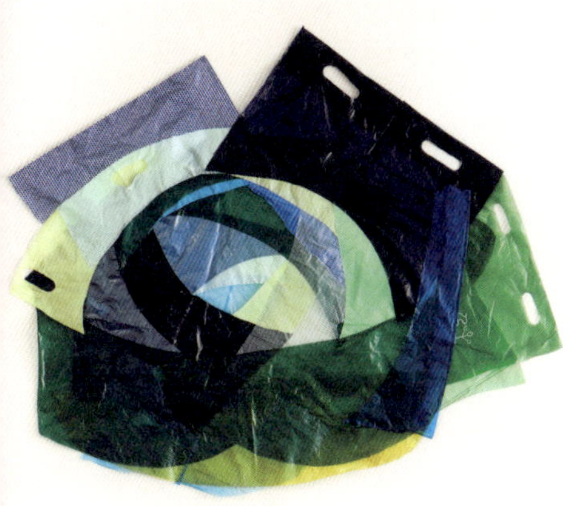

Dianna Cohen, WAVELENS, 2007, plastic bags, handles, and thread. Installation view at the Ionion Center for Arts and Culture, Kefalonia Island, Greece, 2014.
© Dianna Cohen

When, in 2009, a small group of committed environmentalists founded Plastic Pollution Coalition (PPC), little did we know just how important Dianna Cohen's own identity as a visual artist would be in shaping a new, unanticipated career as a global, anti-plastics activist. Over a decade later, she is PPC's CEO and the driving force behind a still-expanding mobilization of hundreds of citizen-activists, 150 celebrity spokespersons and well over 1,200 institutional members, including businesses, research institutes, NGOs, and other influencers and change agents. Among PPC's earliest strategies for building a global coalition to stop plastic pollution and raise awareness about its toxic impacts on human and other-than-human health was the decision to use artwork as a platform to communicate with several distinct publics concerning this accelerating ecological crisis. The material force of plastic itself was the inspiration for Dianna's activist work (FIGURES 1—2) which has included two- and three-dimensional experiments with plastic as a primary medium. Since 2009, PPC has continued to utilize art as a potent form of communication. Its various campaigns and messaging are guided by science and multi-disciplinary research, but the phenomenological resources of art make the difference between indifference and engagement.

Therefore, in our conversations spanning over ten years, Dianna and I have often talked about the strategic conjoining of science and art as a mobilizing force behind PPC's activism. While PPC has actively engaged efforts to encourage governmental policy change related to the steward-ship (or lack thereof) of the oceans and to the stemming of plastic effluence in all its forms, the organization works on the premise that speaking truth to power is only part of the mission. For there to be political will for change, a person has to see not only power but powerlessness—and their own relationship to *both*. That is something which art, in its great variety of forms and modes, can do very effectively. Beyond the paradigms of (bio) power that structure the lives and deaths of human and nonhuman creatures as well as the very ecosystems of water, land and air, is the experience of relationality and interdependency, through both viewing and participating in various forms of artistic work. ___ *Jennifer Wagner-Lawlor*

20

Plastic Reality ___ Plastics are oleophilic: this means that they function like attractors for other organic pollutants. In sample studies, preconsumer plastic nurdles have been found to have more than a million chemicals attached to their surface. Furthermore, the manufacture of plastics—whether the carbon source comes from petroleum or is plant-based (as in bioplastic)—requires chemical additives such as phthalates and bisphenols, including BPA and BPS, which are known endocrine disruptors. *Body Burden: The Pollution in Newborns*[1] is a 2005 U.S. study by the Environmental Working Group (EWG) which found that the umbilical-cord blood from newborns was prepolluted with 220 different chemicals, including BPA, phthalates and flame-retardants. Exposure to these hormone disruptors, *in utero* and *postpartum* throughout childhood, has been linked to shortened anogenital distance, small penis size, feminization of boys, early menses in girls, and lower IQ. EWG's study of adults revealed exposure levels linked to lower sexual function, sterility, infertility, obesity, diabetes, breast cancer, brain cancer, and prostate cancer.[2] The World Wildlife Fund's report *No Plastic in Nature*[3] reveals that adults now ingest five grams of plastic—about the weight of a plastic credit card—*every week*. These chemicals often leach out of plastics, including water bottles and food packaging, impacting our health before they are even thrown away. This leaching generally is not what »recycling« means, and thinking about and working through the life cycle and recycling of plastic began to inform my sense not only of the »life« of my art, but also of the art of living. I had been working with plastic bags since the early 1990s, and I found, ten years on, that some of my work was beginning to change, as I recounted in 2014:

As time went by, I was surprised, looking back at early work, to learn that plastic photodegrades: some of the bags in my artwork were fissuring and flaking, falling apart. I was forced to reconsider the material and my intentions of making art out of it. I thought at first that this degradation was good, that there is an end to plastic. But this was—like so much else about plastic—a false promise. Part of its alchemy. Like alchemists of olden times, we can try magically to produce gold—and in one sense we did; but what we really produced was toxic dross.[4]

Plastic, I learned, is anything but inert—as the history of plastic chemistry, featuring chemical explosions (including a few fatal ones) illustrates. Today, materials conservators at prestigious institutions such as The Smithsonian and the Getty Museum of Art research and debate how to preserve their plastic and rubber artifacts, which deteriorate, fade, fragment, and sometimes spontaneously combust and disintegrate. When and how is hard to say: »Even with knowledge transfer and recent advances in material characterization,« report Odile Madden (Smithsonian Institution) and Tom Learner (Getty Museum of Art), »our understanding of plastics stability remains rudimentary.«[5] While Madden and Learner note the irony of »the seemingly thankless task of trying to preserve a class of material that

FIG. 2
Dianna Cohen, FALDA, 2005,
plastic bags, handles, and thread.
Installation view at the Ionion Center
for Arts and Culture,
Kefalonia Island, Greece, 2014.
© Dianna Cohen

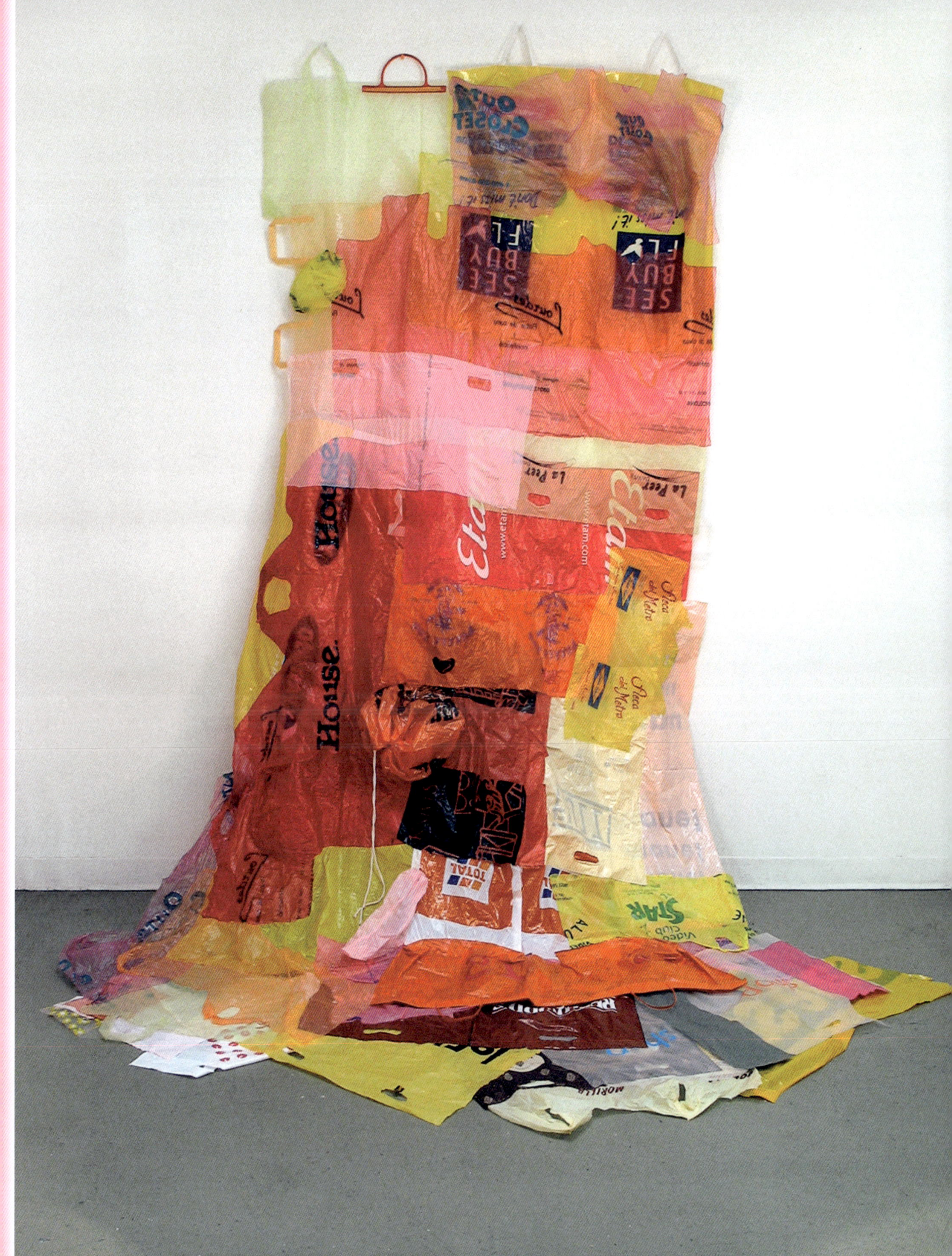

OUTa
CLOSET
SEE
BUY
FLY
SEE
BUY
FLY
Don't miss it!
La Fest
Eta
Eta
House
House
Area
Metro
Area
Metro
TOTAL
STAR
VIDEO CLUB
MORILL

almost defies preservation,«[6] ensuring »a great future in plastics«[7] for the conservators, my own reflections led me to a different conclusion. Even if one could »turn off the faucet« of plastic production in this instant, we would still be left with hundreds of millions of tons of nonbiodegradable plastic waste that we have already tossed »away,« into a waste stream that is ultimately global and interconnected, or entangled. As a visual artist, the transformation and discovery of the ephemerality of plastic bags as my primary material in the creative work motivated me to learn more, and discover just how pressing the issue of plastic pollution had become, especially in the oceans, which I have spent so much time in and around. I learned that just skimming objects off the surface was futile without recognizing that we have to follow the course of plastic »upstream« to the source points. That network of waste streams includes not just the roads

FIG. 3
Pamela Longobardi, SUBMERGENCE / EMERGENCE, 2015, ocean plastic, propeller-fouled driftnet, silicone, and specimen pins.
© Pamela Longobardi

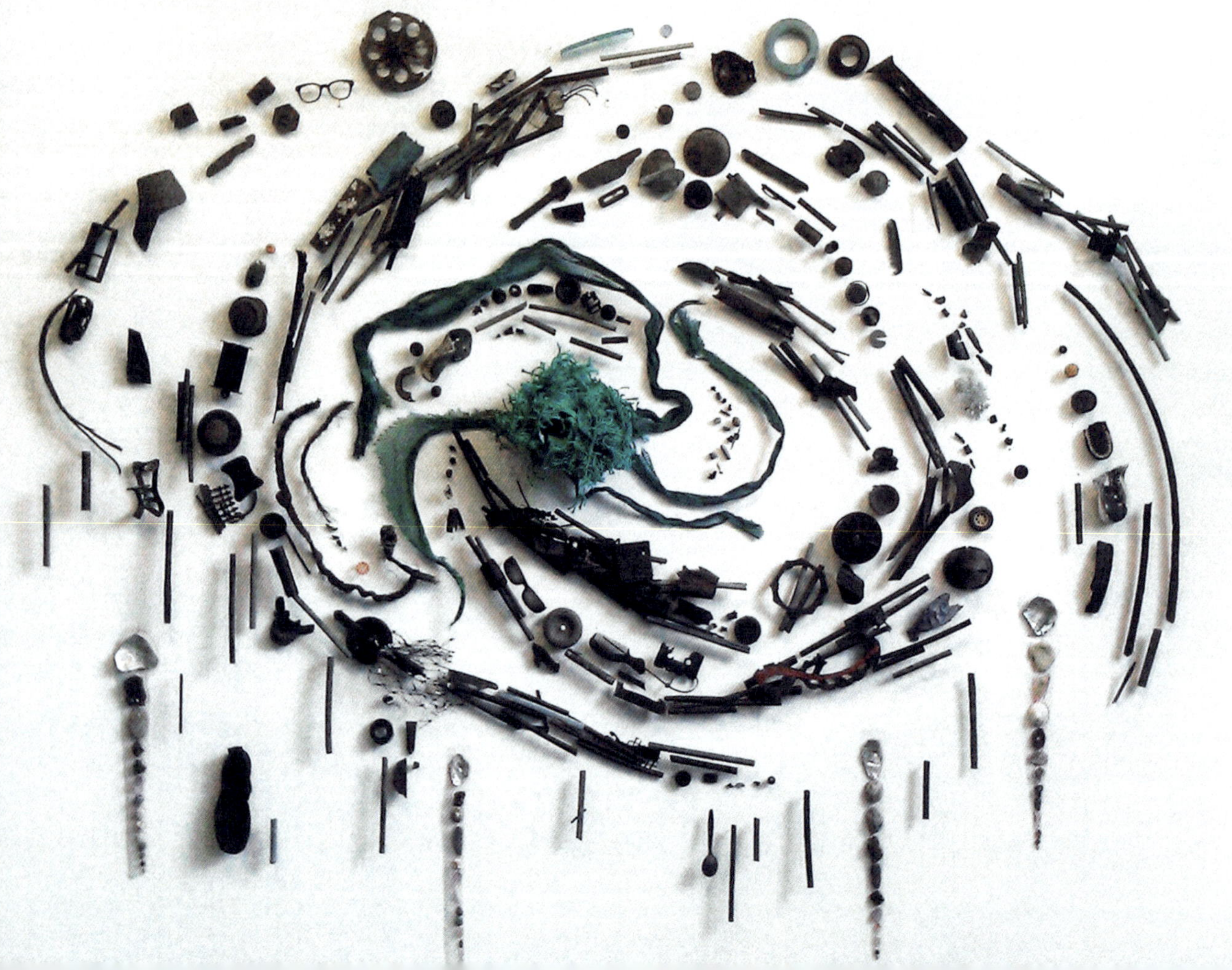

from consumer to landfill, but the paths of our circulating oceans, and of surface and subsurface water flows ultimately headed to the oceans. The waste stream also includes atmospheric currents carrying toxic particulate and gaseous remains of incinerated plastics—and that waste stream is eventually also found in our own bloodstreams. But, as Captain Charles Moore, credited with »discovering« the Great Pacific Garbage Patch, likes to say, the ocean is »downstream from everything,«[8] and with time the so-called plastic soup increasingly resembles a plastic slurry.

Those of us who believe that art can change the world talk about the »work« that art can do, not just about »art works« as objects. Art has been called »forcework,«[9] for example, referring to the kinds of force fields energized by the interchanges of art and environment, artwork and spectator. It also refers to the work going on between *material* and *idea*:

FIG. 4
Pamela Longobardi, OUROBOROS, 2014, ocean plastic removed from beaches and sea caves of Kefalonia Greece, steel wire, mesh. Installation view in the 2014 exhibition PLASTIC FREE ISLAND at Ionion Center for Arts and Culture, Kefalonia Island, Greece.
© Drifters Project, photo: Pamela Longobardi

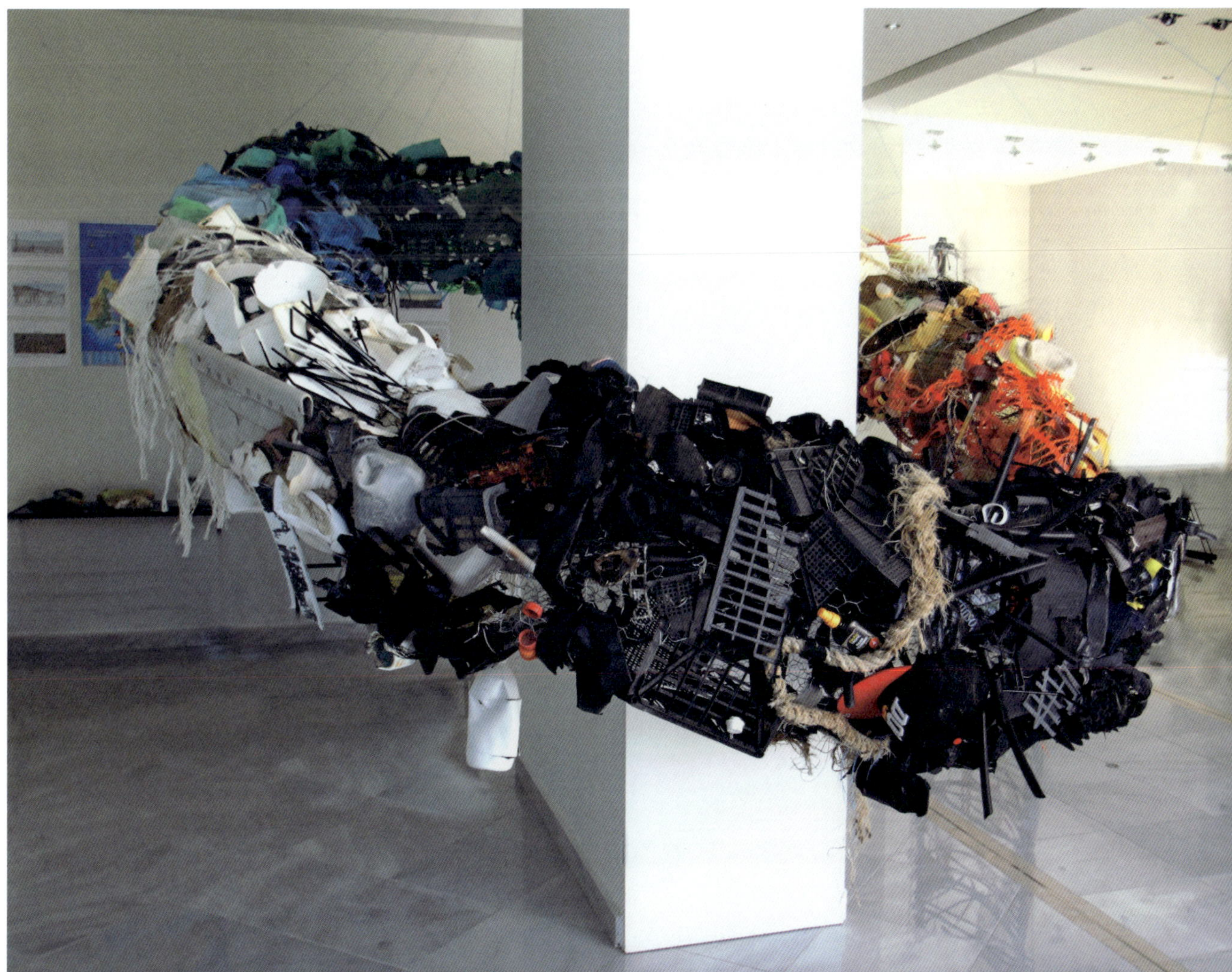

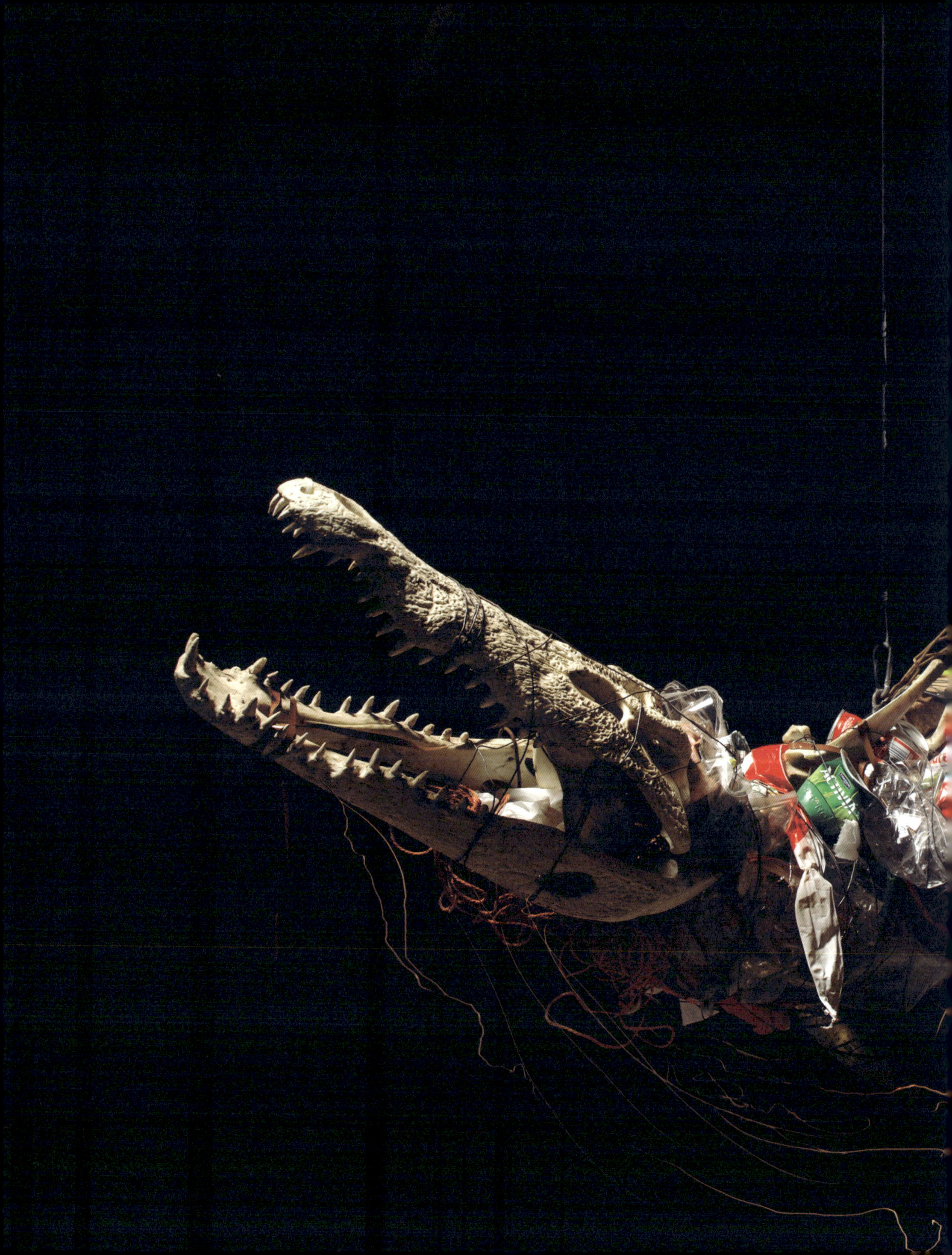

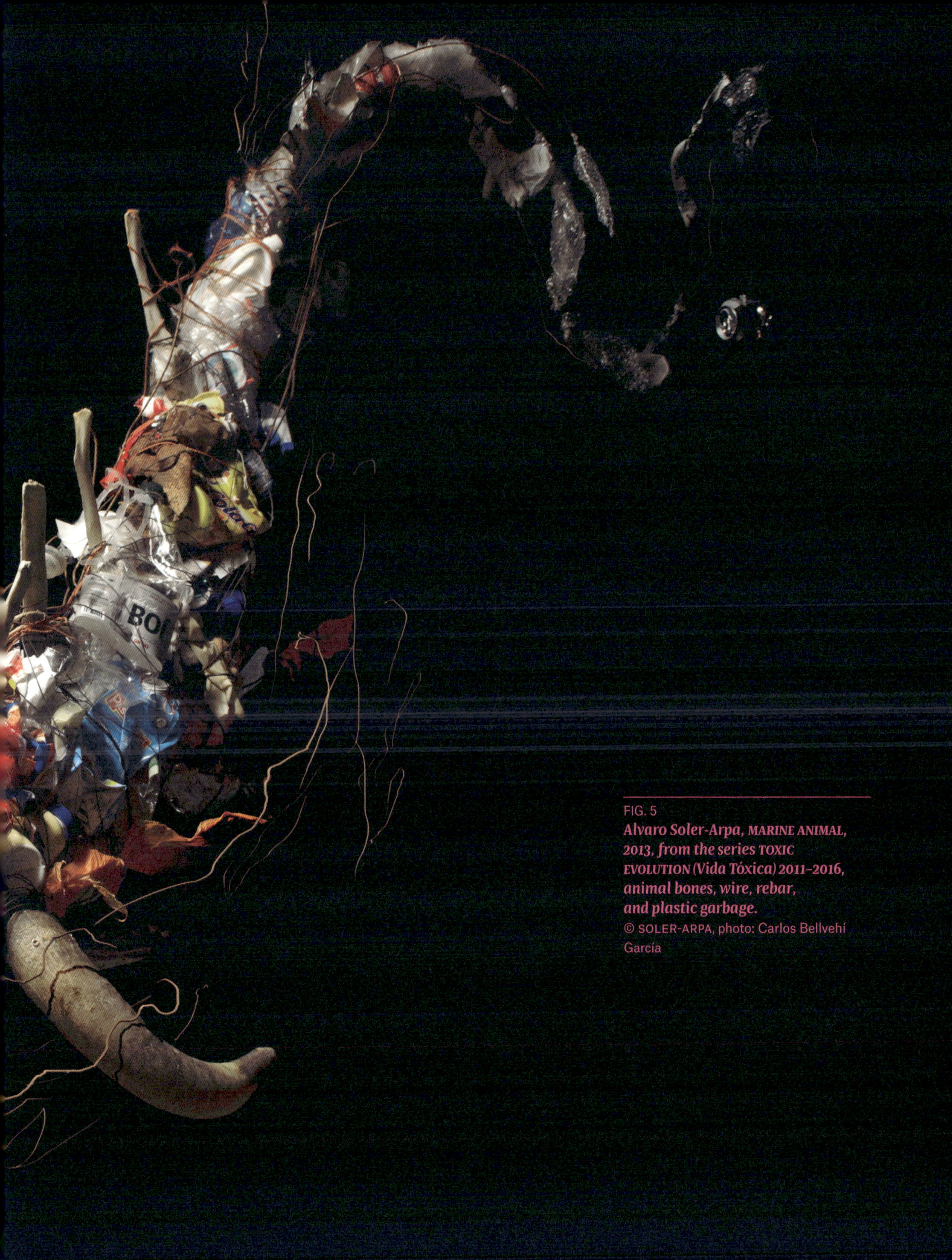

FIG. 5
Alvaro Soler-Arpa, MARINE ANIMAL, 2013, from the series TOXIC EVOLUTION (Vida Tóxica) 2011–2016, animal bones, wire, rebar, and plastic garbage.
© SOLER-ARPA, photo: Carlos Bellvehí García

26

in Greek mythology the sculptor Pygmalion learns about the »reciprocal shaping«[10] of matter and desire. There, as here, the plastic molding of material takes on a life of its own, begins to realize or visualize alternative visions of the future – good and bad. But the best thing art can do – as we move toward PPC's role as an *activist* force – is expose relations of power. Many have written about the capitalist structuring of Western culture, of which plastic, with remarkable efficiency and mythical power, is the symbolic form.[11] More than symbol, though, art literally impresses us, entering our body through our senses, materializing itself in/on us in *that* way – a *reciprocal* shaping of object and subject (the human subject).

Plastic Power ___ In 2010, scientists Dr. Sylvia Earle, Dr. Arlene Blum, and Jeanne Rizzo (Registered Nurse), presented at *TEDxGreatPacificGarbagePatch*, as researchers were just uncovering the effects of plastic on the environment and human health. Today, over ten years later, plastic and chemicals in plastic are part of our body chemistry: it's in our gut, it's the stuff of our clothing, computers, cosmetics, even the cash we carry. PPC's decade-plus life has adopted a diverse set of activist strategies, recognizing that in order to gain any traction in reducing the use of single-use plastics, scientific reports are not enough – because they don't meet people where they are. It's one thing to talk about plastic pollution washed up on coasts where the stuff makes itself visible as waves of flotsam and jetsam, and where sand is increasingly made from degraded plastics (FIGURES 3, 5, 7). It's another to raise awareness inland; and yet another to move urban dwellers for whom the appearance of the plastic shopping bag in the 1970s seemed like a godsend (before tons of them started clogging city sewers and wrapping lamp posts).

As PPC cofounders we wanted to involve business and policy leaders; we wanted to talk to women who are still the primary purchasers of household products; we wanted to talk to children; we wanted to talk to entertainment and sports venues selling millions of drinks a year in plastic cups and bottles, and food in styrofoam; we wanted to reach the managers of the artists and athletes who play there. We did not want to turn people off with relentlessly negative messaging, even as we did and do want to impress on people the scale of plastic's impact on environmental and human health by providing up-to-date scientific facts. To do that, PPC turned consciously to the transformative potential of literature, art and song to communicate both the science *and* the urgency of the problem.

The day-long *TEDxGreatPacificGarbagePatch* happening included poet Ellyn Maybe's reading of a new poem, »Ocean Song,« which imagines the ocean speaking to a »colony« of plastic bags and bottles: »The bags were sullen and drifting deeper into the askew./But the ocean smiled a knowing smile and even laughed and said./Like bags can turn inside out, the mind wanders along in its own dance.«[12] A dance piece by performance

artist Lila Roo enacted a baby bird being born and stretching its wings from
a pile of plastic detritus (FIGURE 6). There was music: PPC Notable Members
Inara George performed »Family Tree« and Jackson Browne performed
»If I Could Be Anywhere,« a contemporary protest song about plastic
pollution in the ocean. Recently, in 2019, Keb' Mo', another PPC Notable
Member, released the song and video »Don't Throw It Away« (with a bluesy
shout-out to »Leo B.« [Leo Baekeland, the chemist who achieved the first
thermoset plastic in 1907]); and in 2020, Browne dropped the rock anthem
»Downhill from Everywhere« on an album of the same name (released
in 2021). Each of these artists addresses the recalcitrance of plastic
and plastic pollution on the one hand and the resistance and resilience
of the ocean itself on the other.

27

FIG. 6
Lila Roo Duncombe-
Lieber, LAYERS,
2013, refuse plastic.
© Lila Roo

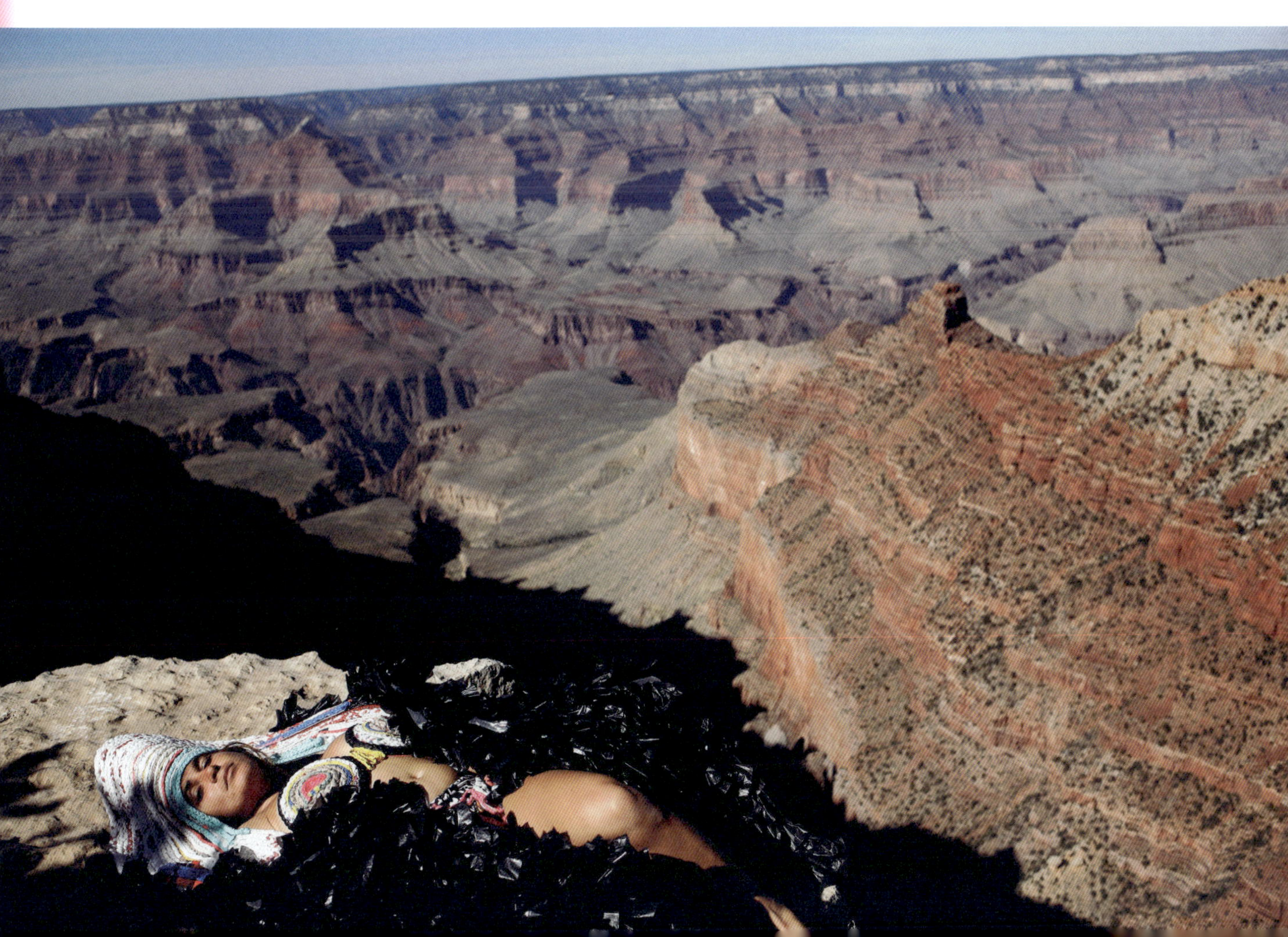

Plastic Revolt ___ Today, it's almost as if plastic is itself a living artist, remaking creation down at cellular and molecular levels, revising the contours of the land- and oceanscapes, reterritorializing the planet through manufacturing product streams as well as waste streams. We are seeing the mutation of human and nonhuman life-forms; we are seeing these unique ecosystems, so-called plastispheres where, according to microbial ecologist Erik Zettler, »at microscopic scales, microbes such as bacteria, algae and other single-celled organisms gather around and colonize plastic and other objects floating in water. [...] We call this community of microbes growing as a thin layer of life (a biofilm) on the outside of plastic the ›plastisphere,‹ analogous to the layer of life on the outside of planet Earth called the ›biosphere.‹«[13] Also, there is *plastiglomerate*, a rock-like material so named by the American Geological Society. These new forms are all markers of the »Plasticene Age.« And PPC can be regarded as the most elaborate and collaborative work of art I have contributed to; like these new forms, PPC is a mobile, hybrid practice of environmental activism. We want people to *see* the problem, but also to *feel* it. We want the visual *and* the visceral. The ocean is downhill from everywhere. Many people first realize the impact of plastic pollution by witnessing firsthand the endless waves of plastic debris washing up on shorelines of islands, coastal communities, lakes and rivers around the world, and the response is »We need to clean this thing up!« But the truth is that as you begin to follow the path of plastic production back upstream, you discover the source points of its toxicity at every major juncture. Plastic pollutes from extraction at the source, or the wellhead, through manufacturing and production, through use (particularly single-use) for a very short amount of time. At this point, its useful life over, plastic becomes a waste management issue on a global scale, whether we find it as litter on streets, rubbish in the bins, as manufacturing detritus, or marine debris. One way to manage the waste is simply to burn it: incineration is euphemistically referred to as waste-to-energy, pyrolysis or chemical recycling. But these are really all just forms of burning. This burning releases particulate pollution into the air, which impacts us all, aggravates the greenhouse effect and releases dioxins into the air. Plastic pollutes us and life and the planet at every stage of its existence. Our use of plastic directly contributes to climate change and global warming. Not just downhill from everywhere but downwind from everywhere.

For all these reasons, one of PPC's earliest projects involved participatory community art projects. PLASTIC FREE ISLAND (2014) is a collaboration between artist Pamela Longobardi and myself to engage island communities in addressing plastic pollution and solutions locally, particularly those areas with fragile ecosystems upon which both islander livelihoods and tourism depend. Proposing a program of forensic beach cleaning trainings, education and large-scale art production, we involved the entire Kefalonia community, including its municipal government and businesses, as it

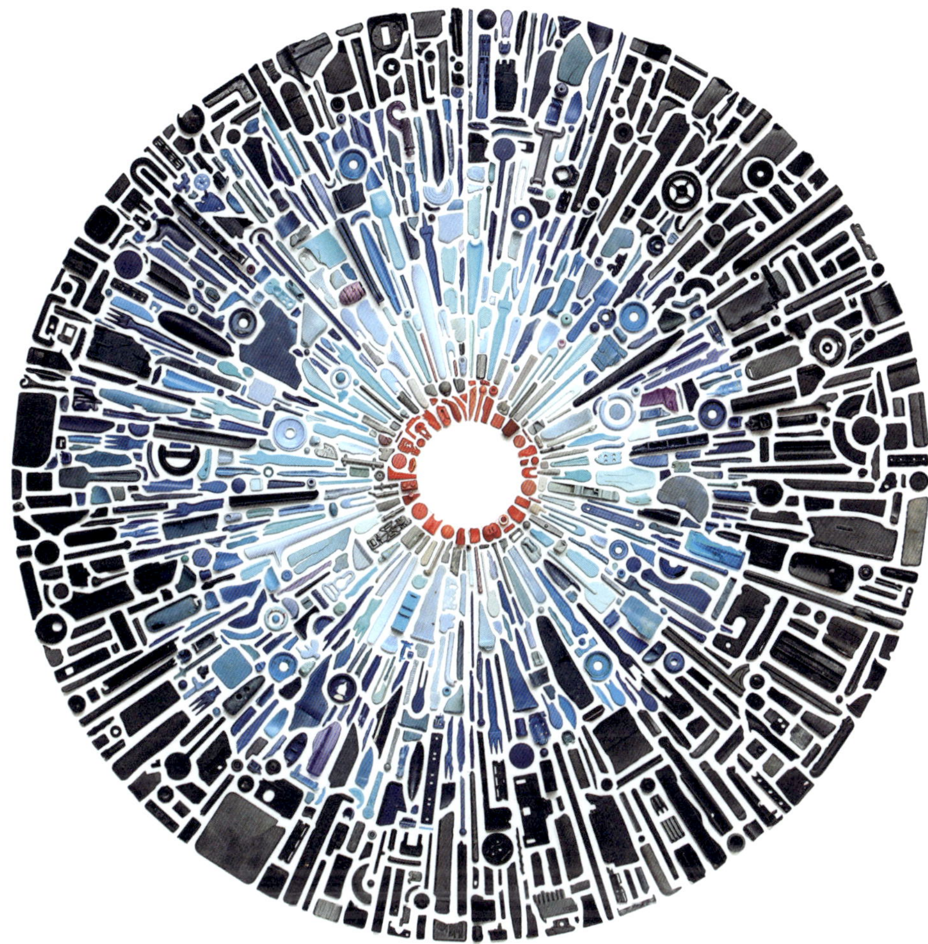

FIG. 7
Steve McPherson, WAVELENGTHS, 2013,
unaltered marine plastic objects
found along the UK coast from
1994–2013.
© Steve McPherson

completed »plastic audits,« learned about the sources and risks of plastic
pollution and mobilized islanders to remove plastic pollution from beaches. Community participants of all ages created art and sculpture from the
detritus cleared from their own shores and from the island's fragile lime-
stone sea caves. A culminating event celebrated our work with community
engagement: a powerful, 44-foot sculpture called OUROBOROS (FIGURE 4)
lay at the center of a public plaza. The name alludes to the ancient symbol
of infinity and the soul of the world; in addition, however, snakes under
duress will eat their tails.

The involvement of PPC meant thinking beyond a single island
or specific site; after all, the scale of plastic pollution runs from the
microscopic to the global to the planetary. PLASTIC FREE ISLAND was an early
example of the notion of scalability that has become essential to many
of PPC's campaigns. One of PPC's most ambitious advocacy projects is

30

REFILL REVOLUTION, which was launched at the Bonnaroo Music and Arts Festival in 2014.[14] Rock festivals are not exactly known for their »leave it cleaner than you found it« approach: a post-festival visitor is likely to find scenes of trash fields and mountains, such as those photographer Andy Hughes memorializes in his book-length photo essay *GLASTONBURY OPUS 1*.[15] Photos like these (FIGURES 8—9) reflect the aftermath of music festivals since Woodstock; but PPC's mission was to change that. I had already convinced various artists to start taking commercial water filters with them on their tours, which eliminated thousands of bottles of water that bands would normally go through. But then PPC worked with the organizers of Bonnaroo to pilot a program that would offer branded,

FIG. 8
Andy Hughes, GLASTONBURY OPUS SERIES, 2015, digital C-Type. Plastic footprint left by hundreds of thousands of attendees at the 2014 Glastonbury Festival, UK.
© Andy Hughes

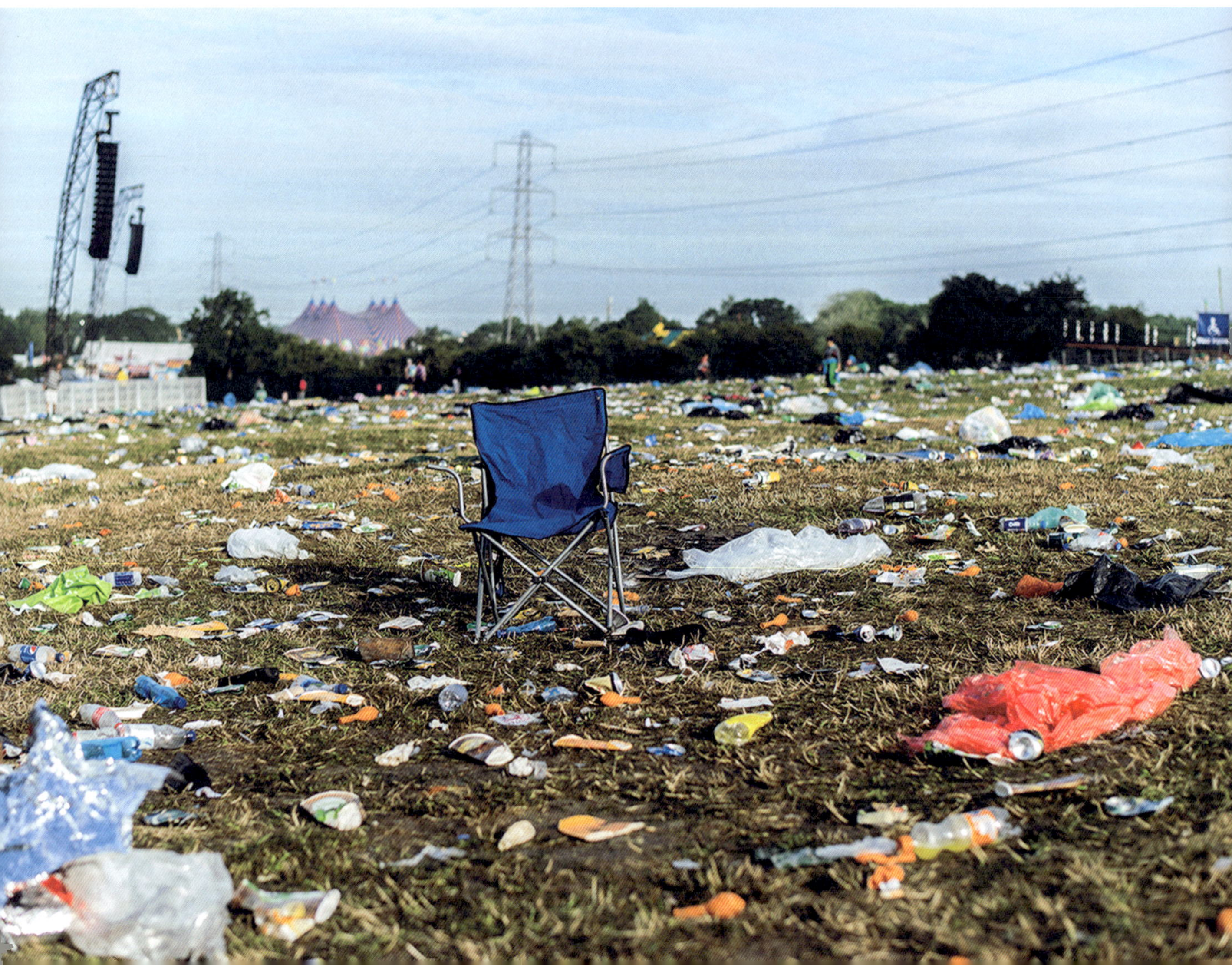

reusable, stainless steel cups to replace plastic beer and water cups. The first year, 7,500 cups were sold at the point of sale for beer and the owner of each cup then received a one-dollar discount on every beer refill purchased during the multiday festival. The festival site already had free water refill stations on site. We aimed to »normalize« reusables and refillables by making them hip. In the first day and a half of the festival, the reusable cups sold out and people began to steal them, coveting the new cool thing. Fans have been collecting them since each year's cup is a different color. The program continues to expand: PPC has since joined forces with the producers of large US festivals including OutsideLands, Coachella and Lollapalooza to Wonderfruit in Thailand and Byron Bay Blues Festival in Australia, along with LiveNation Entertainment and Kim and Jack Johnson to create a resource called BYOBottle.org (Bring Your Own Bottle). PPC creates tools that people can use to reduce their plastic footprint at such events.

PPC marshals every form of creative communication to further its goals: short videos and public service announcements, music, comic books, film, animation, video, the visual arts—and in its latest project, virtual reality technology. »360° Sea Plastic« is a five-minute video produced by Jetlagged and PPC. For its 2020–2021 »challenge« period, the Conrad Foundation, in collaboration with The Winsor Foundation, Prince Albert II Foundation and PPC, has created a new special challenge, *Oceans: The Plastic Pollution Problem*. PPC contributed an informational short video, »Open Your Eyes,« featuring PPC supporter, actor Jeff Bridges; in addition, »360° Sea Plastic« gives teams entering the challenge an underwater VR view of plastic pollution.

As PPC enters its second decade in 2021, the forcework of art will continue to join science in informing the shape of PPC advocacy—all this to a higher purpose, as the world awakens to the plastic pollution crisis and seeks solutions and alternatives. We continue the vital work to create a systems shift towards a world free of plastic pollution. Utilizing passion and creativity, PPC connects our coalition members as we reach out to the world at large. We engage our community and educate, inspire and activate by speaking from heart to heart. As artists, scientists, youth activists, conscientious businesses, and policy makers, we have the power to create change. Onward to a plastic-free world!

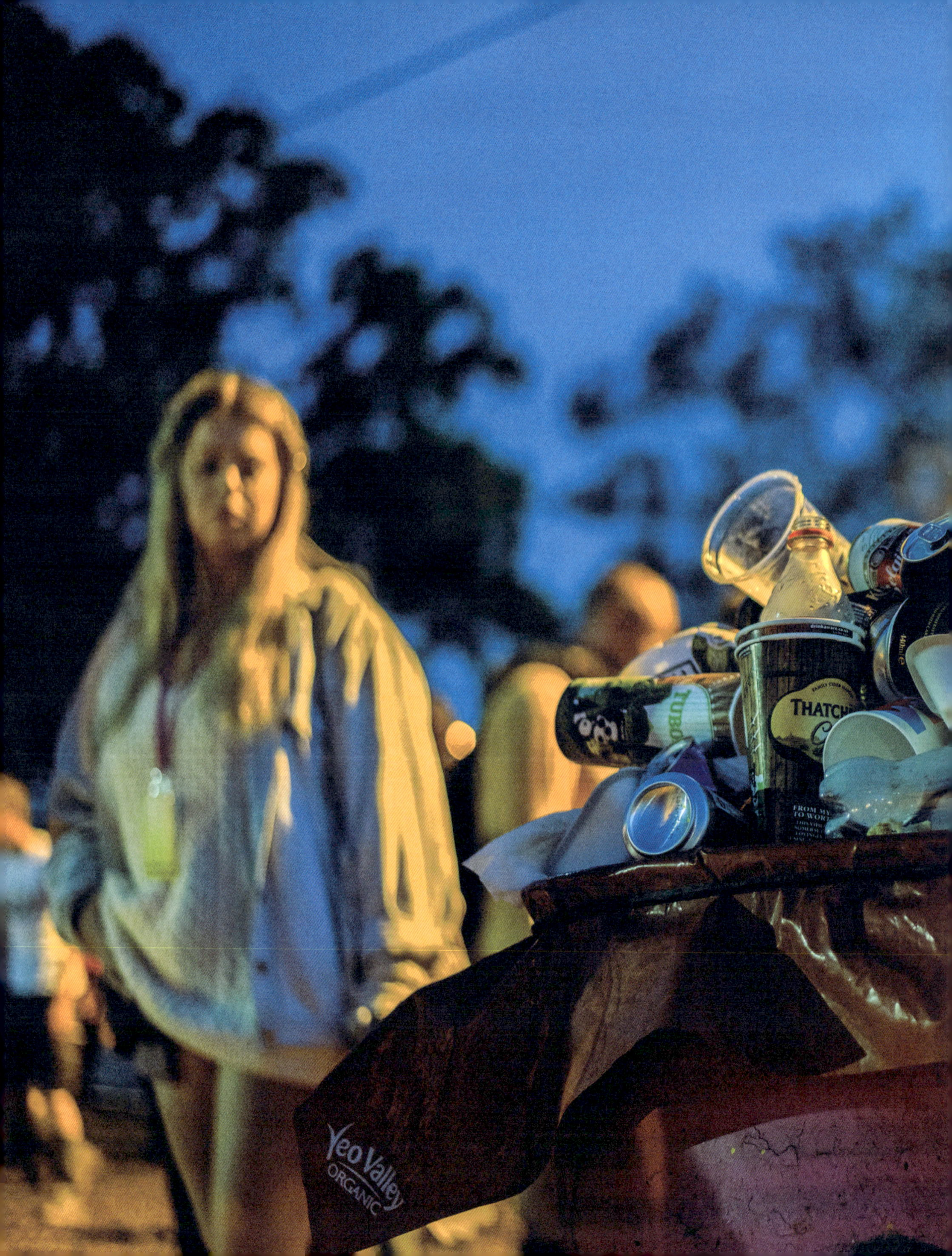

THATCHE
FAMILY CIDER MAKERS
FROM MY
TO WOR
Tubo
Yeo Valley
ORGANIC

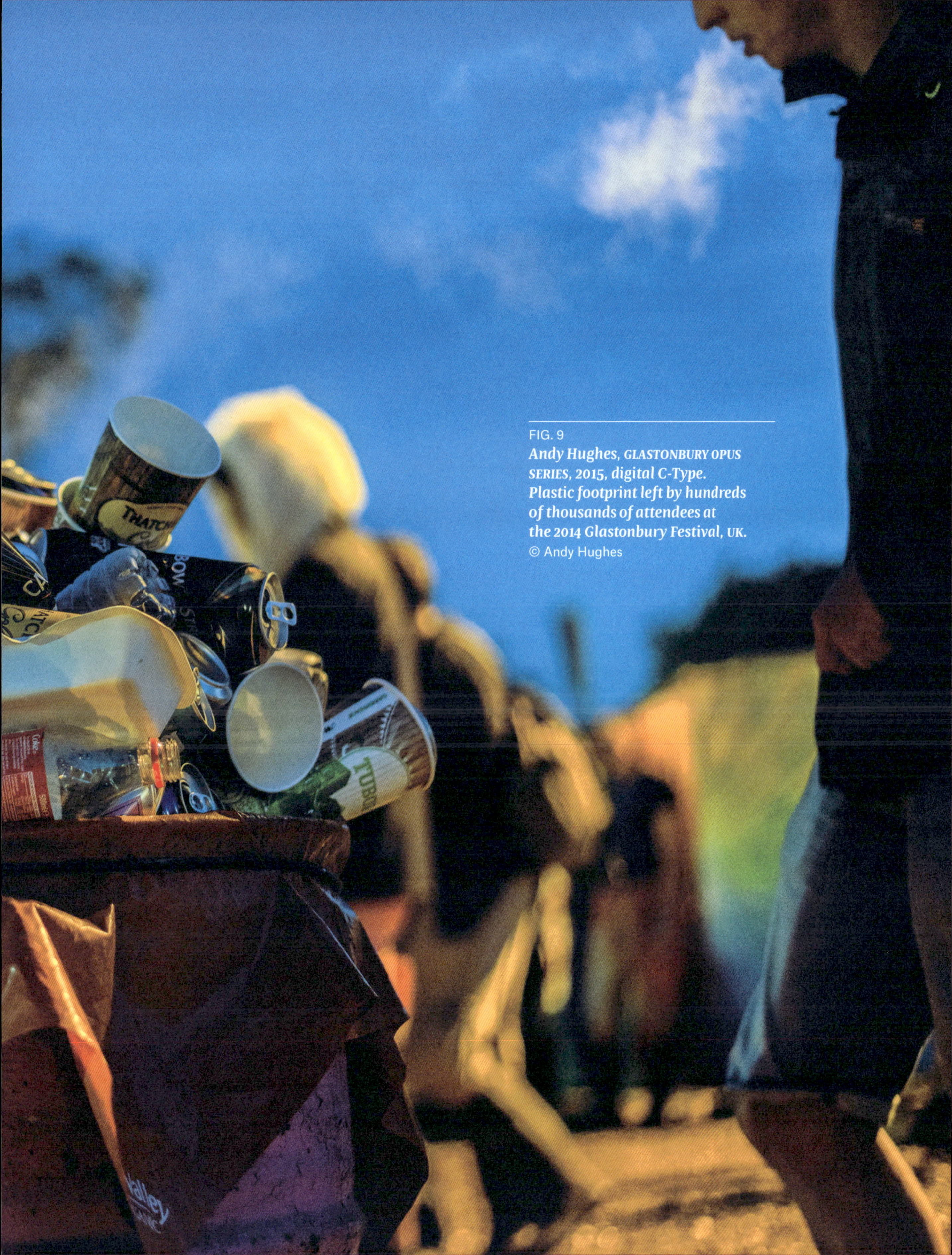

FIG. 9
Andy Hughes, *GLASTONBURY OPUS SERIES*, 2015, digital C-Type. Plastic footprint left by hundreds of thousands of attendees at the 2014 Glastonbury Festival, UK.
© Andy Hughes

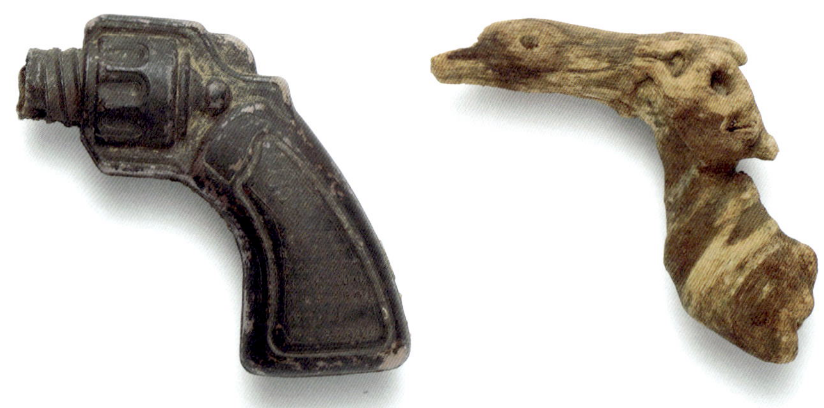

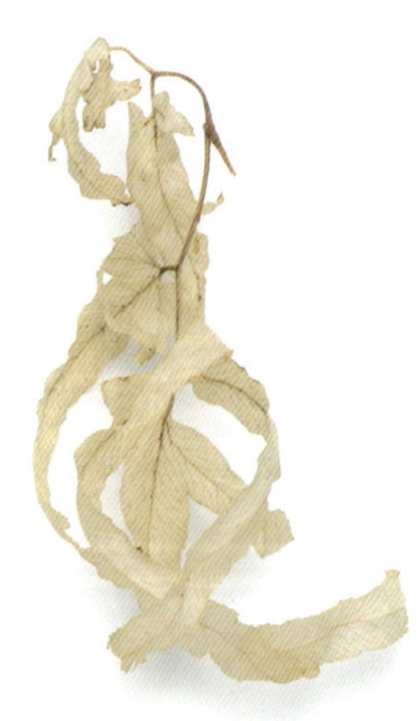

FIG. 10
liina klauss, PLASTIC GUN –
DRIFTWOOD from the INVOLUNTARY
PAIR series. A collection of
plastic artifacts and natural
specimens picked up on
Asian beaches, 2013–2020.
© liinaklauss

FIG. 11
liina klauss, PLASTIC FILM –
FERN LEAF from the INVOLUNTARY
PAIR series. A collection of
plastic artifacts and natural
specimens picked up on
Asian beaches, 2013–2020.
© liinaklauss

FIG. 12
liina klauss, PLASTIC FRAGMENT –
CRAB CLAW from the INVOLUNTARY
PAIR series. A collection of
plastic artifacts and natural
specimens picked up on
Asian beaches, 2013–2020.
© liinaklauss

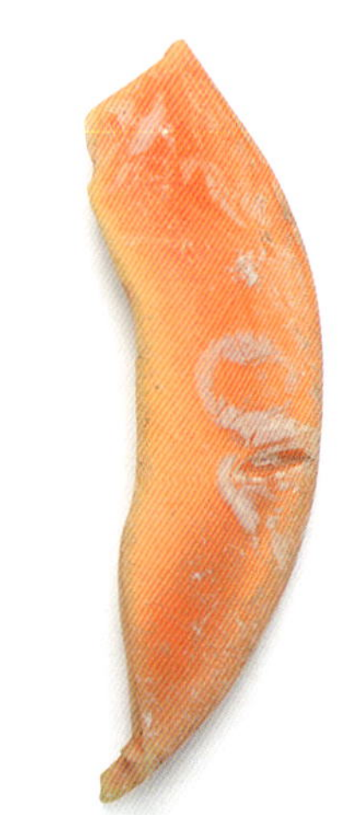

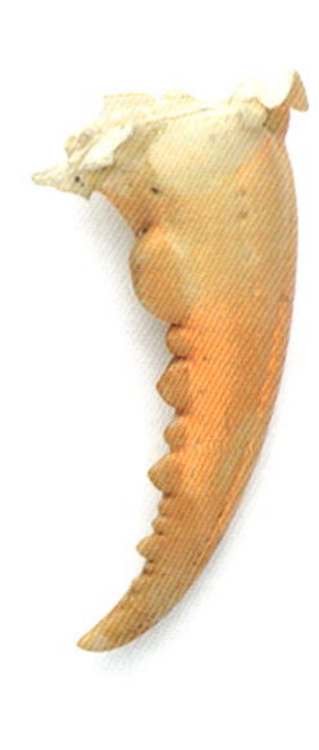

1   Jane Houlihan, Timothy Kropp, Richard Wiles, Sean Gray, and Chris Campbell, Environmental Working Group, *Body Burden: The Pollution in Newborns: A Benchmark Investigation of Industrial Chemicals, Pollutants and Pesticides in Umbilical Cord Blood*, July 14 (2005). Online: www.ewg.org/research/body-burden-pollution-newborns (last accessed September 23, 2020).

2   Ibid.

3   World Wildlife Fund, *No Plastic in Nature: A Practical Guide for Business Engagement*, Report, February 25 (2019). Online: www.worldwildlife.org/publications/no-plastic-in-nature-a-practical-guide-for-business-engagement (last accessed June 2, 2020).

4   Dianna Cohen with Jennifer Wagner-Lawlor, »Plastic is Washed Up,« *Antennae: The Journal of Nature in Visual Culture* 29 (2014): 6–30.

5   Odile Madden and Tom Learner, »Preserving Plastics: An Evolving Material, A Maturing Profession,« *Conservation Perspectives: The GCI Newsletter* 29, no. 1 (2014): 4–9. Online: www.getty.edu/conservation/publications_resources/newsletters/29_1 (last accessed June 17, 2020).

6   Ibid., 9.

7   Ibid., 5.

8   Thomas Morton, »Oh, This is Great! Humans Have Finally Ruined the Ocean,« TOXIC: Garbage Island, *Vice*, April 8 (2008). Online: www.vice.com/sv/article/zn3qky/toxic-garbage-island-1-of-3 (last accessed July 3, 2020).

9   Krzysztof Ziarek, *The Force of Art*, Stanford: Stanford University Press (2004), 29.

10 Ibid., 11.

11 See Heather Davis, »Life & Death in the Anthropocene: A Short History of Plastic,« in *Art in the Anthropocene: Encounters Among Aesthetics, Politics, Environments and Epistemologies*, eds. Heather Davis and Etienne Turpin, London: Open Humanities Press (2015): 347–358; Jennifer Wagner-Lawlor, »Plastic's ›Untiring Solicitation‹: Geographies of Myth, Corporate Alibis, and the Plaesthetics of the Matacão« (forthcoming 2021); Jennifer Wagner-Lawlor, »Regarding Intimacy, Regard, and Transformative Feminist Practice in the Art of Pamela Longobardi,« *Feminist Studies* 42 (2016): 649–688; and Jennifer Wagner-Lawlor, »Poor Theory and the Art of Plastic Pollution in Nigeria: Relational Aesthetics, Human Ecology, and ›Good Housekeeping,‹« *Social Dynamics* 44 (2018): 198–220.

12 Quoted with permission of the artist, Ellyn Maybe.

13 Erik Zettler, »The ›Plastisphere‹: A New Marine Ecosystem,« Smithsonian Institution, OCEAN: Finding Your Blue, July (2013). Online: https://ocean.si.edu/ocean-life/plastisphere-new-marine-ecosystem (last accessed July 5, 2020).

14 See Dianna Cohen, »The Music Industry's Battle Against Plastic Junk,« *RollingStone*, May 19 (2016).

15 Andrew Hughes, *Glastonbury Opus 1* (2015). Online: https://itunes.apple.com/us/book/id1022093606 (last accessed June 21, 2020).

With our essay we would like to introduce some new artistic formats in the context of Hybrid Art, which have emerged in recent years in connection with our activities at the art and research platform Art Laboratory Berlin as well as elsewhere: performative »doing it with others« (DIWO) actions by artists in the form of participatory and hands-on workshops. Since Art Laboratory Berlin opened in 2006 we present, research, and publish on contemporary art at the interface of art, science, and technology.

In recent years we have focused on the fields of art, biology, and biotechnology as well as artistic research. We will introduce three international artists, whose artistic focus is dedicated to the consequences of climate change and the crisis our oceans are facing due to massive plastic pollution: Robertina Šebjanič, Kat Austen, and Mary Maggic. Over the past five years we have presented their latest art projects to audiences in Berlin and internationally with exhibitions, performances, workshops, and conference presentations.

Artistic Research and Ecology: Pollution, Plastic, Water

Regine Rapp and Christian de Lutz

While awareness of pollution and its negative ecological effects on our environment has been around for over a century, the last few decades have brought together various strands of environmental concern and action due to the severe climate and biodiversity crises that can no longer be ignored. This unprecedented crisis caused by humanity has led to some scientists proposing a new geological epoch dating from the commencement of significant human impact on Earth's geology, ecosystems, and climate: the Anthropocene. At the same time, in part due to these concerns and in part due to new technologies and their accessibility, artists have begun to develop new artistic strategies to confront these crises. These new approaches are ways to bridge the epistemic gaps between science and the broader public, and these bridges are used as tools for activism and finding common solutions.

Aquatic Artistic Research for Post-anthropocentric Awareness ___ The element of water – marine ecosystems in particular – and the importance of sound in marine communication systems, as well as the effects of human interventions in acoustic environments of aquatic ecosystems are central to the modes of artistic research of Ljubljana-based artist Robertina Šebjanič. Her work involves close collaboration with marine biologists from around the world and she utilizes a multitude of artistic media, but primarily interactive sound installations and live performances.

In her AURELIA 1+HZ series of works, Robertina Šebjanič engaged with both biopolitical and technological attempts to prolong life as well as advancing new critical reflections on interspecies cohabitation. With AURELIA 1+HZ she explored agency and sentience in one of the »simplest« of all multicellular creatures, the jellyfish, bringing it into a relationship with a machine. The artist chose to work with jellyfish because of their usefulness in aging research and also because she is fascinated by the fact that these creatures have existed on Earth for over 500 million years. This timescale undoubtedly provides us with a yet more differentiated view on the phenomenon of the Anthropocene.

Her interactive installation AURELIA 1+HZ/PROTO VIVA GENERATOR (2014) proposes the mutual coexistence of animal and machine (FIGURE 1). In contrast to modern robots, which are mostly driven by artificial

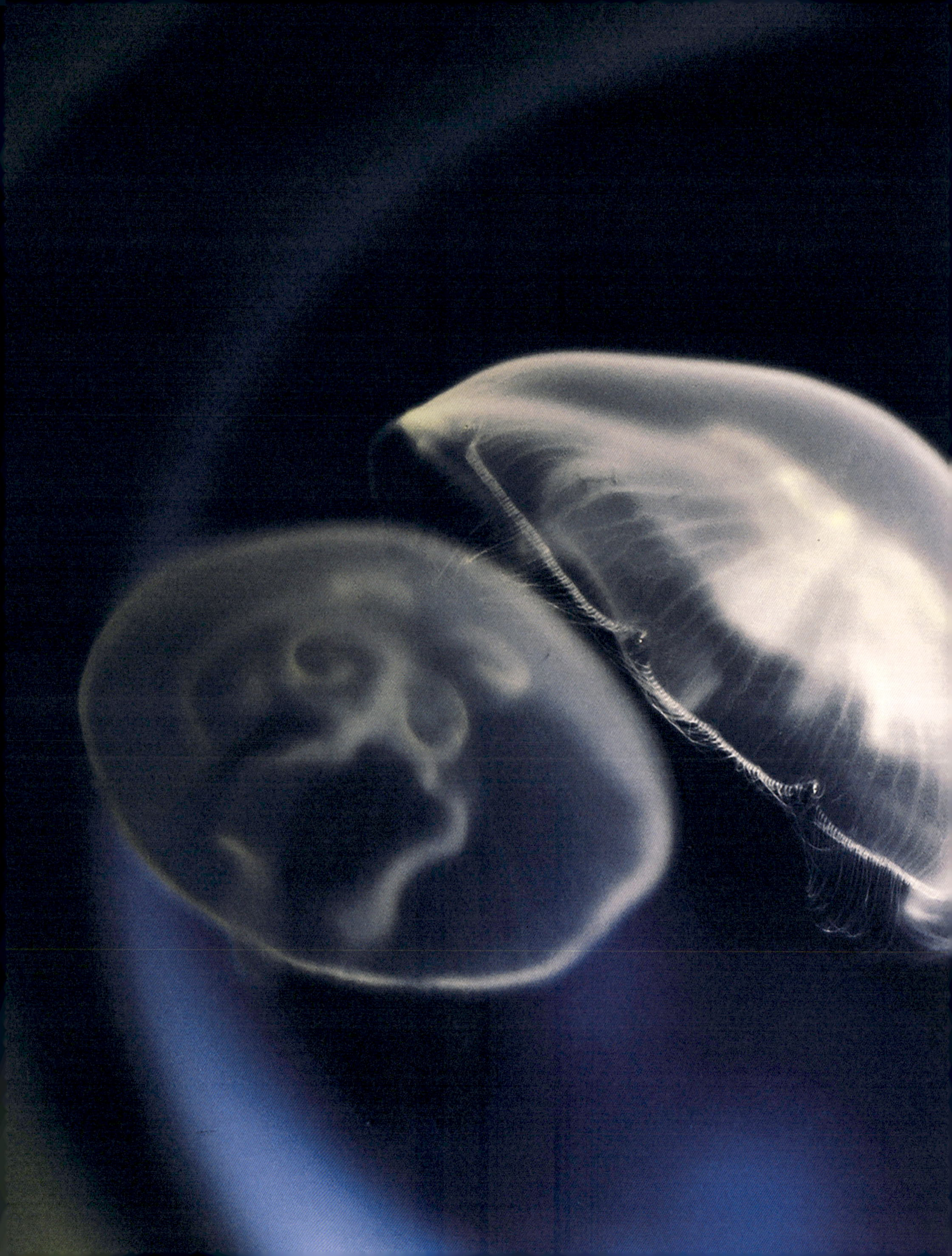

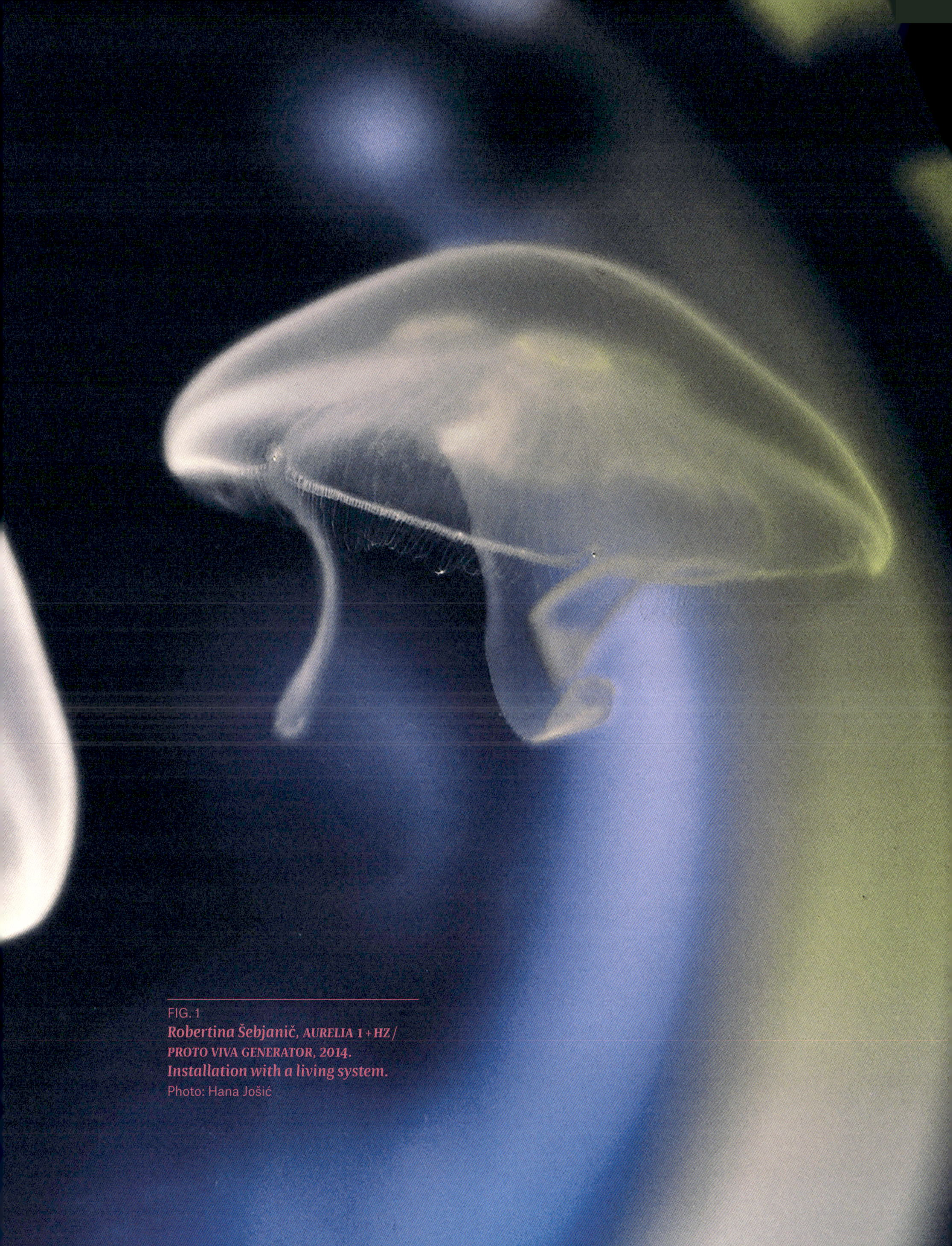

Robertina Šebjanič, *AURELIA 1+HZ/ PROTO VIVA GENERATOR, 2014. Installation with a living system.*
Photo: Hana Jošić

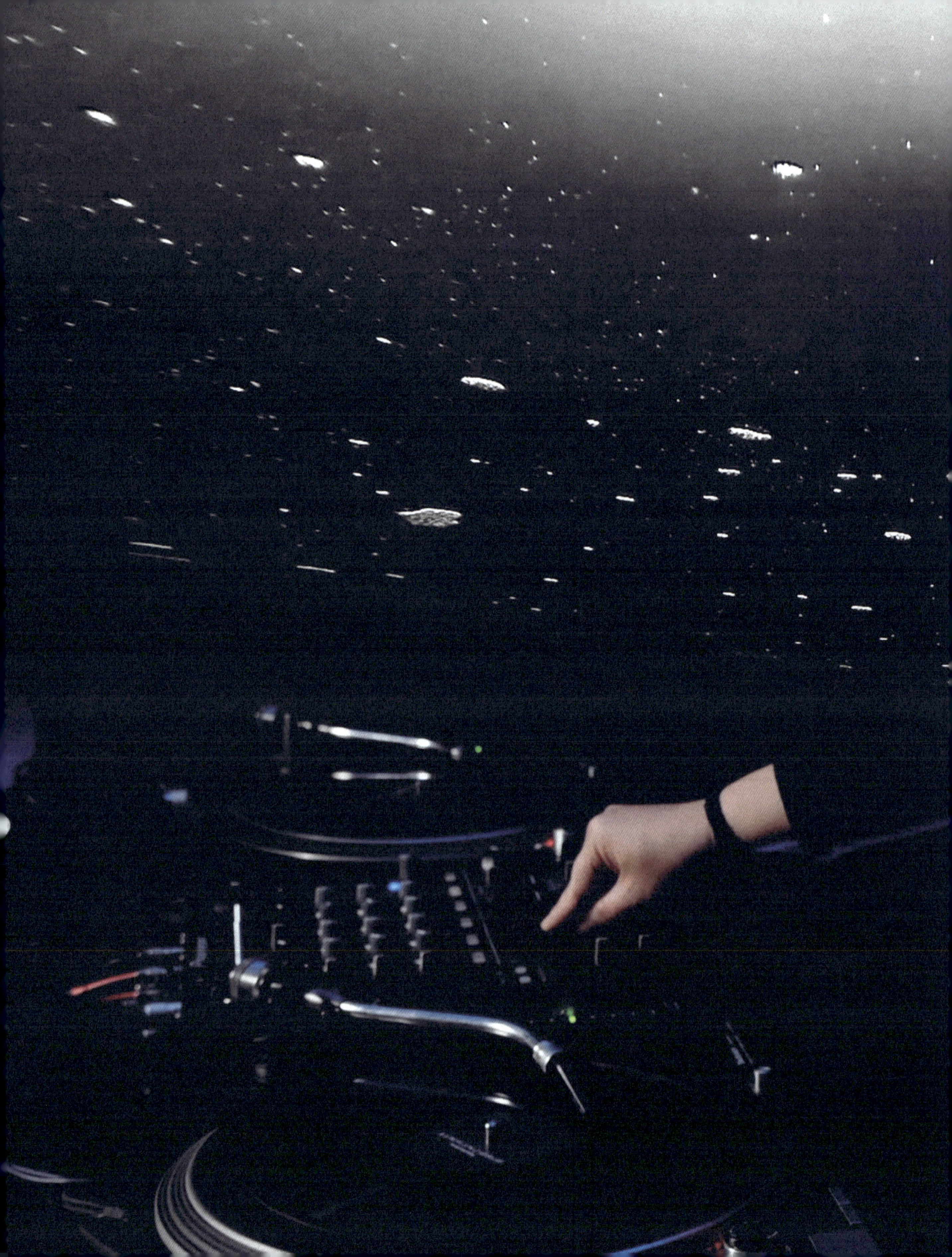

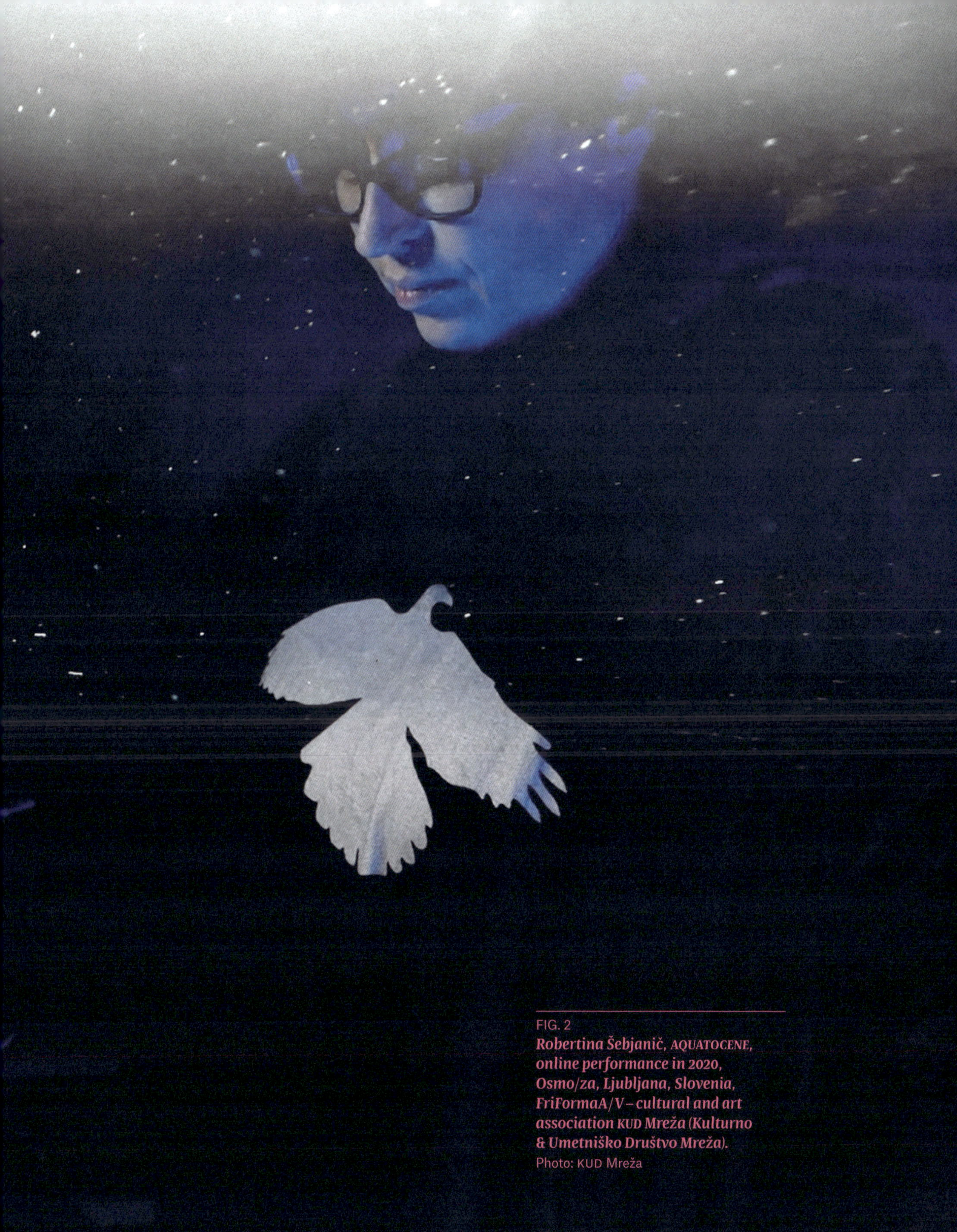

FIG. 2
Robertina Šebjanič, AQUATOCENE,
online performance in 2020,
Osmo/za, Ljubljana, Slovenia,
FriFormaA/V – cultural and art
association KUD Mreža (Kulturno
& Umetniško Društvo Mreža).
Photo: KUD Mreža

intelligence, this project uses a living organism to bring life to a simple machine, and in a way to express itself through the machine. Living jellyfish actually run the installation: its mechatronic parts, video and sound. The movements and contractions of the jellyfish are recorded by an HD camera. The captured data is then transformed in real time into code, which in turn navigates the mechanisms of the installation.[1]

44

Noise Pollution and Aqua-Performative Art ___ Robertina Šebjanič's long-term project AQUATOCENE/SUBAQUATIC QUEST FOR SERENITY is concerned with human underwater sound pollution and is based on a large number of underwater sound recordings made at different European sites along the Atlantic Ocean and the Mediterranean Sea. The artistic media chosen are various—from interactive sound installations to live performances and vinyl albums. In this project the artist's approach is most innovative, she spotlights a nonhuman perspective in which the anthropocentric dilemma fully unfolds: »Sound is the main communication tool for most marine animals and plants that dwell in the deep darkness of the world's oceans and seas,« says Robertina Šebjanič, »Despite widespread knowledge of certain aquatic sounds, like those produced by animals such as whales, the public is predominantly not aware that the underwater soundscape is as rich as our terrestrial one.«[2]

There are many in-depth scientific investigations of the severe and harmful impact of anthropogenic noise on fragile marine acoustic environments. Noise pollution, which is caused by human activity, has increased intensely over the last century.[3] Whereas naturally occurring environmental noise includes water movement and weather conditions, as well as the noise made by marine fauna and tectonic movement, the sources of anthropogenic underwater noise are numerous: »Anthropogenic noises include the sound of watercraft (from jet skis to supertankers); offshore oil/gas exploration and production noise; sonar—especially military high-power equipment,« says marine bioacoustics specialist Michael Stocker, »underwater telemetry and communication for mineral exploration and research; fish ›bombing‹ and other underwater explosives; civil engineering projects, and overflying aircraft.«[4]

Robertina Šebjanič's art projects are a noteworthy example of emerging Hybrid Arts practice: She rigorously researches marine ecosystems, collaborates with marine scientists, and collects underwater sound recordings using both scientific and hacked hydrophones. She then creates spatially and temporally intense artworks in which the audience is able to immerse, acoustically and aesthetically, in underwater soundscapes and reflect on topics related to the issues while experiencing the project.

SUBAQUATIC/AQUATOCENE SOUND SCAPE was presented for the first time at Art Laboratory Berlin in 2016 as a minimalist audiovisual installation with five audio pieces with headphones and a subtle black-and-white projection of flowing water. From her accumulated archive she shared underwater recordings from coastal waters off places like Roscoff in France, Bergen in Norway, Izmir in Turkey, Dubrovnik in Croatia, and Koper in Slovenia with gallery visitors.[5] »Looking at a video showing water currents, associative cross-references are created between image and sound,« says Berlin-based art theoretician Carola Hartlieb about the sound installation, »allowing ships to pass by in the mind's eye, tourists to splash around in the water, or rhythmic sounds of sea creatures to appear vividly.«[6]

While in the 2016 exhibition at Art Laboratory Berlin, the visitors could perceive the sound of each on-site recording in its authentic mode and reflect on the phenomenon of human sound pollution and the negative effects of underwater noise, the piece later evolved into a live performance with mixed recordings.[7]

This follow-up project challenges the audience of Robertina Šebjanič's performance with an auditory palimpsest: she has collected and recorded underwater sound and condensed it into a single coherent experience. Different units of recorded subaquatic data from sites on different oceans, at different times, and in different conditions, are melded together. The sounds are bubbling, clicking, squealing, splashing, gurgling, rushing, chattering, hissing. Using DJ aesthetics the artist transforms the familiar perception of underwater field recordings into another level of sound experience, mixing the sound of marine environments with the help of turntables and electronics (FIGURE 2). The artist de- and recontextualized her field recordings as the subaquatic sounds were finally pressed on vinyl.

Robertina Šebjanič presented the AQUATOCENE performance at Art Laboratory Berlin in March 2019 in the context of the group show WATERY ECOLOGIES together with artist Kat Austen, who performed a DJed version of her own work on seawater with THE MATTER OF THE SOUL, which we shall elaborate on in the following section.[8]

Plastics, Art, and Workshopology ___ Berlin-based artist Kat Austen, who holds a PhD in Chemistry and has a background in science journalism, combines scientific knowledge, hacked equipment, and ethnographic research. Crucial to her artistic work is the experience of living in a time of dramatic climate change, such as the melting of the Arctic ice shield. Her approach to artistic research is driven by the desire to develop more empathy on our part for the effects of climate change. In her project THE MATTER OF THE SOUL (2017–ongoing), exhibited at Art Laboratory Berlin in 2019, she examines the impact of climate change in the Canadian High Arctic through a multimedia sound work, sculpture, and performance (FIGURES 3—5).[9]

During the last few years Kat Austen has also been working on the subject of microplastics in the wild, first in sea life, then in urban waterways, and more recently in trees. An important part of her artistic research and practice has been an evolving series of workshops in collaboration with scientists, artists, and DIY science practitioners. Kat Austen's artistic hybrid practice of chemistry, open science, environmentalism, and activism finds a strongly engaging performative format in these workshops. While the workshop format is prominent in DIY and open science as a format for education and the exchange of tools, protocols, and information, since the early 2000s it has also played an important role in hybrid art practice.[10] From props drawn from artistic practice to a performative structure intended to draw out discourse, we can look at workshops such as these as a further extension of performance art, but one that includes all participants (human and nonhuman) in developing new social structures for exchange and action. Workshops also act as a hinge between artistic research and production. And the workshop itself is a place for gathering various forms of input from participants, who range from qualified scientists in various fields to citizen scientists and DIY enthusiasts, hackers from various disciplines to inexperienced lay people, who may be newbies to the process, but who often have a lot to offer from their own varied life experiences.

FIG. 3
Kat Austen, THE MATTER OF THE SOUL, 2018, audiovisual installation (back), and audio cassettes relief sculpture and canvas cassette case (front). Installation view during the 2019 exhibition WATERY ECOLOGIES at Art Laboratory Berlin, Germany.
Photo: Tim Deussen

FIG. 4
Kat Austen, THE MATTER OF THE SOUL, 2018, video screen shot of the audio-visual installation during the 2019 exhibition WATERY ECOLOGIES at Art Laboratory Berlin, Germany.
Photo: Kat Austen

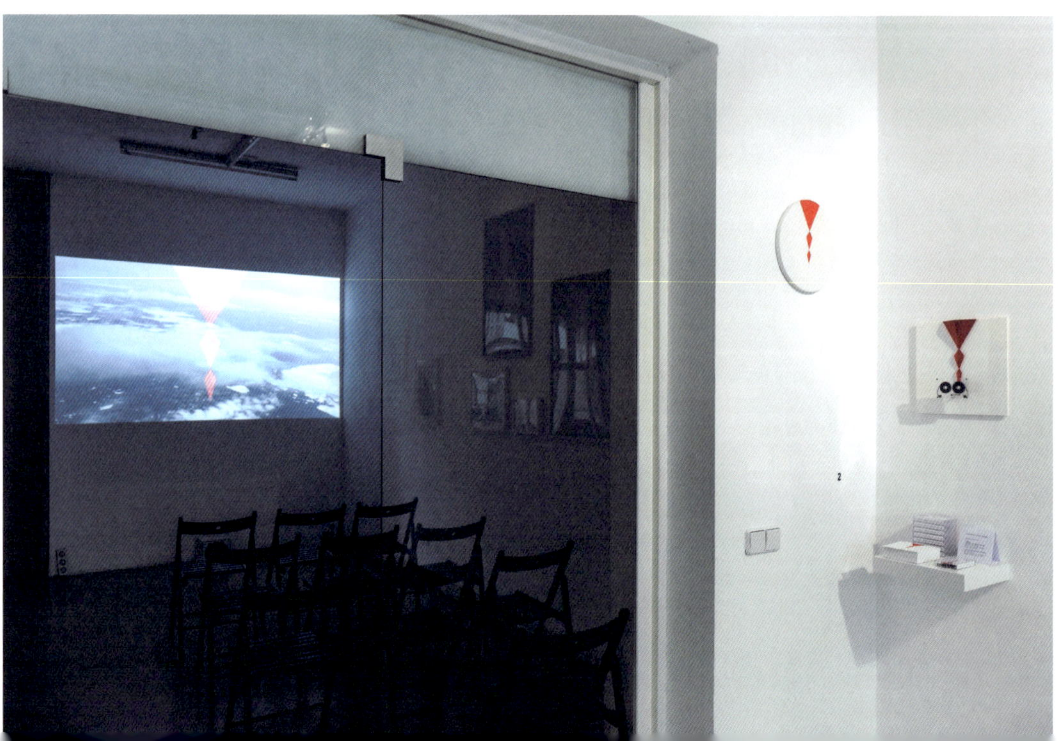

(Un)Real Ecologies and Microplastics __ Kat Austen has conducted several workshops on microplastics: Together with scientist, DIY practitioner, and intermedia artist Gjino Šutič[11] she developed a series of workshops to isolate microplastic particles. In 2018 she developed another microplastics workshop together with Berlin-based microbiologist Joana MacLean who works at the German Research Centre for Geosciences in Potsdam. (UN)REAL ECOLOGIES explores the prevalence of microplastics in the Panke, a small river that flows through Brandenburg and Berlin. This workshop was the first event developed by the art and science collective DIY Hack the Panke, based at Art Laboratory Berlin, whose purpose is »to explore the Panke River for living organisms and critically examine its complex history of human use. Members of DIY Hack the Panke plan public workshops on topics such as river flora, fauna and microbiology, plastic waste, and other pollutants.«[12]

(UN)REAL ECOLOGIES was a two-day workshop to which Kat Austen and Joana MacLean invited a dozen participants to explore a »new understanding of the reality of the Panke's ecosystem, with plastic present and wholly a part of it – a microcosm that allows us to ask: What is nature?«

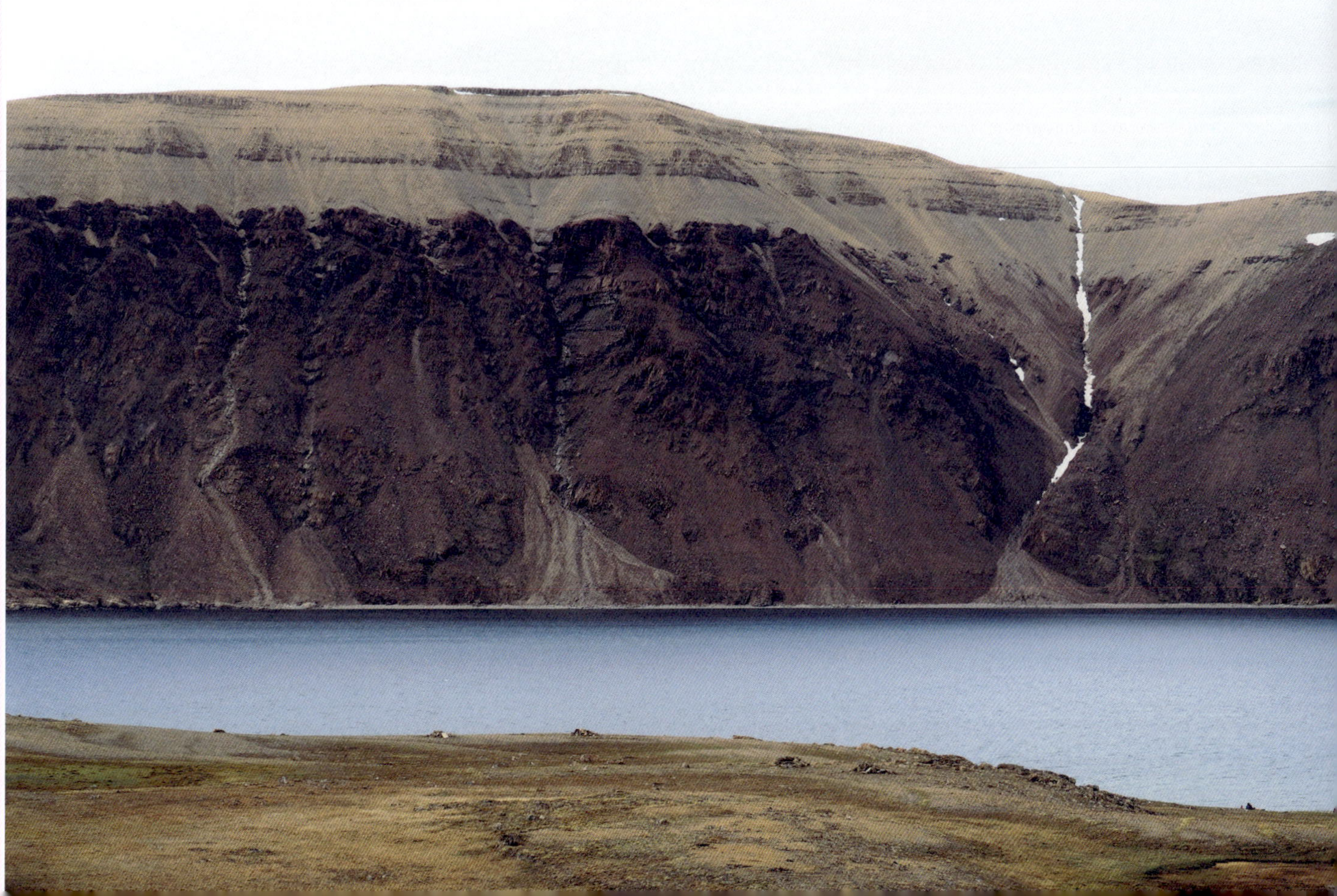

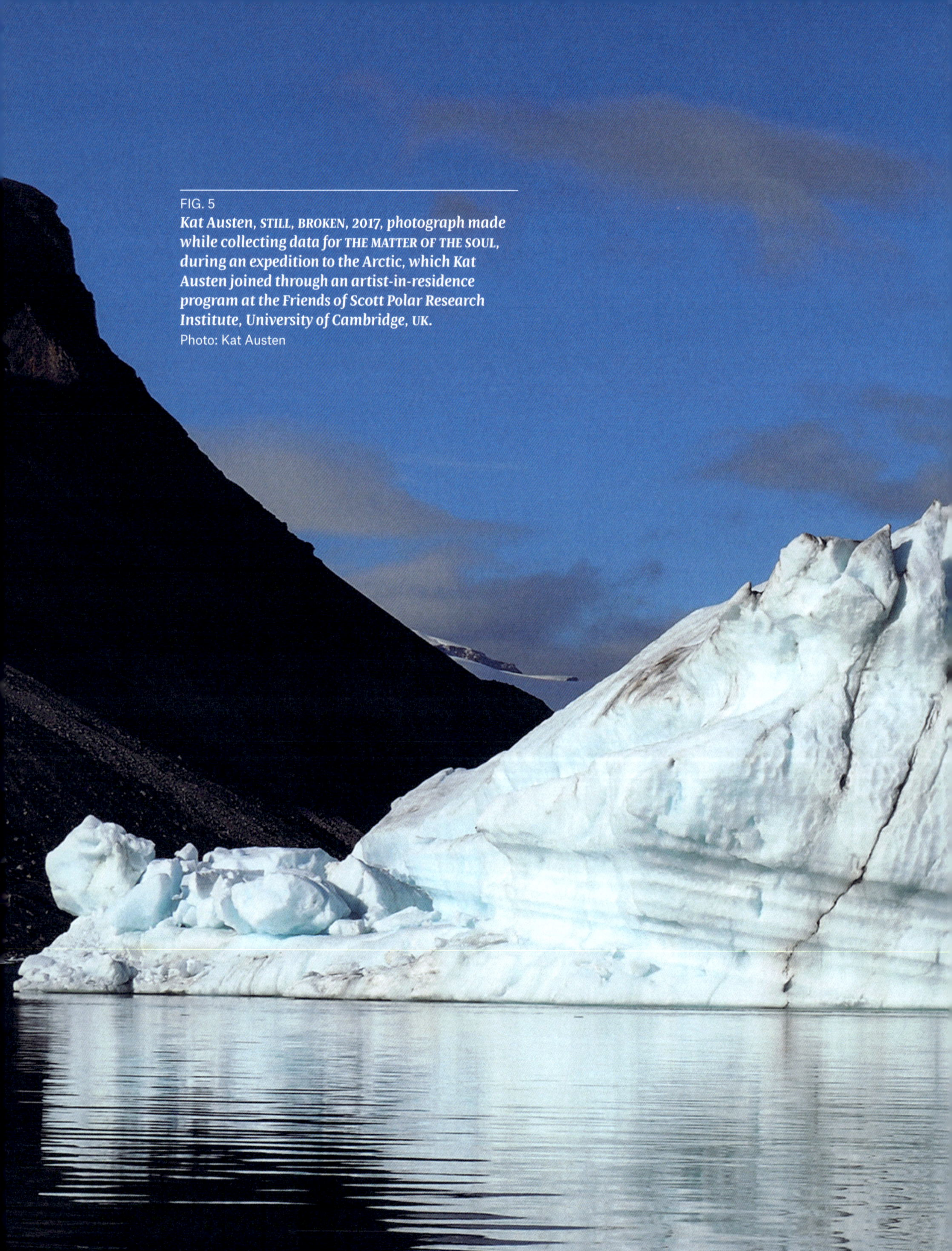

FIG. 5
Kat Austen, STILL, BROKEN, 2017, photograph made
while collecting data for THE MATTER OF THE SOUL,
during an expedition to the Arctic, which Kat
Austen joined through an artist-in-residence
program at the Friends of Scott Polar Research
Institute, University of Cambridge, UK.
Photo: Kat Austen

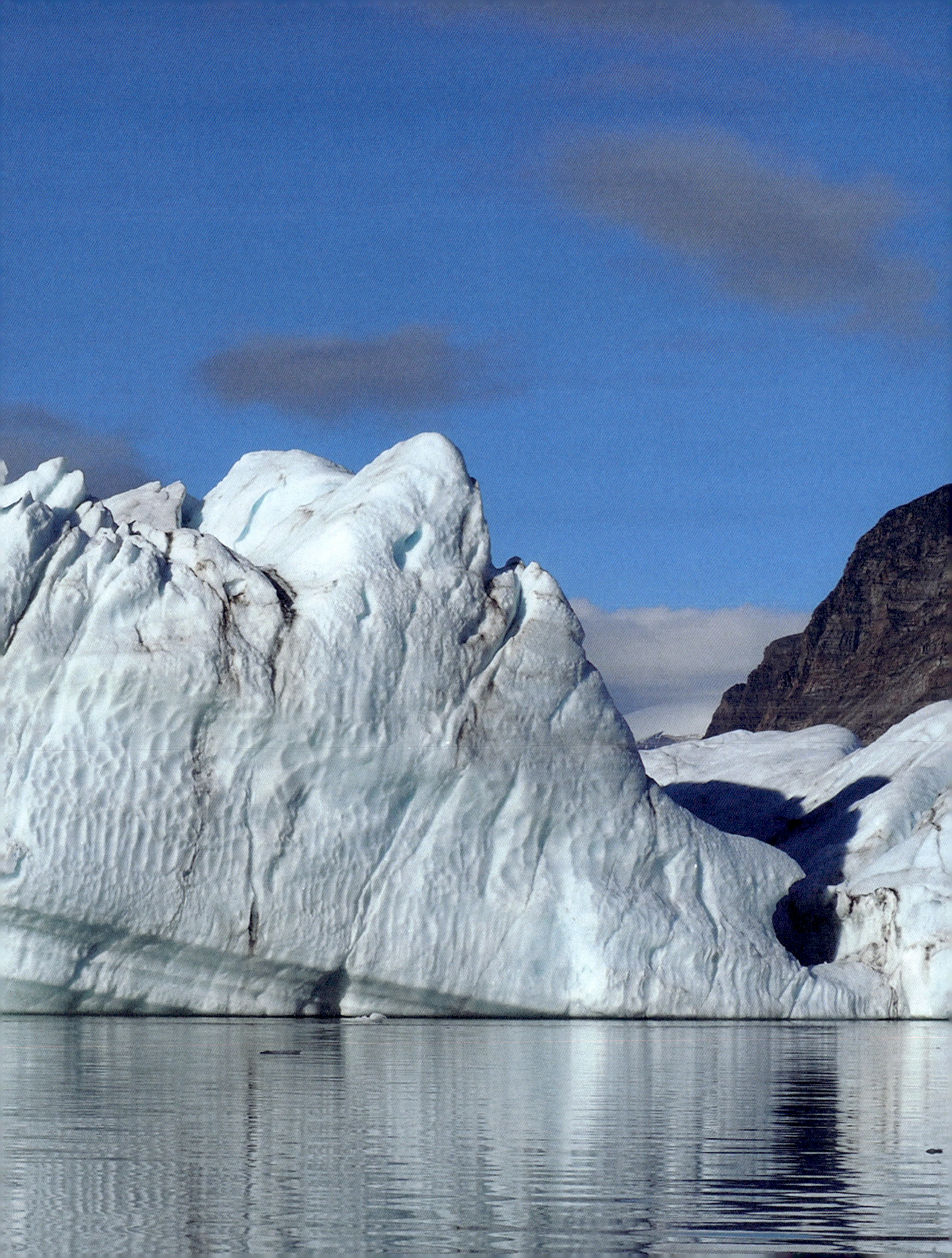

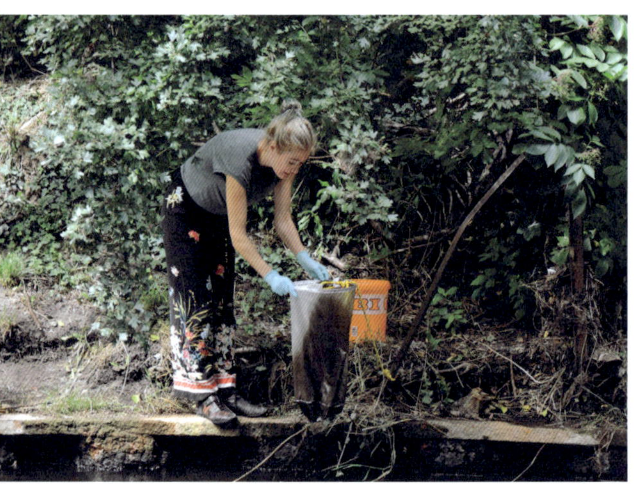

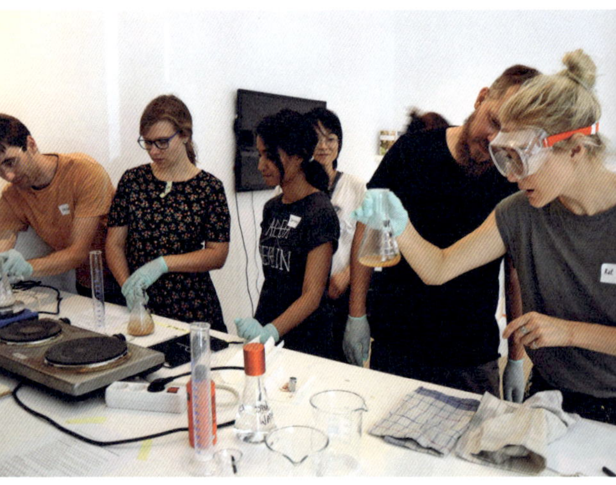

as well as how do »organisms and microorganisms exist with and construct with these human-made materials?«[13]

The group investigated two sections of the Panke in the Berlin borough of Wedding: upstream was a stretch of the river near the border of the borough of Pankow; downstream was near Gerichtstrasse, not far from where the Panke goes underground. Before the workshop started, fine nets, used for DIY beer fermentation, were placed in the river to collect drifting debris (FIGURE 6). At the beginning of the workshop these nets were collected, along with several buckets of small and medium-sized discarded objects ranging from plastic toys to phone cables and rusty iron bars. Back at a pop-up DIY chemistry lab at Art Laboratory Berlin, the participants were divided into two groups. One team examined the larger plastic and other items for biota—algae and water plants, as well as microscopic life—using a digital microscope. Meanwhile, the other team sieved the samples from the beer nets. Larger objects and even some organisms, such as small freshwater shrimp, were discovered.

Objects larger than the net mesh but smaller than all but one sieve, were taken by the second team for a two-step chemical protocol: subjected to wet hydrogen peroxide oxidation, where hydrogen peroxide and iron were added and heated to 75° C to dissolve organic matter. The solution was

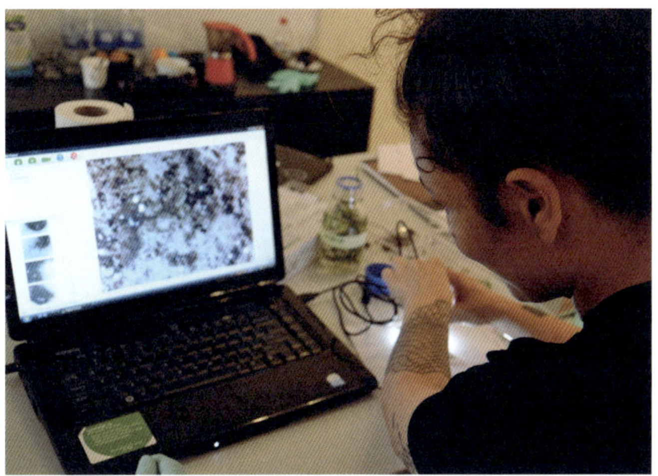

again sieved and washed. What remains is either minerals or microplastics, and can be stained for differentiation under a microscope. Finally, determining the source and type of the microplastics was done using a density column, in this case a DIY column made of vegetable oil, honey, agave syrup, and dishwashing soap. As each liquid has a different density, exactly where the specific plastic particles settle determines its composition.[14] The teams changed roles so that all participants could undertake both chemical and microscopic investigations. In this way Kat Austen and Joana MacLean were able to try wet hydrogen peroxide oxidation with both lab quality peroxide (30%) and a 9% solution from hair dye on samples from both upstream and downstream (FIGURES 7—8).

However, this description omits a more complex set of activities and social interactions: the participants came from a variety of fields and included designers, artists, open science enthusiasts, and also an employee of a local sanitation firm who was working with a team to recycle plastic waste. Finally, in-depth discussions took place about the long-term implications of the Anthropocene, and new definitions of our common understanding of »nature« and »culture.« Although the workshop investigated a specific locality, during the workshop, participants were made aware of the ubiquity of microplastics in the environment, from the deepest oceans to the highest mountains. »Even in geology rocks bear traces, or an inscription, of their history, determining and being determined by the activities of the creatures that reside, pass through, live and die within particular environments,« New York based scholar Heather Davis notes in her essay »Plastic: Accumulation without Metabolism,« »This sense of ecology, tied to the notion of place-making, is defied by a material such as plastic. There is no ›local‹ for plastic. Instead, plastic exists everywhere and anywhere.«[15]

The following year (UN)REAL ECOLOGIES was repeated several times with
students from the Gustav-Freytag secondary school in Berlin[16] and as
a public workshop.[17] In April 2020, during the first COVID-19 lockdown,
Art Laboratory Berlin invited Kat Austen and Joana MacLean to organize
an online workshop on »Microplastics and Coexistence,« which took the
topics of previous live workshops to a new discursive level. Kat Austen
presented previous projects and her ongoing project *Stranger to the Trees*,[18]
which explores the interaction between birch trees and microplastics
through controlled artistic and scientific experiments, both in situ in forests in Germany and in Poland. The question was whether living organisms
and systems and microplastics can coexist constructively. Joana MacLean's
research is concerned with the question of how microorganisms live
on plastics and microplastics, and whether naturally occurring microorganisms can remediate plastic.

 During the online event we again became aware of the potential
of the open format of DIY approaches, as Kat Austen notes, »[They]
allow us to rediscover our agency in the world, the ability to research
and make sense of the world is to have agency within it, and when you
are aware of your agency you are more able to and likely to act, and to
act in a constructive way that will change the problem you are looking at.
And so for me, the development of these DIY techniques is a political act.«[19]

 The complex combination of DIY science, community building,
making, discussion, and exchange are all vigorously organized by Kat
Austen, whose workshops are always performative as well as informative,
interactive, and didactic. The act of doing (and making) together, stresses
both aesthetic and ethical approaches to our interaction with the world
around us, with implicit political goals. Her workshops and installations,
sculptures and sound pieces form a synthesis of what otherwise might be
considered dueling epistemologies from the natural sciences and the arts.

Queering the River and Watery Entanglements __ Another artist whose
artistic approach strongly and rigorously combines workshops with a
DIY science character is Vienna-based nonbinary Chinese-American artist
Mary Maggic. Key aspects in their works of recent years are issues like
hormone biopolitics and environmental toxicity – and the intention to
de-hierarchize and share specific scientific insights with a broader,
non-specialized public. Mary Maggic contributed important aspects
about their artistic discourse in the talk »From Molecular Colonization
to Molecular Collaborations« at Art Laboratory Berlin's symposium
»Nonhuman Agents« in fall 2017.[20] The artist also played an integral role
during the exhibition WATERY ECOLOGIES at Art Laboratory Berlin in 2019
and in the exhibition and conference »*THE CAMILLE DIARIES*: New Artistic
Positions on M/otherhood, Life and Care« in summer 2020, presenting the
multimedia installation *MILIK BERSAMA REKOMBINAN (RECOMBINANTS COMMONS)*,

which reflects on the symbiotic web in which humans, plants, animals, and the (polluted) environment are entangled on molecular, organic, ethical, and biopolitical levels (FIGURES 9—10).[21]

The installation is an outgrowth of an artistic research project *River Gynecology*, which was the result of a ten-month Fulbright residency in Yogyakarta, Indonesia, in 2019. Both the residency and the artistic research project have their origins in the artist's participation in Hackteria Lab 2014,[22] hosted by the art and science collective Lifepatch and the international DIY biology platform Hackteria. Mary Maggic's artistic research and practice explores the hormonal outcomes of water pollution through diverse media, not least via hybrid workshops that utilize both artistic strategies and DIY science protocols. The artist stresses the importance of the Lifepatch collective who utilize art, science, and technology workshops to involve their local communities in finding solutions for environmental and social problems. Mary Maggic was introduced to both Lifepatch's practice and their involvement with projects along the Code (cho-deh) river at the 2014 DIY science festival. The 2019 residency was located at Lifepatch and the artist collaborated with members of the collective, as well as scientists at Gadjah Mada University in Yogyakarta.[23]

The installation includes living biomaterial and consists of several elements: Two vast wall pieces on both exhibition walls present *objets trouvés* at eye level with the visitors — numerous pieces of garbage (such as tins, paper, cups, rusty nails) taken from the Code river and attached to latex (the piece's olfactory output is one of several multisensorial challenges for the viewer). Central to the installation is a bamboo sculpture in the form of the river, constructed at knee height on metal stands, it meanders through the long exhibition space. It is filled with agar dyed with Remazol Blue (RBBR), a textile dye referencing the role of the textile industry in both Yogyakarta's economy and the pollution of the Code River. On this »blue river« there are also 20 Petri dishes containing both agar and RBBR as well as live oyster mushroom mycelium, which literally remediates the dye, turning the Petri dishes from blue to light pink during the exhibition. The use of oyster mycelium as fungal bioremediation becomes a natural form of river purification. The chosen placement of the Petri dishes on the bamboo sculpture refers to places along the river where Mary Maggic and Lifepatch carried out field research.

At the far end of the installation, a large bright and colorful video projected on the back wall becomes visible: a circle rotating in both directions, showing both images of found plastic waste and diagrams referring to the molecular fusion between plastic waste and the human body, in an aesthetic style often used by the artist. This mandala-like spinning disk of constantly rotating plastic packaging (and its kaleidoscope-like effect) point to the artist's concept of »molecular queering.« The symmetrically set-up installation is accompanied by an omnipresent audio piece:

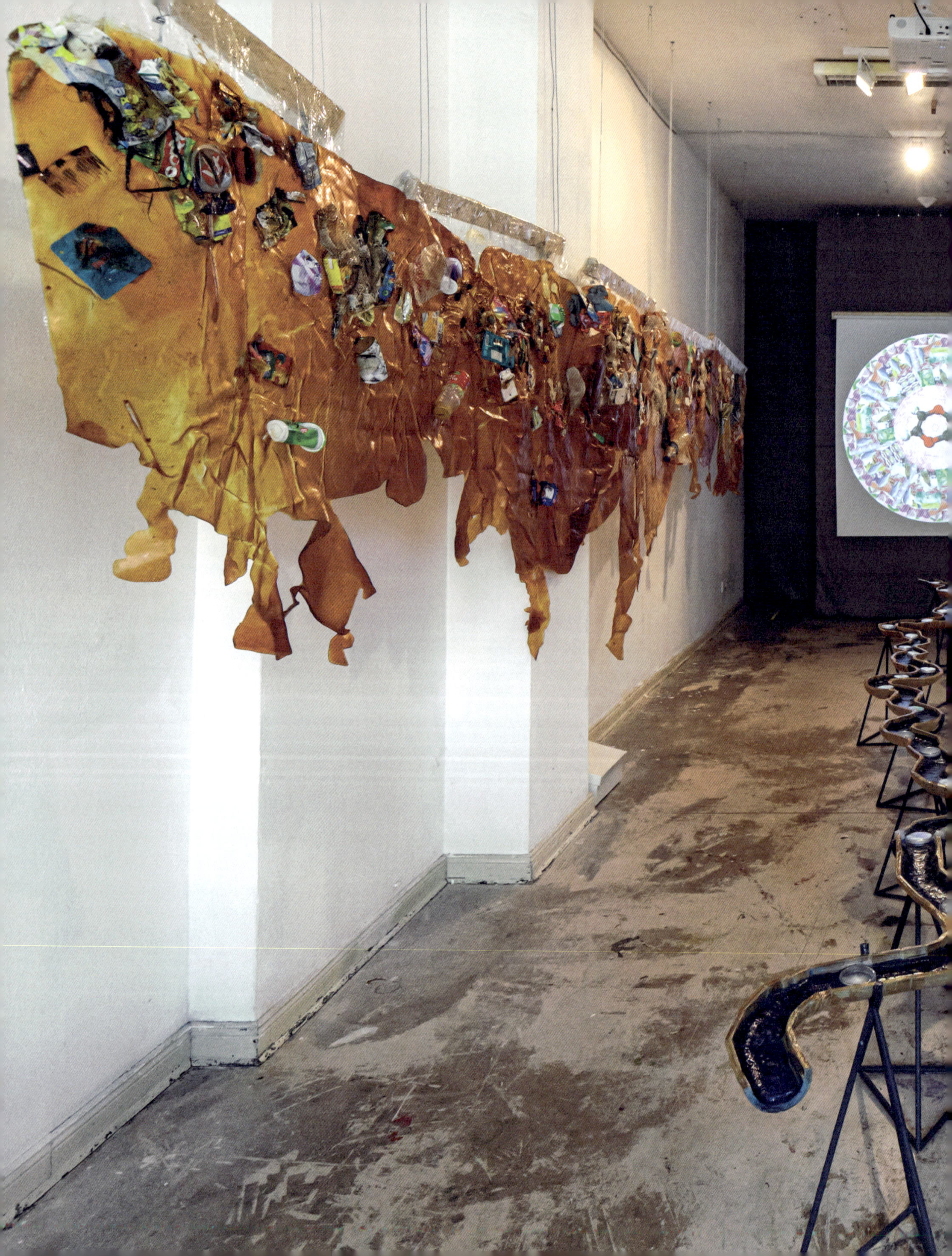

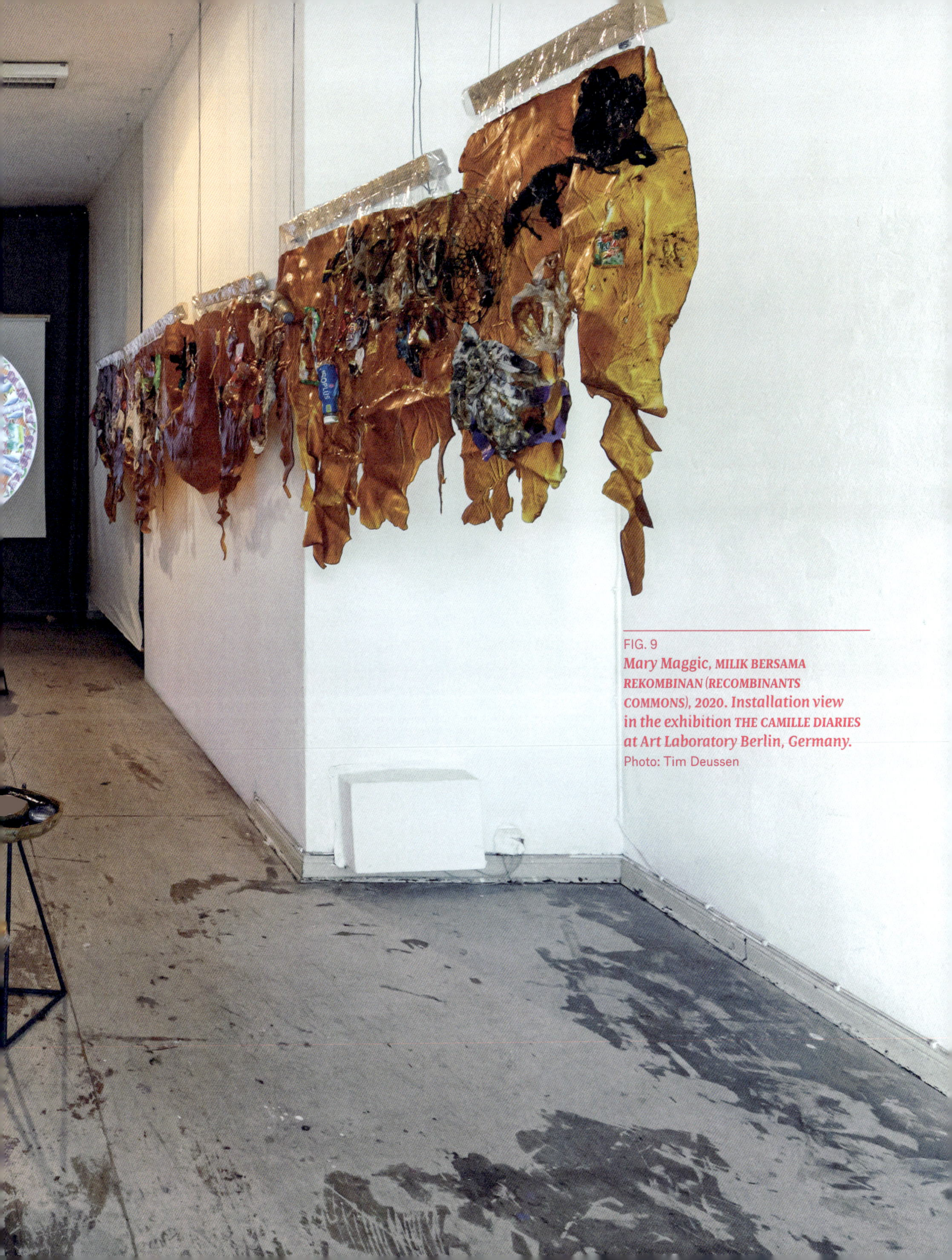

FIG. 9
Mary Maggic, MILIK BERSAMA REKOMBINAN (RECOMBINANTS COMMONS), 2020. Installation view in the exhibition THE CAMILLE DIARIES at Art Laboratory Berlin, Germany.
Photo: Tim Deussen

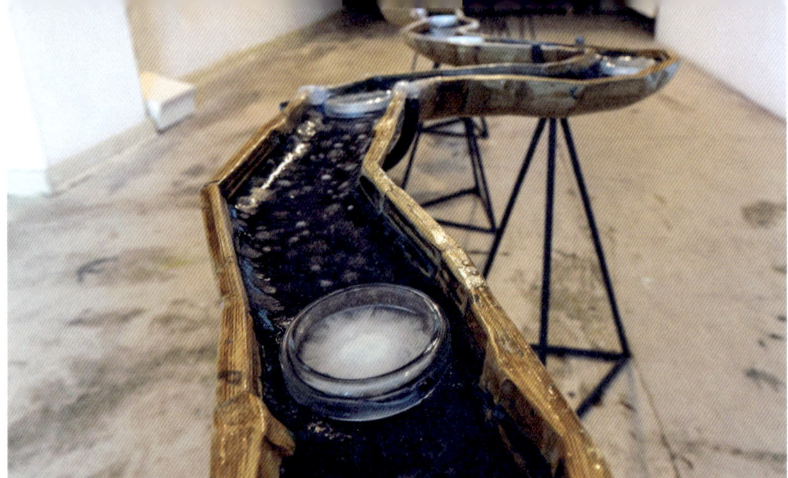

FIG. 10
Mary Maggic, MILIK BERSAMA REKOMBINAN (RECOMBINANTS COMMONS), 2020. Detail of the installation, river sculpture with live oyster mycelium at Art Laboratory Berlin, Germany.
Photo: Tim Deussen

»How the World Stopped Moving« resembles a tale of the Anthropocene in which the river speaks to a little girl about the pain of »digestion« because of all the plastic in the river's belly.

The artist explains the complexity of this art project based on research in situ as follows: »At first glance, River Code in Yogyakarta is a surreal landscape colonized by plastic, with its marginalized citizens tied intimately with the natural water sources for daily life, i.e., cooking, cleaning, fishing, and playing.« It also refers to the interconnectedness of urban ecology and biopolitics: »While the root of the problem is complex and multifaceted (lack of government infrastructure, pollution as colonialism), it can be seen as a cultural issue that requires new strategies for soliciting empathy in toxic-becoming.«[24]

Mary Maggic's concept of »molecular queering« resonates with the writings of Heather Davis who in her essay »Toxic Progeny: The Plastisphere and Other Queer Futures« notes that toxicity »forces us to reveal the ways in which we are multiply composed — of plastic, of toxins, of queer morphologies. [...] In seeking to refashion the molecular structure of organic and inorganic compounds, we believed so much in our own hubris that we seemed surprised to encounter negative consequences. But so many of us already know that this is a fantasy that can no longer be sustained.«[25]

Conclusion ___ The artists we introduced use technology to connect art and science, society, and ecology. Their projects combine artistic research and myriad forms of artistic practice: multimedia installations, performances, and performative DIWO actions in the form of workshops. These workshops can be thought of not only as educational, but also as vehicles for mutual exchange and development between the artist and the public, in which the participants also become performers (and in some cases so do natural objects, subjects, and materials). The multimedia installations produced by these artists combine new techniques and media from science and technology, but use them *critically* to ask complex questions about our world today through the deft use of a mixture of aesthetics and ethics, philosophy and politics, knowledge sharing and activism.

1   From a conversation between the artist and Regine Rapp, August 2016.

2   https://robertina.net/aquatocene/ (last accessed January 1, 2021).

3   Michael Stocker, »Fish, Mollusks, and Other Sea Animals' Use of Sound, and the Impact of Anthropogenic Noise in the Marine Acoustic Environment,« *Journal of the Acoustical Society of America* 112, no. 5 (2002); DOI: 10.1121/1.4779979 (last accessed January 22, 2021).

4   Ibid.

5   Robertina Šebjanič's solo exhibition »Nonhuman Subjectivities: Aural Aquatic Presence« ran from September to October 2016 at Art Laboratory Berlin, see www.artlaboratory-berlin.org/html/eng-exh-41.htm (last accessed January 22, 2021).

6   Carola Hartlieb, »*Auratische Unterwasserklänge: Robertina Šebjanič bei Art Laboratory Berlin*,« Art-in-Berlin online portal, September 9, 2016, www.art-in-berlin.de/incbmeld.php?id=4040 (last accessed January 23, 2021). English translation by the authors.

7   The artist presented the *Aquatocene* performance at Art Laboratory Berlin in March 2019 and in the context of a FriForma event in December 2020, later as an online performance.

8   www.artlaboratory-berlin.org/html/de-event-archive.htm (last accessed January 22, 2021).

9   www.katausten.com/portfolio/the-matter-of-the-soul/ (last accessed January 20, 2021).

10 The workshop as performative practice and activist exchange goes back at least as far as 2002 with the tactical media workshops of Critical Art Ensemble (CAE): http://critical-art.net/halifax-begs-your-pardon-halifax-nova-scotia-tactical-media-workshop-2002/ (last accessed January 20, 2021). More interesting for the current text was CAE's installation *Molecular Invasion* at the Corcoran Gallery of Art, Washington, D.C., 2002–2004, created along with Beatriz da Costa and Claire Pentecost, which was developed together with biotechnology students in a precursor of present-day hybrid art workshops. See http://critical-art.net/molecular-invasion/ (last accessed January 20, 2021).

11 www.artscatalyst.org/artist/gjino-sutic or Gjino Šutič's website, currently under construction: http://gjino.info/ (last accessed January 20, 2021).

12 http://artlaboratory-berlin.org/html/eng-research.htm and http://artlaboratory-berlin.org/html/eng-DIY-Hack-the-Panke.htm (last accessed January 20, 2021).

13 http://artlaboratory-berlin.org/html/eng-event-41.htm (last accessed January 20, 2021).

14 http://artlaboratory-berlin.org/assets/pdf/MICROPLASTICS_Workshop_Description_and_Protocols.pdf (last accessed January 20, 2021).

15 Heather Davis, »Plastic: Accumulation Without Metabolism,« in *Placing the Golden Spike: Landscapes of the Anthropocene*, eds. Sara Krajewski and Dehlia Hanna, Milwaukee: Inova (2015), 69. Online: https://monoskop.org/images/8/81/Placing_the_Golden_Spike_Landscapes_of_the_Anthropocene_2015.pdf (last accessed January 20, 2021).

16 http://artlaboratory-berlin.org/html/eng-DIY-Hack-the-Panke.htm and https://www.gustav-freytag-schule.berlin/wp-content/uploads/2019/06/Mikroplastik-NaWi-Ag.pdf (last accessed January 20, 2021).

17 http://artlaboratory-berlin.org/html/eng-event-44.htm (last accessed January 20, 2021).

18 https://www.katausten.com/portfolio/stranger-to-the-trees/ (last accessed January 20, 2021).

19 »Microplastics and Coexistence« was an online workshop, which took place at Art Laboratory Berlin on April 22, 2020, http://artlaboratory-berlin.org/html/eng-event-54.htm and https://youtu.be/FcmRQOEAF0k (last accessed January 20, 2021).

20 http://artlaboratory-berlin.org/html/eng-event-40.htm (last accessed January 20, 2021). Mary Maggic's talk »From Molecular Colonization to Molecular Collaborations«: www.youtube.com/watch?v=YKe4rhiKJZ4&feature=emb_logo (last accessed January 23, 2021).

21 www.artlaboratory-berlin.org/html/eng-Camille-Diaries.htm (last accessed January 24, 2021).

22 For documentation of the 2014 Hackteria Lab, see www.youtube.com/watch?v=kS6qmB98PCl&feature=youtu.be (last accessed January 22, 2021). For more information on Lifepatch: https://lifepatch.org/main_page/ (last accessed January 20, 2021).

23 Details of the project at the artist's website: https://maggic.ooo/Milik-Bersama-Rekombinan (last accessed January 20, 2021).

24 See https://maggic.ooo/Milik-Bersama-Rekombinan (last accessed January 20, 2021).

25 Heather Davis, »Toxic Progeny: The Plastisphere and Other Queer Futures,« *philoSOPHIA* 5, no. 2 (2015), 244.

*This chapter introduces the interdisciplinary
art practices of Canadian artist Max Liboiron
and Belgian artist Maarten Vanden Eynde,
both concerned with the consequences of plastic
pollution on marine ecosystems. While Max
Liboiron offers community-based citizen science
strategies for monitoring plastic pollution in
marine animals and develops innovative research
approaches with discard studies and anticolonial
scientific practices, Maarten Vanden Eynde
travels the world's oceans to collect marine plastic
debris to raise awareness about the impact of
mass consumerism and environmental injustice,
from which countries of the Global South are
suffering disproportionately.*

Toxic Plastic Politics: Rethinking Plastic Pollution through Art and Activism

Ingeborg Reichle

Art and Environmental Justice ___ Politics of ecology and environmental activism have found increasing resonance in the contemporary art world, giving rise to a wide range of artistic responses, especially when it is a question of ecological emergencies like climate change and other forms of environmental destruction driven by the violence of contemporary fossil-fuel-based capitalism.[1] At the intersection of art, activism, and marine plastic pollution new communities have evolved that want to raise awareness about the urgency and magnitude of this global challenge. With a diverse set of new artistic approaches they created a new sphere to foster environmental justice that became particularly attractive to critical hybrid practitioners, who often have a background in art and activism as well as in science.

Reorienting within a World of Plastic The art practice of Canadian artist Max Liboiron has long been concerned with the materiality of waste and the endeavor to open up a wider debate about systems of waste and their sociocultural and economic implications, with a strong focus on plastic pollution. Moreover, with a PhD from the Department of Media, Culture, and Communication at New York University (and active in various academic communities), Max Liboiron finds it important to articulate new critical frameworks by developing interdisciplinary research approaches with *discard studies* and anticolonial scientific practices with the aim of introducing environmental justice work to academia.[2] Her references to scientific knowledge, scientific research methods, and laboratory bench work, are a major strategy in her artistic practice, but at the same time she wants to do science differently. She therefore critiques established scientific approaches, especially with regard to their inherent colonialist worldviews that obviously make scientific communities prefer and privilege certain topics and articulate certain questions while suppressing and avoiding others. Rethinking (non)participation in science activities and what is recognized in general as »doing science« is also a question of the historicity of the process that forms scientific communities, science identities, research methods, or academic standards.

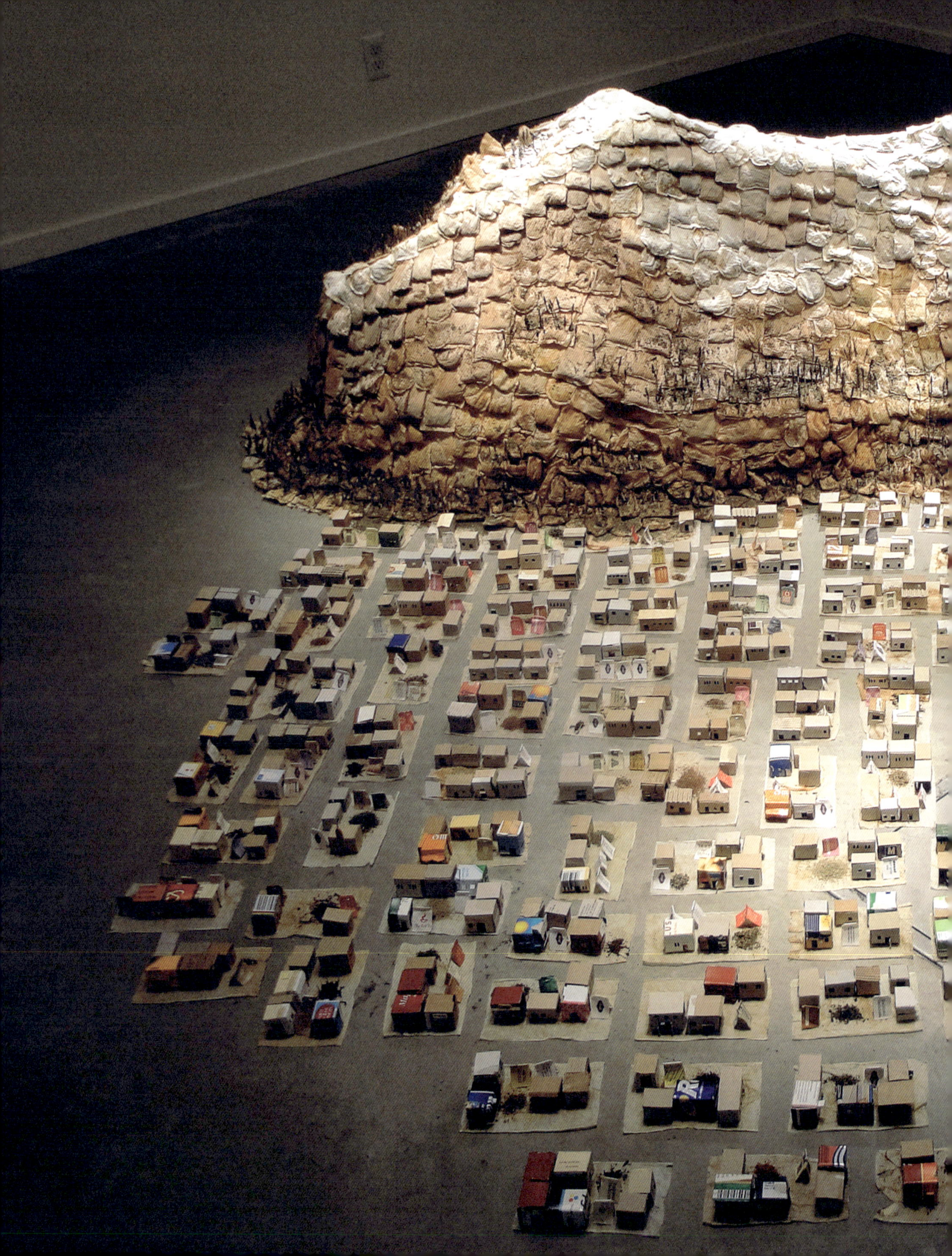

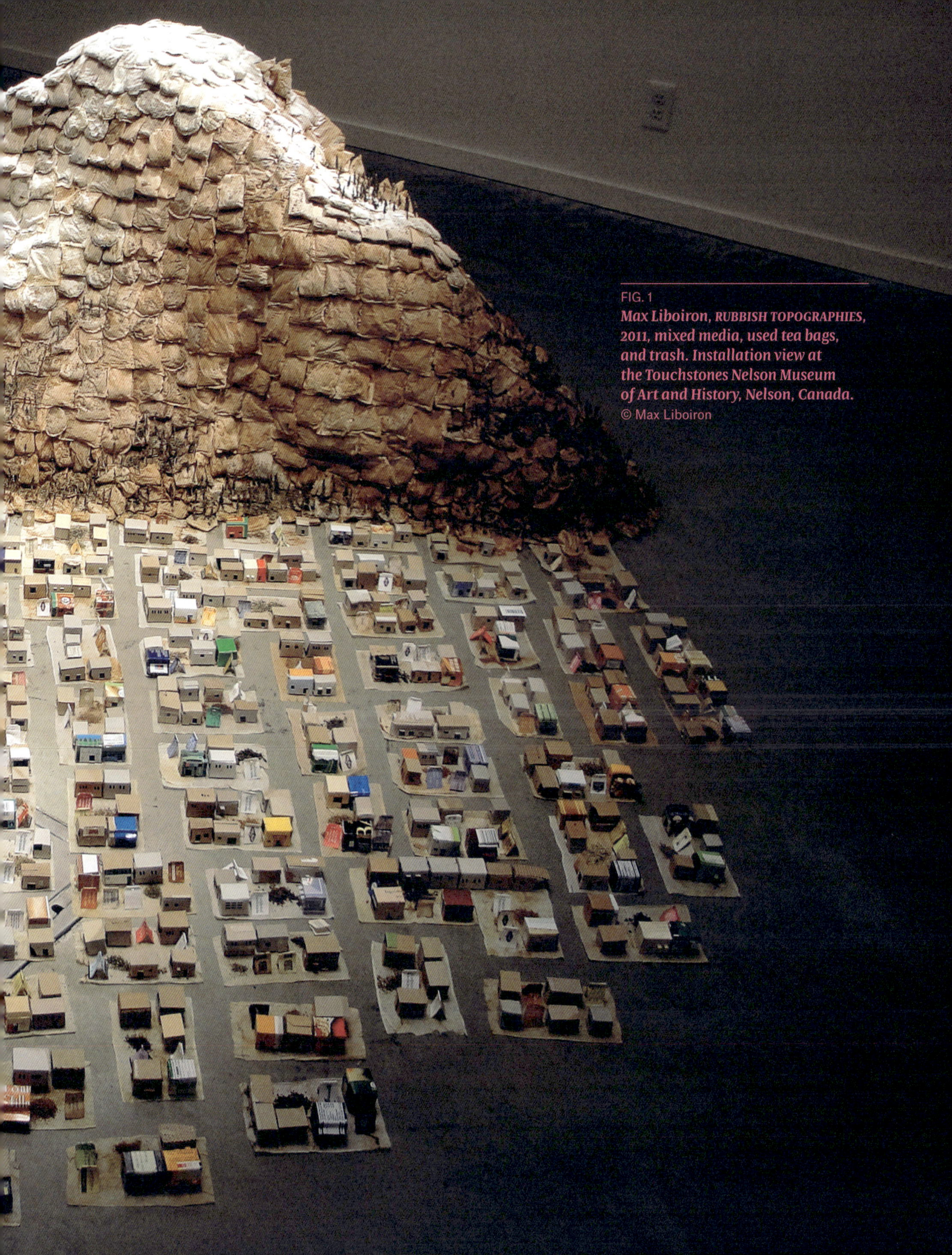

FIG. 1
Max Liboiron, RUBBISH TOPOGRAPHIES,
2011, mixed media, used tea bags,
and trash. Installation view at
the Touchstones Nelson Museum
of Art and History, Nelson, Canada.
© Max Liboiron

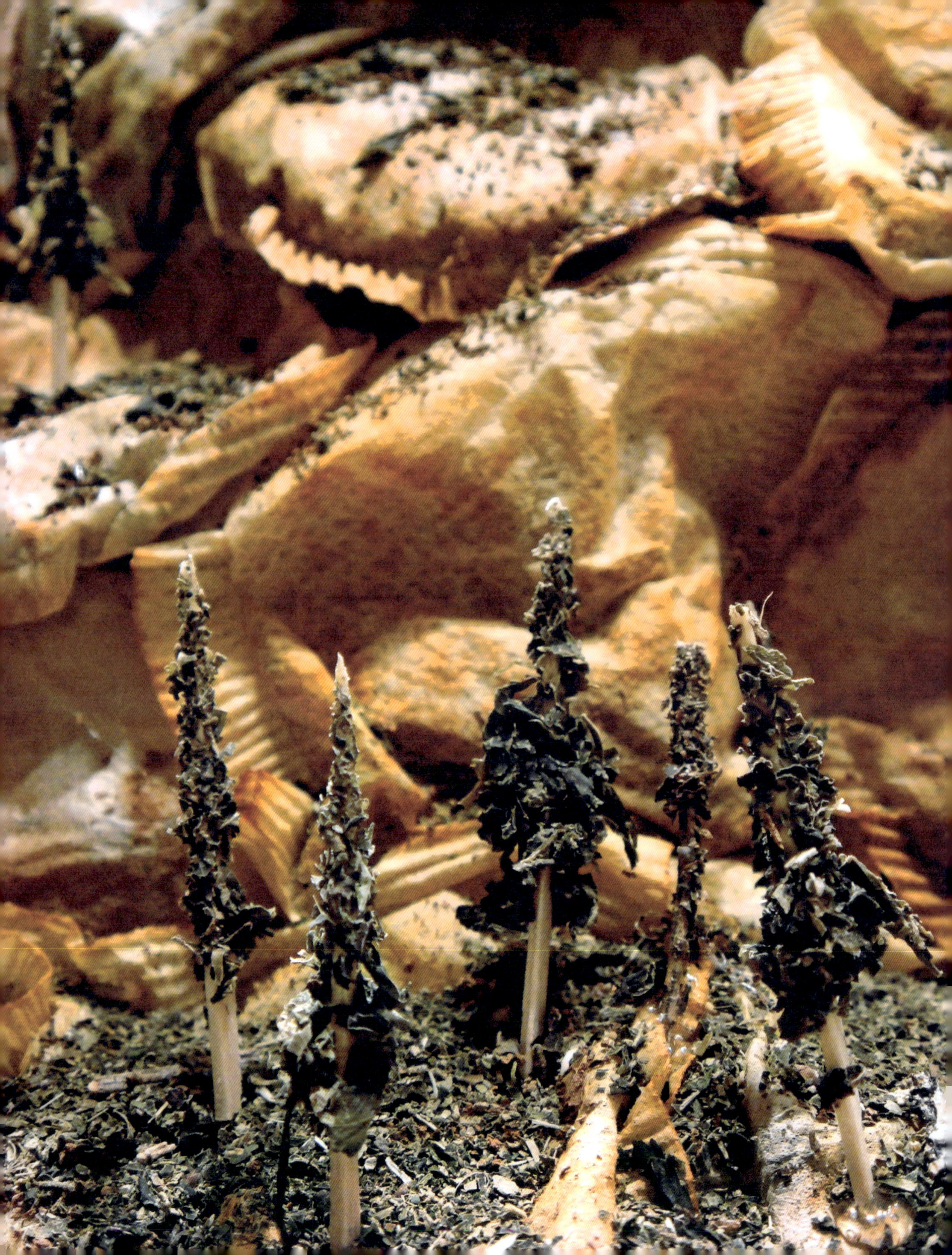

With art projects DINNER PLATES (NORTHERN FULMAR) and COD OBJECTS (INGESTION STUDY) (2016), part of the larger project SEEING LIKE A SCIENTIST, Max Liboiron looked for marine plastics in the guts of fish and birds. COD OBJECTS (INGESTION STUDY) shows digital microscopic images as the result of plastic ingestion studies from the guts of Atlantic cod taken with a camera built into the laboratory microscope at CLEAR, Civic Laboratory for Environmental Action Research, which she runs. Based at the Department of Geography in Memorial University of Newfoundland, St. John's, CLEAR explores marine microplastics and wild food projects as well as food security and food sovereignty with a focus on the community-based, and citizen science, monitoring of plastic pollution.

During the 2015 Newfoundland food fishery, Max Liboiron analyzed 205 Atlantic cod with a group of students, searching for marine microplastic particles in their guts, while developing a citizen science dissection and analysis protocol. The difficulty here was, as with all plastic ingestion studies, to detect the microplastic particles among the other pieces of digested food because under the microscopic landscape of the gastric contents the microplastic particles seemed indistinguishable from the rest. In 2015 the cod stocks had recovered somewhat after the total collapse of the Atlantic northwest cod fishery in 1992, due to massive overfishing since the beginning of the 1970s. This collapse did irreversible damage to the Atlantic cod population, which was brought to the brink of extinction and had a devastating socioeconomic impact on Newfoundland communities.

With plastic ingestion studies, Max Liboiron explored, with her community, the effects of fish eating plastics because of the vast distribution of micro- and nanoplastics in marine environments. Fish eat plankton but also microplastic particles, which have about the same size as many planktonic organisms. Fish are ingesting plastic directly but also indirectly through feeding on zooplankton—the basis of marine food webs—which also eat microplastic particles. The damage plastic does to fish eating microplastic particles is twofold: on the one hand it can physically damage the digestive tract, and on the other there is a high risk of potential uptake of toxic pollutants by the organism, because microplastic particles are very efficient absorbent surfaces for pollutants.

For DINNER PLATES (NORTHERN FULMAR), the artist chose the Northern fulmar (from the Labrador Sea): not an endangered species like the Atlantic cod, but an abundant marine animal. Because of its wide distribution, the Northern fulmar is a key indicator for monitoring the level of exposure of marine birds to plastic debris. Again the artist was dissecting the gastrointestinal tracts of a marine animal searching with her community for plastic-like objects, which were then removed from the bird's guts, put into a petri dish, and analyzed under a microscope before finally being photographed (FIGURE 3).

FIG. 2
Max Liboiron, RUBBISH TOPOGRAPHIES, 2011 (detail), mixed media, used tea bags, and trash. Installation view at the Touchstones Nelson Museum of Art and History, Nelson, Canada.
© Max Liboiron

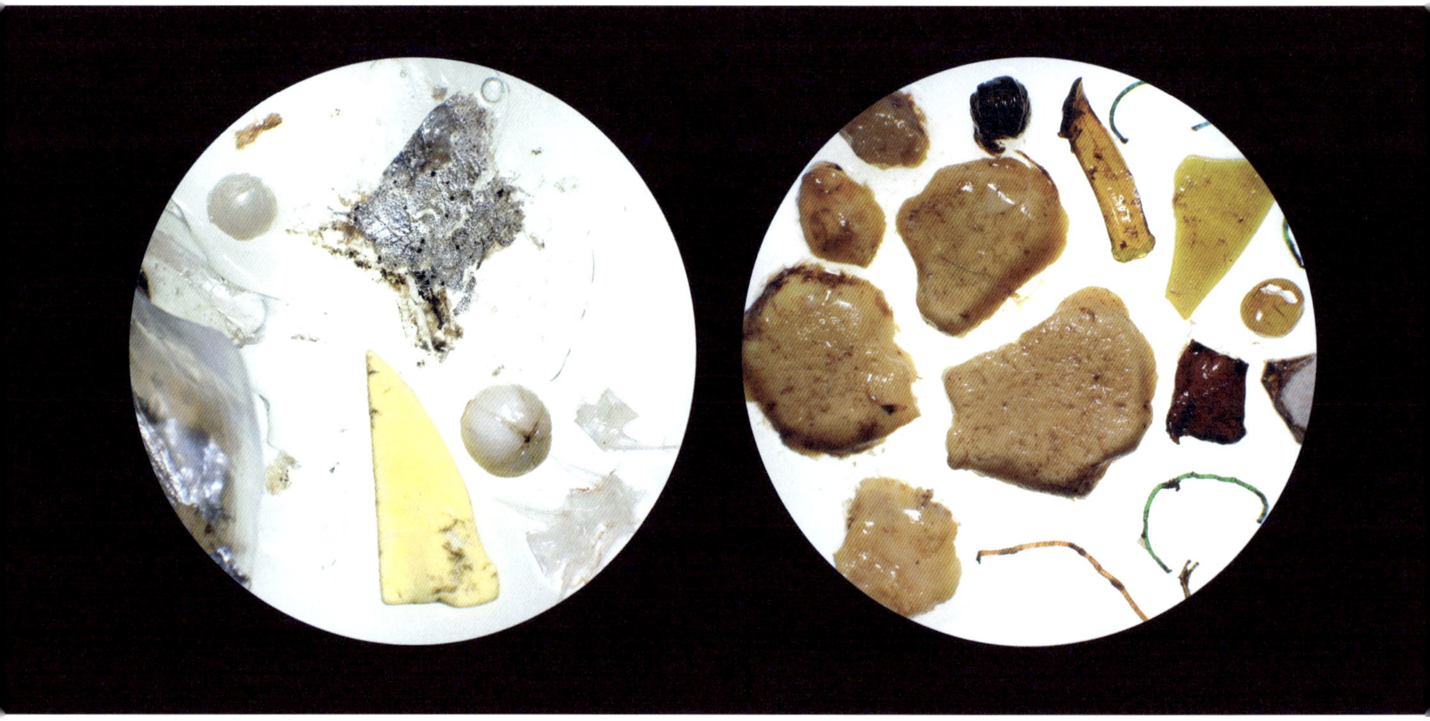

FIG. 3
Max Liboiron, DINNER PLATES
(NORTHERN FULMAR), 2015, *digital
photographs of items taken
from the guts of* NORTHERN FULMAR.
*These images are part of the
larger series* SEEING LIKE A SCIENTIST.
© Max Liboiron

FIG. 4
Max Liboiron, SEA GLOBES, 2013–2014,
*ocean plastics, historical landfill,
New York City kitsch globe. View
of the piece in the 2014 exhibition*
GYRE: THE PLASTIC OCEAN *at the
Anchorage Museum, Alaska*, USA.
© Max Liboiron

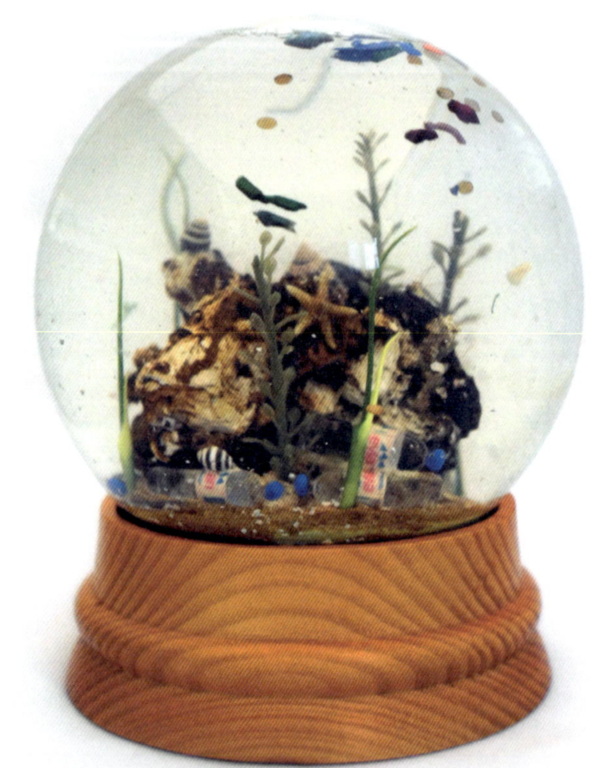

Rethinking plastic pollution and making art from waste was already
on Max Liboiron's mind when she was living in New York City, and with
SEA GLOBES, 2013–2014, she took a close look at the degree of plastic
pollution in the Hudson River (FIGURE 4). The artist took water samples
from the Hudson River, south of Brooklyn, collected bituminous coal
from a landfill that closed in the 1930s, and bought some snails from
a SoHo taxidermy shop in downtown Manhattan. She then arranged all
the parts and snails in typical holiday souvenir style, like a snow globe.
While snow globes mostly show idealized, miniaturized scenes of
landscapes in a kitschy way, in SEA GLOBES the artist carefully presented
the waterfront environment of New York City, accurately and the
way it looks today, using also tiny miniatures of plastic bottles, still
the number-one item found in shoreline trash at the Hudson River.

Trash Transformation __ With participatory art installations like *Material Afterlife:
Circulation* (2009–), *New York Trash Exchange* (NYTE) (2010), ELOCATION (2010),
RUBBISH TOPOGRAPHIES (2011), *Steady-State: Development Without Growth*
(2011), *Founder/Worker* (2011), or *Trash Transformation* (2013) and others,
Max Liboiron was making art from trash while at the same time looking at
entire systems of waste, giving gallery visitors a structural system defined
by a set of rules offering ways to interact with objects considered trash.
ELOCATION (2010), for example, was based on a model-scale display of an

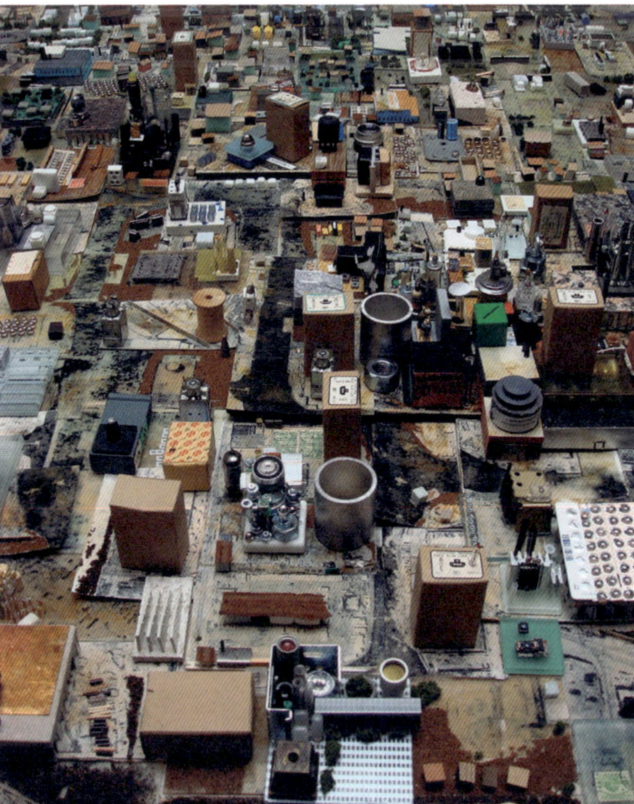

68

area of East Brooklyn, where the local communities at the time were going through a challenging and disrupting transformative process because of major gentrification, followed by cultural displacement. Built from tiny pieces of trash, some of the architectural objects depicting people's neighborhoods were glued on the table while others were not (FIGURE 5). Gallery visitors were free to interact with the movable objects: anything that was not glued down on the table people could take away with them. People were also invited to glue down anything they wanted or had brought with them, adding their perspective on the transformation process of this area of East Brooklyn. At the end of the exhibition the shape and appearance of the miniature neighborhood on the display had changed because many visitors had taken objects with them, exchanged parts of the installation, or even made major modifications with objects they brought to the gallery.

With *RUBBISH TOPOGRAPHIES* (2011), Max Liboiron again involved gallery visitors, but also her friends, community, and even strangers. The rule of interaction was to send the artist clean and dry used tea bags as raw material for her installation. Each and every donated tea bag was carefully inserted by the artist into the design of the construction of the installation, which consisted of a pile of tea bags and about ten dozen miniature houses made of discarded cardboard and other sorts of trash. All objects were meticulous arranged to imagine a kind of miniature

FIG. 6
Maarten Vanden Eynde,
HOMO STUPIDUS STUPIDUS, 2008,
human skeleton, clay. Installation
view at the 2012 exhibition
THE MUSEUM OF FORGOTTEN HISTORY,
Museum of Contemporary Art
Antwerp, Belgium.
Photo: Maarten Vanden Eynde

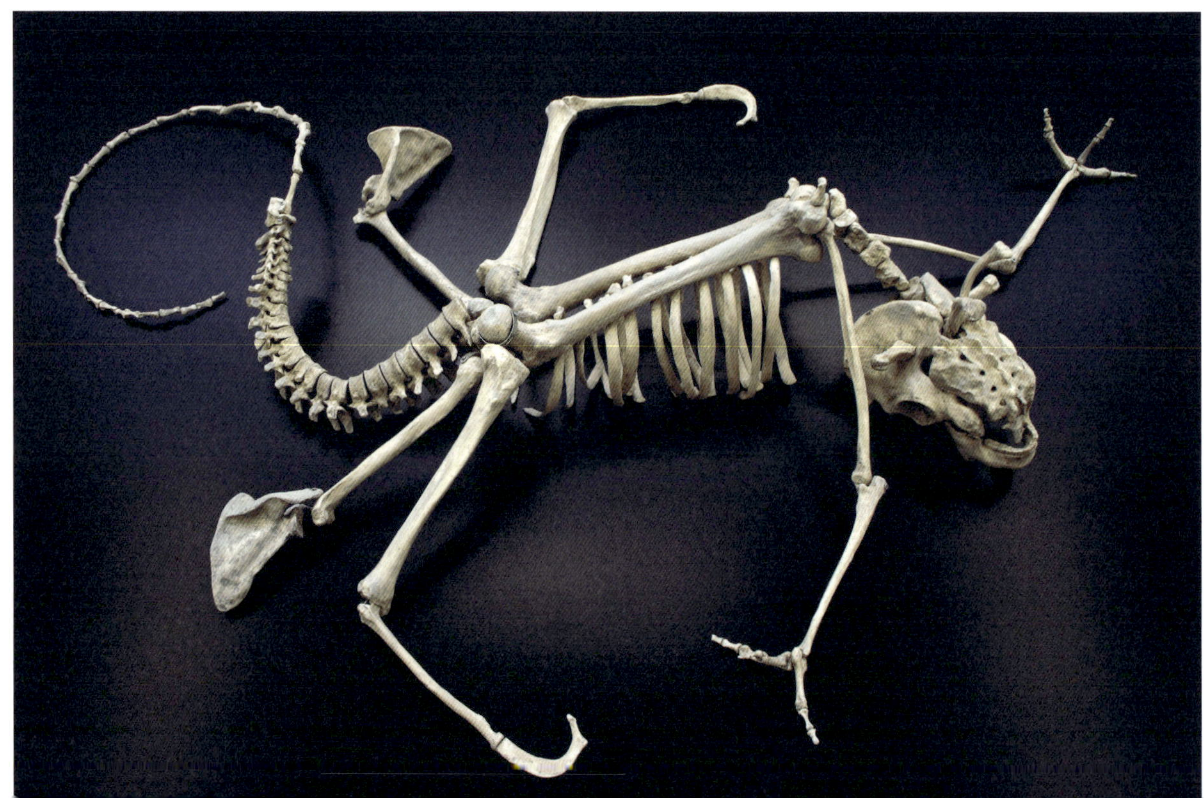

city at the foot of a mountain, a mountain made of landscapes of tea bags in all possible shapes and shades of color (FIGURES 1—2). By asking to donate used tea bags, which are usually thrown away after one use, the artist wanted to make people care about their waste and rethink the systems of waste in which they are locked in rather than moralizing about the trash practices of each and every individual involved. Turning trash into an aesthetic resource was intended to help people develop a different perspective on their waste, especially when having the option to reuse it in a meaningful way while participating in an art project.

Inserting non-art objects or even trash, thus expanding the language of art in multiple ways, has a long history in art. At the beginning of the twentieth century, Western avant-garde artists began to blend their art with objects that were not considered art. By inserting scraps of newspaper and small pieces of wallpaper and other discarded items onto the canvasses of their cubist paintings, artists like Georges Braque (1882–1963) and Pablo Picasso (1881–1973) crossed the demarcation line between the everyday world and the world of fine arts because any fragment of the ordinary world would say more about reality at the dawn of the twentieth century than any painting could. A few years later Dadaist artist and writer Kurt Schwitters (1887–1948) found the greatest pleasure in turning discarded items through art's transformative power into meaningful artistic materials for his collages and *Merz-Bilder*, fostering an approach towards art, which earned him great deal of criticism and contempt. Although his art was widely rejected, Schwitters remained deeply fascinated with turning trash into an aesthetic resource because there was no underlying meaningful history of its material use that would limit his artistic expression, quite similar to how artists evaluated the use of synthetic materials like Bakelite at the time.[3] Max Liboiron's artistic objectives seem not so much to set out to rethink the value of trash as an artistic medium, but to offer a model–conceived as a system defined by a set of rules–to challenge our existing ideas and attitudes toward our systems of waste, which seem so familiar to us that we hardly notice them or simply overlook them. With plastic ingestion studies from the guts of marine animals like the Atlantic cod, which could end up on our dinner plates in one form or another, the perspective on systems of waste come full circle and the consequences become clear because the artist makes them visible and understandable.

The First and the Last Things ___ Taking gallery visitors to the man-made plastic waste of the world's oceans is also the concern of Belgian artist Maarten Vanden Eynde, who began exploring marine plastic pollution immediately after he learned about the Great Pacific Garbage Patch in 2008 and subsequently that much of the world's ocean debris is floating in five gyres. These five gyres are large systems of rotating ocean currents, located in

69

the North and South Atlantic, the North and South Pacific, and the Indian Ocean. The artist responded with art projects like PLASTIC REEF (2008–2012), *1000 Miles Away From Home* (2009–2013), CONTINENTAL DRIFT #2 (2014), and others. His great interest in plastic pollution is based on the main characteristics of this material made of petrochemicals: the long polymer strands that make up plastic are not biodegradable and therefore this material lasts forever.

Objects made of plastics will be the fossils of the future to come. With his long-term research project, which Maarten Vanden Eynde calls *Genetology or Science of First Things*, he developed, from 2003 to 2014, an oppositional concept to the established concept of *Eschatology* (The Science of Last Things) in order to reconstruct civilizations from the past. With Genetology the artist explored his interest in diverse methodologies of academic disciplines—his enthusiasm for archaeology, history, the humanities and natural sciences as well—articulating a fictional scientific methodology as a novel way to frame our perception of the world.[4]

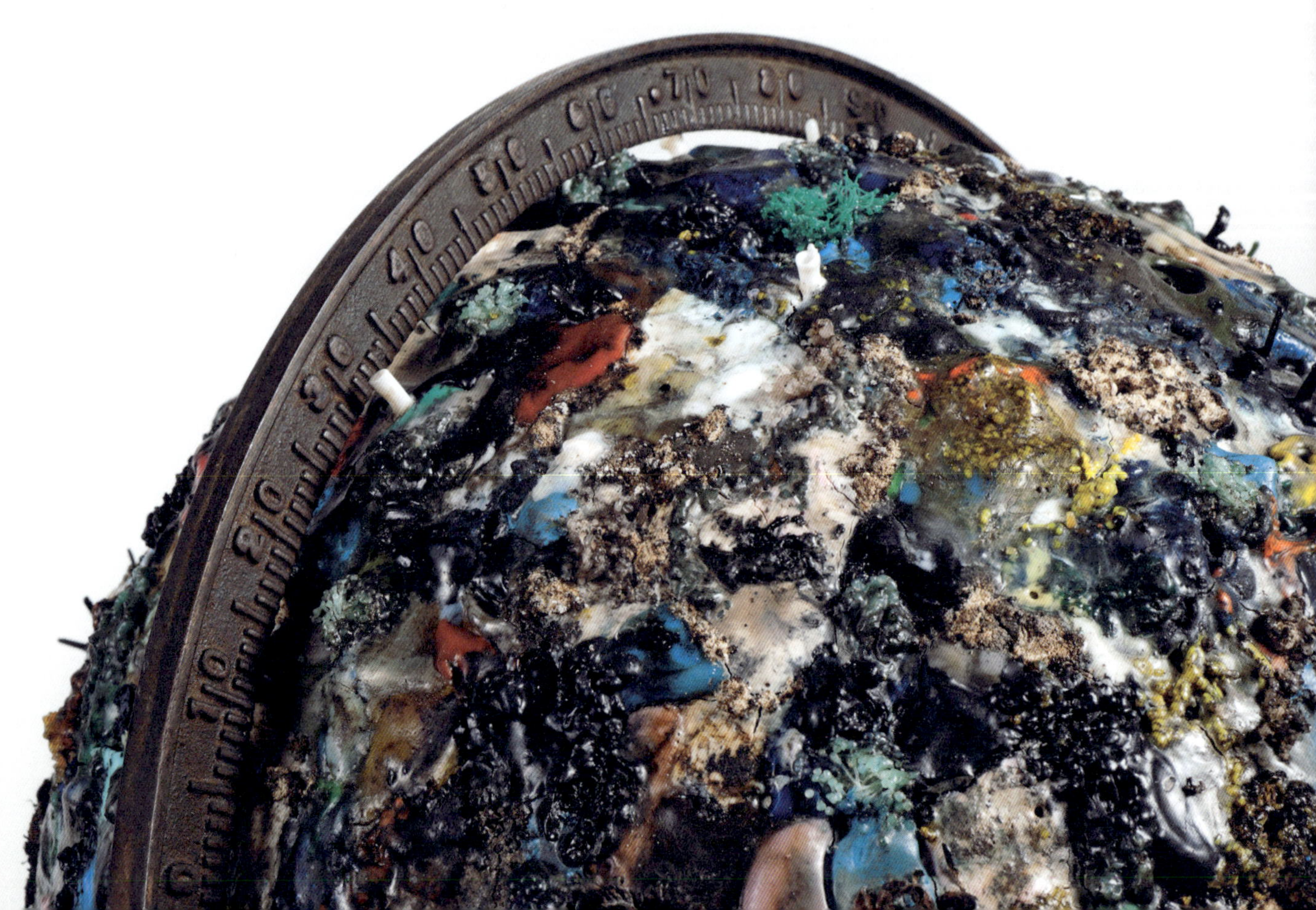

The ubiquitous present of plastic has left an irrevocable footprint on the planet, whereas the toxicological responses to plastic are not yet fully understood. When all other evidence of the existence of our civilization will have vanished from the face of the Earth, plastic objects will still exist and continue to poison ecosystems. With HOMO STUPIDUS STUPIDUS (2008), the artist makes a critical visual statement about the denomination of modern man as *Homo sapiens sapiens*, the wise and sensible man as 18th century Swedish botanist Carl von Linné (1707–1778) called his own species, when introducing his system of nature through the three kingdoms of nature, according to classes, orders, genera and species, still known today as the Linnaean taxonomy. Maarten Vanden Eynde took a human skeleton apart and put it back together in an unsystematic and rather senseless way to symbolize that the traditional denomination of modern man as *wise and sensible* no longer justifies for a failed species, a species that knows so little about its past and present and the world at large; that it destroys the planet to an extent which no other species ever has.

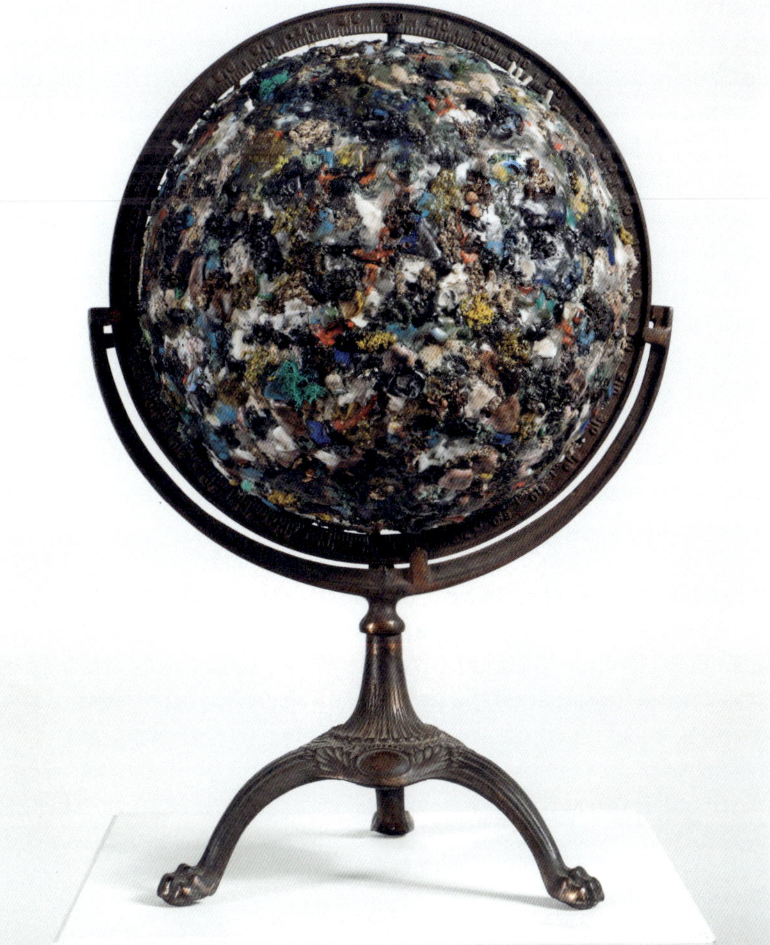

FIG. 7
Maarten Vanden Eynde,
CONTINENTAL DRIFT, 2014 *(detail),*
vintage globe, melted plastic
debris from the world's oceans,
variable sizes.
Photo: Philippe De Gobert

FIG. 8
Maarten Vanden Eynde,
CONTINENTAL DRIFT, 2014,
vintage globe, melted plastic
debris from the world's oceans,
variable sizes.
Photo: Philippe De Gobert

The *Theater of the World* and Plastic Globes __ The sculptural installation PLASTIC REEF is made of melted plastic debris Maarten Vanden Eynde extracted from about 1000 kilograms of collected plastic in the years from 2008 to 2013 while travelling to the five gyres located in the Pacific, Atlantic, and Indian Oceans (FIGURE 11). The sculpture was constantly growing over the course of time, resembling the impression of a colorful coral reef as a kind of substitute in perspective to the dying of the world's coral reefs: The title of the artwork makes reference to a phenomena occurring across the globe caused by a process known as »bleaching«: the world's coral reefs are dying because they have, amongst other things, difficulty adapting to rising water temperatures due to climate change. The title also reflects possible responses of fragile coastal ecosystems, such as coral reefs, to their massive exposure to microplastic and nanoplastic particles.

CONTINENTAL DRIFT #2 is a globe also made of melted plastic debris (FIGURES 7—8). The title of this artwork refers to the Flemish cartographer and geographer Abraham Ortelius (1527–1598), the creator of the first modern atlas, the *Theatrum Orbis Terrarum (Theater of the World)*, and considered to be the first scholar who imagined the continents were once a single landmass before drifting apart, albeit his idea about continental drift went unnoticed until the 20th century. The shape Maarten Vanden Eynde gave his artwork also recalls the oldest surviving terrestrial globe (earth apple), which was produced under the direction of the German textile merchant and cartographer Martin Behaim (1459–1507) in the years 1490 to 1492, shortly before the New World became known to Europeans.

With his monumental piece GLOBE (2013), the artist again refers to the motive of the globe, this time to articulate the phenomenon of planned and progressive obsolescence (FIGURE 9). The intention to shorten the lifespan of consumer goods or make them go out of fashion after a certain period of time to accelerate the rate of consumption and thus economic growth, increased dramatically after World War II due to the enormous capacity of the production system and the logic of Western markets, characterized by innovation and accelerating production and sales cycles. To build the GLOBE 8.5 meters in diameter, Maarten Vanden Eynde used different kinds of scrap and rubbish found in and around the small village of Saint-Mihiel, France, setting up the installation in an old rubbish dump.

Inequity of Waste: Plastic Pollution and the North-South Divide __

To rethink the way wealthy countries are externalizing their plastic crisis by shipping their contaminated or mixed plastic waste to countries of the Global South, the artist takes a de-colonial perspective with MAMAMUNDI (2010), showing the continued presence of colonial logics of current hazardous waste politics (FIGURE 10). The artist put a pile of recycled scrap plastic objects on top of the head of a wooden statue from Africa depicting a pregnant nude woman standing upright and holding a large bowl with both of her arms, probably to collect wood. The female statue represents the origin of life and the sustaining of the family, whereas today the only thing left for her to do is to collect and sort the garbage from the Global North. The idea the sculpture articulates is not so far from the reality in some parts of Africa, where pregnant women and children, regardless of the toxicity of the materials and fumes, have to sort garbage to survive. According to the artist, this artwork refers also to a female version of the representation of Atlas, a character from Greek mythology who was punished by the gods after the titans were defeated in battle, forced to carry the vault of heaven. With this sublime sculpture, Maarten Vanden Eynde raises awareness of the fact that the countries of the Global South suffer disproportionately from environmental injustice.

73

FIG. 9
Maarten Vanden Eynde,
GLOBE, 2013, various materials.
Permanent installation at Vent
des Forêts, Lorraine, France.
Photo: Marjolijn Dijkman

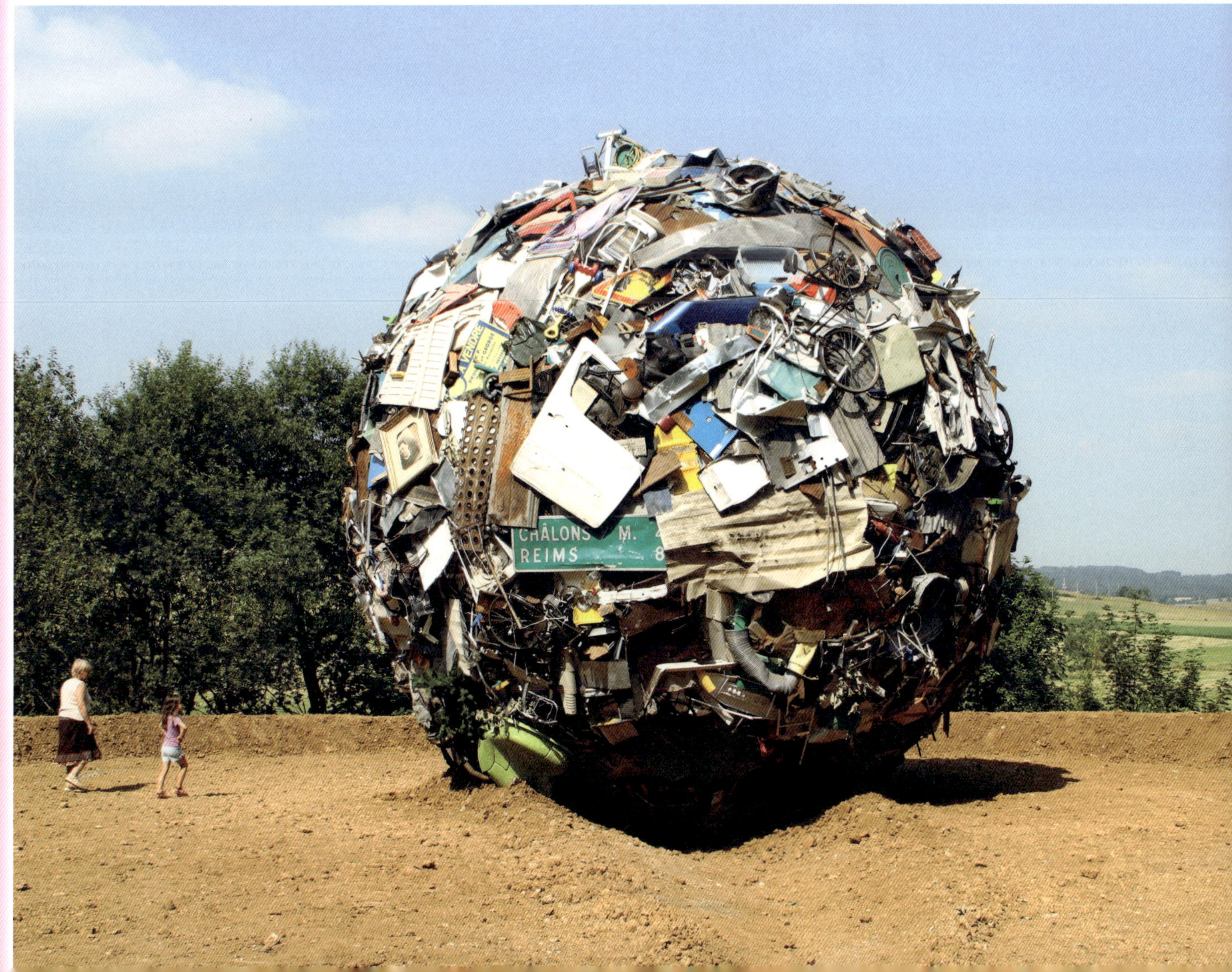

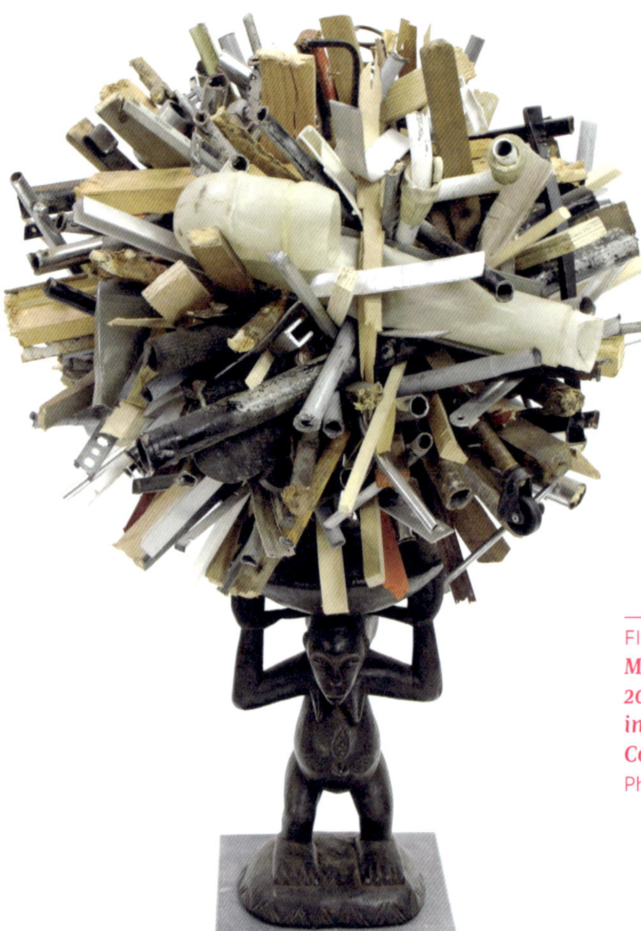

FIG. 10
Maarten Vanden Eynde, MAMAMUNDI, 2010, wood and mixed media. 2012 installation view at the Museum of Contemporary Art Antwerp, Belgium.
Photo: Maarten Vanden Eynde

Plastic recycling is very profitable and a global billion dollar business. Countries of the Global North like the USA or the European Union export their plastic waste to the Global South for more than three decades, including African countries such as Ethiopia and Senegal. Since China closed its doors to most plastic imports in early 2018, the handling of plastic recycling moved progressively to poor countries, located especially in regions like East Asia and the Pacific (particularly China's neighboring countries: Indonesia, the Philippines, Malaysia, and Vietnam). Over-whelmed by the sheer magnitude of waste volume, many of these countries have not yet developed an appropriate basic waste management infrastructure and show a high degree of waste mismanagement and practice limited environmental regulation. As a consequence, plastic waste is inadequately disposed—disposal in dumps or open, uncontrolled landfills—plastic waste enters the ocean easily via inland waterways or transport by wind, therefore the effects of plastic pollution in the world's oceans becomes worse.

Toxic Plastic Politics and Living in the United States of Plastic __

The current global plastic crisis motivates artists like Max Liboiron and Maarten Vanden Eynde to explore interdisciplinary collaborative models at the level of artistic production and activist intervention to engage their communities in collectively rethinking the way we want our future lives to be and what kind of aspirations will be possible or even probable under the auspices of fossil-fuel-based capitalism and mass consumerism. Both artists welcome, in their communities, potential allies, artists, scientists, scholars, and hybrid practitioners from diverse backgrounds and disciplines to engage in global concerns including ocean plastics and the global North-South divide, and join forces effectively to foster positive change.

 The perspective on plastic pollution offers major insights into the deep interconnectedness of the social and the ecological questions involved, and discloses the urgency of fostering systemic change, especially because the heavy toxic burdens associated with plastic are contributing to a massive global health crisis affecting human health as well as all other organisms, because basic features of biology are shared across all life forms. The magnitude of the impact of ocean pollution is only beginning to be understood, but it is obviously a highly complex phenomenon that needs global cooperation as a response as well as a holistic approach because potential ecological collapses are interrelated. Rebuilding *greener* economies and fostering *greener* science after the COVID-19 crisis could be a turning point in giving rise to a more sustainable future. Whether this opportunity (which would only be a first step) will be taken or not, or if the responsible decision-making elites in industry and government bodies decide to make us carry on living in the United States of Plastic remains open.

75

1   See T. J. Demos, *Decolonizing Nature: Contemporary Art and the Politics of Ecology*, Berlin: Sternberg Press (2016), and Amanda Boetzkes, *Plastic Capitalism: Contemporary Art and the Drive to Waste*, Cambridge, MA: MIT Press (2019), Jennifer Gabrys, Gay Hawkins, and Mike Michael, eds., *Accumulation. The Material Politics of Plastic*, London, New York: Routledge (2013), and Lea Vergine, ed., *Trash. From Junk to Art*. Exhibition catalog of Museo di Arte Moderna e Contemporanea di Trento e Rovereto, Mailand (1997).

2   Max Liboiron, *Redefining Pollution: Plastics in the Wild*, PhD dissertation, New York University (2012). In 2010 Liboiron and Robin Nagle cofounded *The Discard Studies Blog* to foster the new interdisciplinary field of discard studies, https://discardstudies.com (last accessed January 3, 2021). Her recent book focuses on plastic pollution and models of anticolonial scientific practice aligned with indigenous concepts of land, ethics, and relations: Max Liboiron, *Pollution Is Colonialism*, Durham, NC: Duke University Press (2021).

3   On the use of plastics and other synthetic materials in art from around 1850 to the period after World War II, see Esther Leslie, *Synthetic Worlds: Nature, Art, and the Chemical Industry*, London: Reaktion Books (2005).

4   Katerina Gregos, Nav Haq, and Jan Zalasiewicz, *Maarten Vanden Eynde: Digging up the Future*, London: Yale University Press (2021).

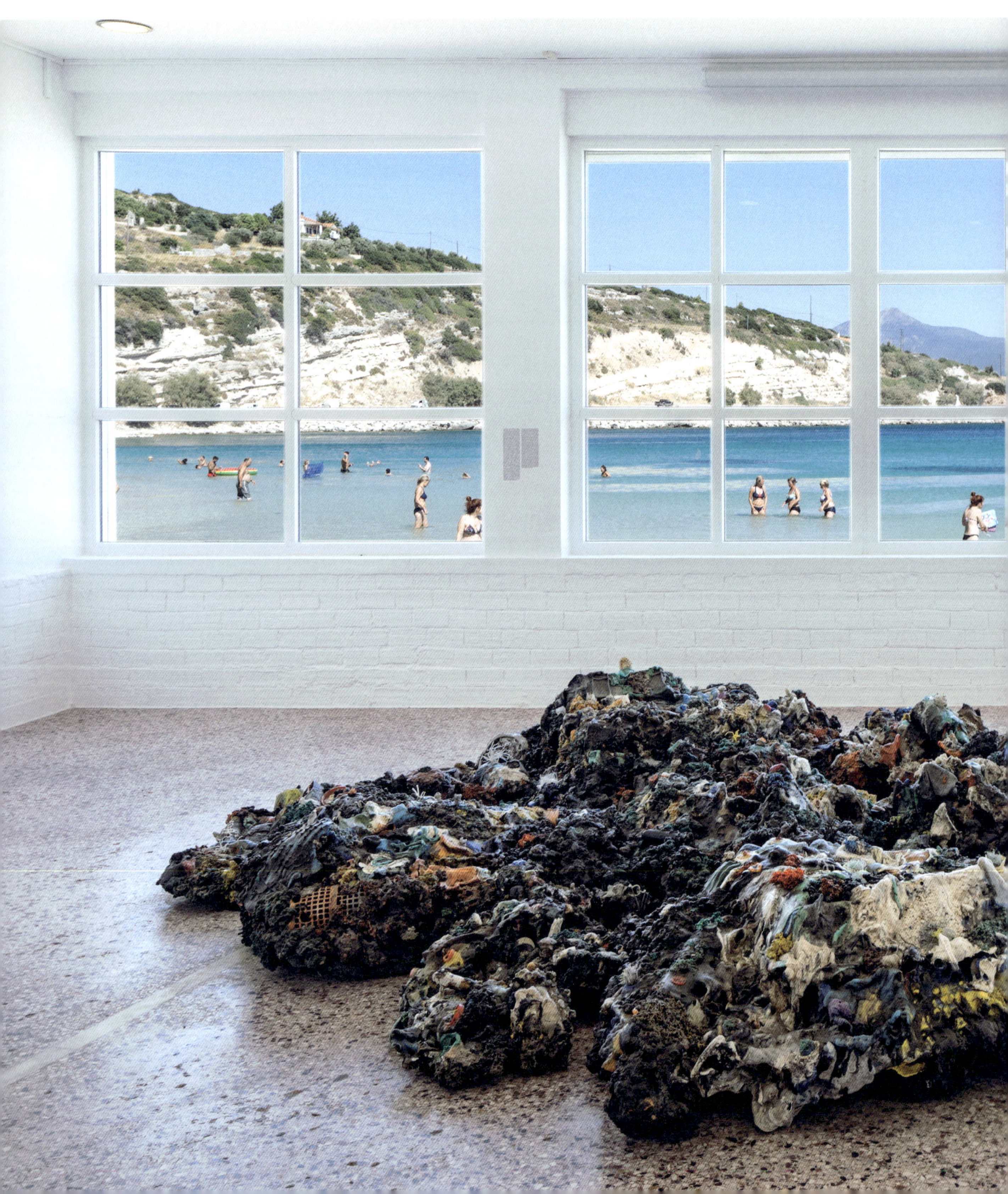

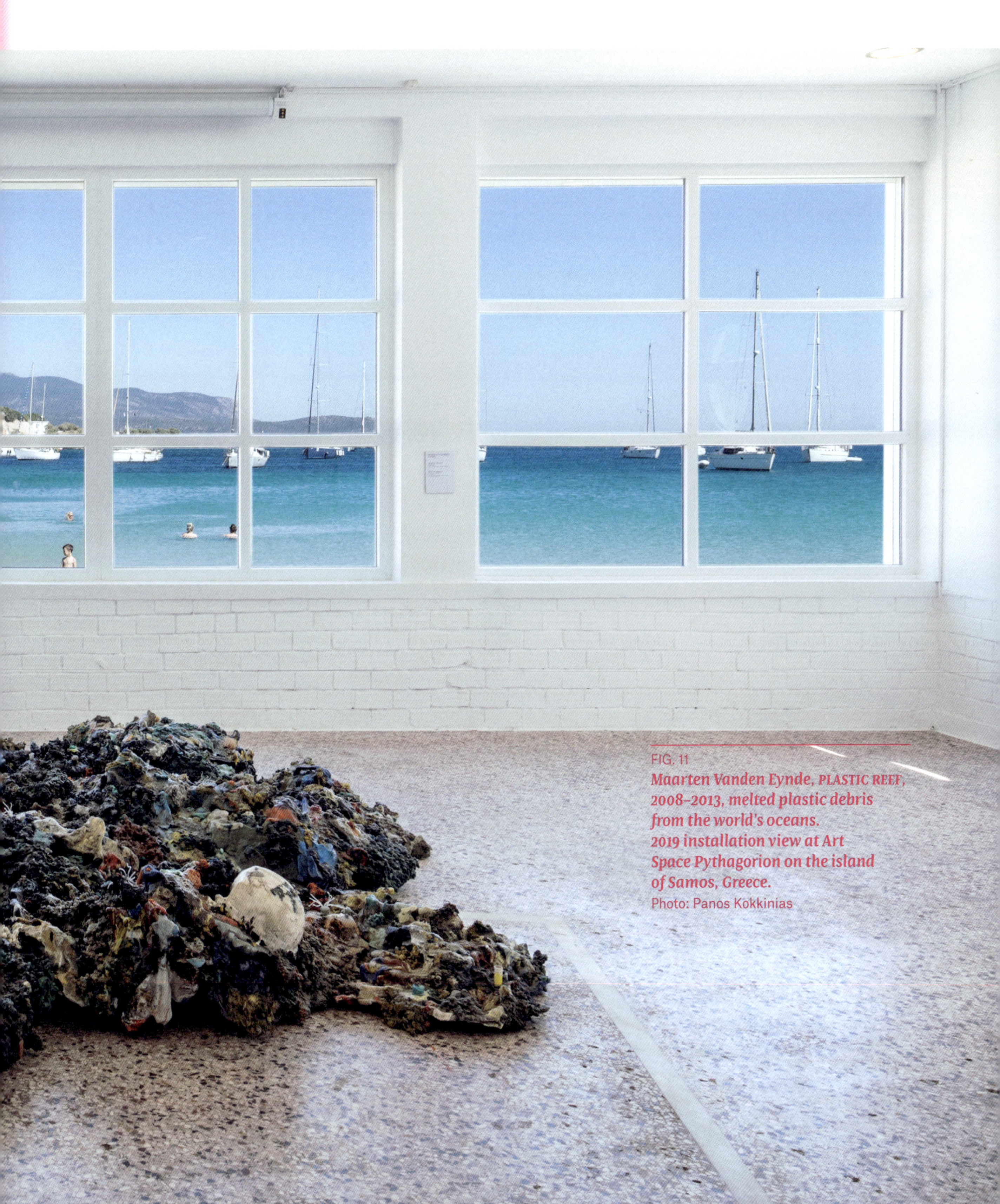

FIG. 11
Maarten Vanden Eynde, PLASTIC REEF,
2008–2013, melted plastic debris
from the world's oceans.
2019 installation view at Art
Space Pythagorion on the island
of Samos, Greece.
Photo: Panos Kokkinias

Through a personal journey to the heart of
American consumerism, this essay takes
the viewer into the thinking and making processes
of two artworks, both of which are reflections
upon the current state of the oceans.
AN ECOSYSTEM OF EXCESS (2014) focuses on plastic
pollution while HOLLOW OCEAN (2021) expands
its scope to the most pressing issues such
as climate change, overfishing, and acidification.

From *An Ecosystem of Excess* to *Hollow Ocean*: Affective Learning in the Service of EcoActivism

Pinar Yoldas

My first encounter with plastic pollution as a severely life-threatening problem occurred when I moved from Istanbul, Turkey, to study in the United States at the University of California, Los Angeles (UCLA). As a young mind who followed her passion for design and science, my plan was to look for a suitable job while adjusting to the new continent during my master's degree. I was on a very limited budget because my studies were solely supported by fellowships. My life in Los Angeles was quite different from my life in Turkey specifically because of this. I had down-graded from being middle-class to a much lower class. Thankfully, the novelty and excitement of a completely new academic environment was enough for me. As months passed and while I was observing the American way of life more closely, I started to notice subtleties that bothered me to my core. I couldn't really cook the few Turkish dishes I knew because I did not have a car and the only grocery store within walking distance in Westwood was Whole Foods, where fresh produce was extremely pricey. The budget solution was to eat processed food which came in layers of packaging that annoyed the heck out of me, since I knew that more than half of the money I had reluctantly paid was going into transport, market-ing, and packaging of the product—packaging all of which was designed to be disposable.[1] Being constantly broke made me realize how my transac-tions were equally idiotic actions. I could only afford what I did not want. Moreover, even for simple pleasures such as a bottle of water, I was creat-ing piles of trash on a daily basis.

One balmy night in my dorm staring at the glowing surface of my most expensive possession, an old MacBook, I discovered where all that trash I had been generating was destined to end up: The Great Pacific Garbage Patch. On the news it was the Pacific Trash Vortex (PTV). Scientists sited it within the North Pacific Subtropical Gyre (20 million km^2). Being an obsessive Google-r, I spent the entire night learning about this fluid site of swirling horror. Some said it was the size of Texas, some said it was bigger than France, one of the largest countries in Europe covering an area of about 551,500 km^2. Plastic trash came mainly from the land (80%) and completed its salty swim to the gyre in about six years. A particular

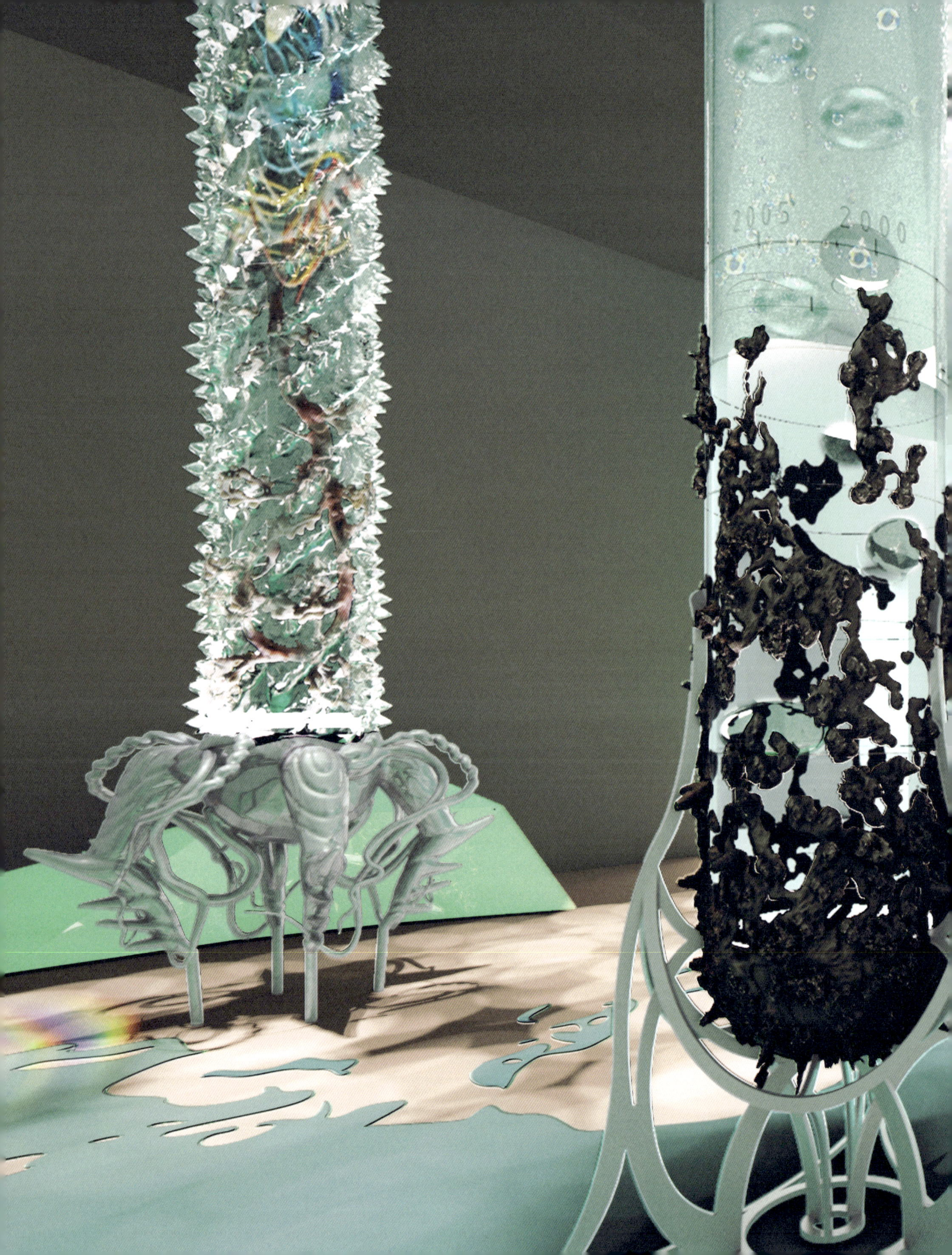

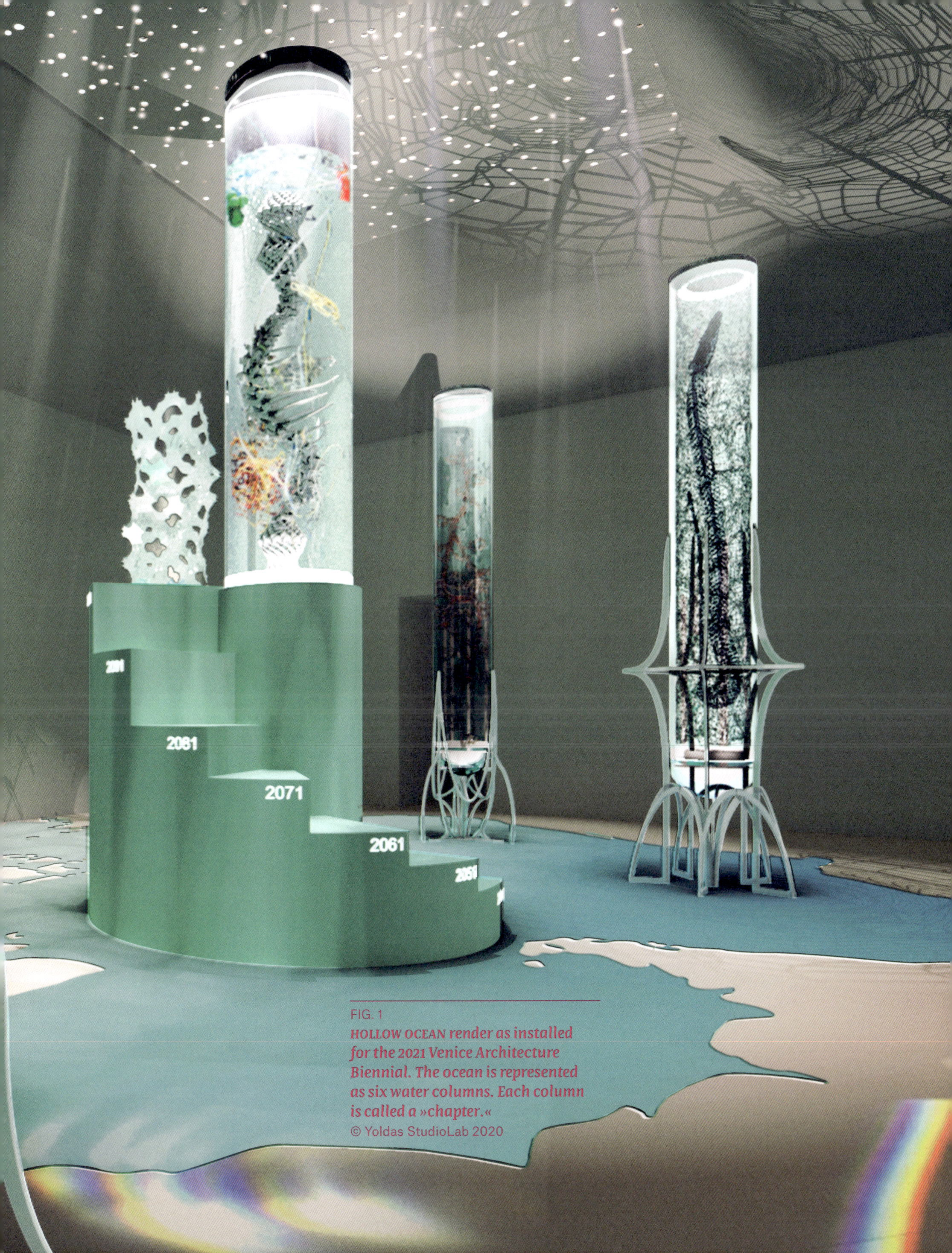

FIG. 1
**HOLLOW OCEAN render as installed
for the 2021 Venice Architecture
Biennial. The ocean is represented
as six water columns. Each column
is called a »chapter.«**
© Yoldas StudioLab 2020

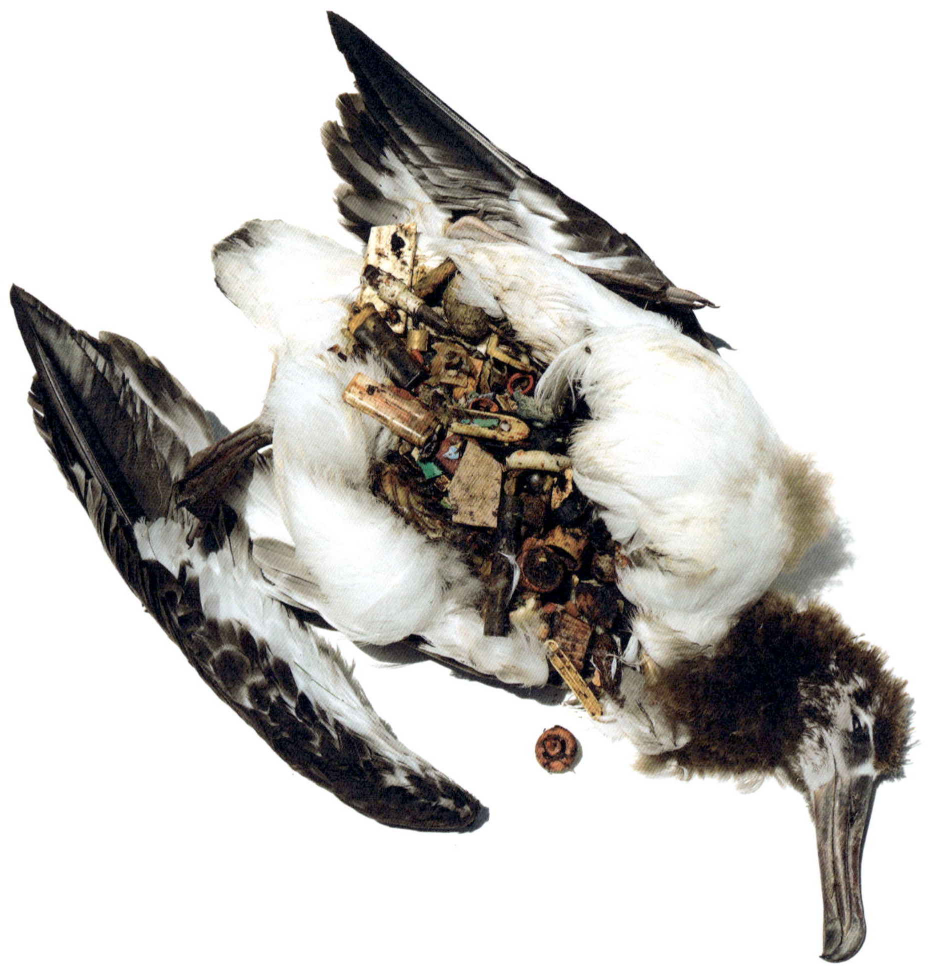

FIG. 2
This photo of a dead Laysan albatross chick from the Kure Atoll, Northwestern Hawaiian Islands, was taken in 2004 by American photographer David Liittschwager, revealing graphically how the seabird's stomach is filled with mainly plastic debris.
Photo: David Liittschwager © 2004

component of sunlight, UV light, broke down plastic polymer into microplastics through a process called photodegradation. Microplastics enter the food chain and are stored in fat and muscle. The North Pacific Subtropical Gyre is one of the largest ecosystems in on Earth.

In the middle of the gyre, thousands of miles away from cities and continents, there is a small island called Kure Atoll. On this northwestern Hawaiian island nests a noble marine bird called the Laysan albatross. There is so much plastic around that the birds ingest plastics and feed plastics to their chicks. It was when I saw a photograph by American photographer David Liittschwager of a decomposing body of a dead Laysan albatross, and another photograph by American artist and photographer Susan Middleton showing the complete content of the stomach of the dead Laysan albatross chick—that reveal the ultimate fusion of nature versus culture—a strange pain sat in my throat (FIGURES 2—3). What a bizarre death, getting fed plastics by your mother who doesn't know it is your death she is feeding you, because she is confusing food with plastics. Being crammed full, but remaining hungry for days on end until finally

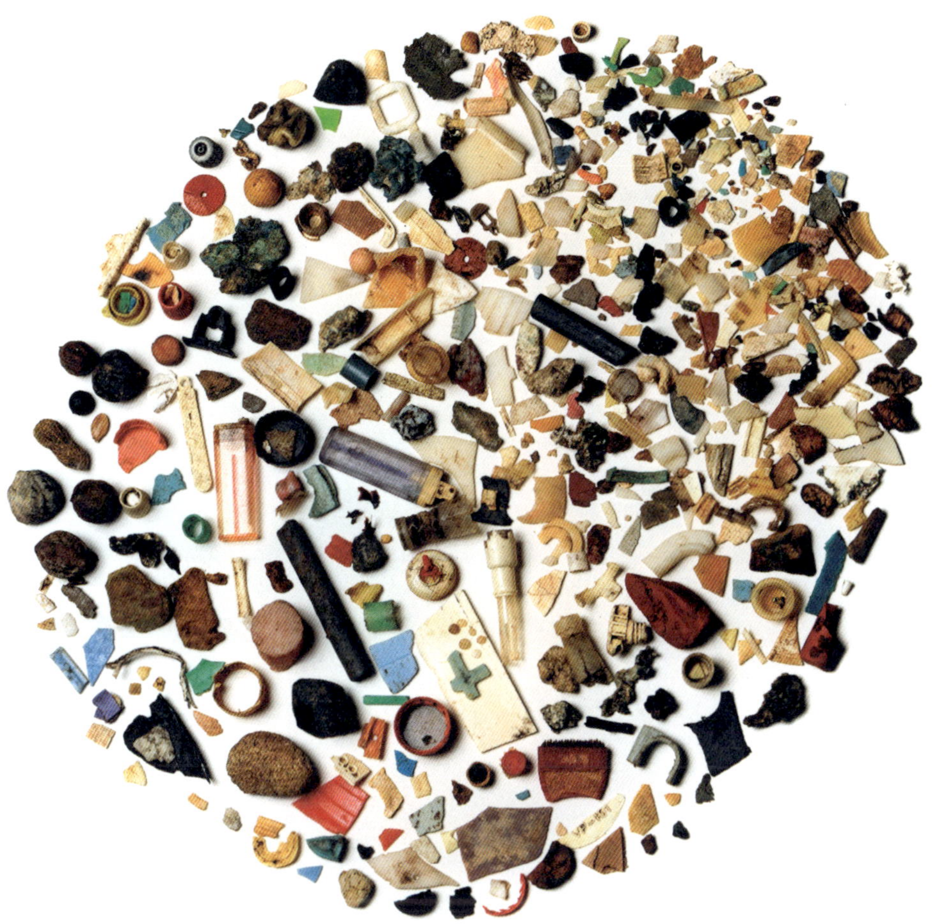

FIG. 3

This photo was taken by American photographer Susan Middleton, and shows the stomach contents of the Laysan albatross chick, laid out on a piece of white plastic.

Photo: Susan Middleton © 2004

your digestive tract gives up, unable to metabolize any food. Your intestines are probably damaged and bleeding internally, blood mixed with stomach acid tearing you apart from inside. The death that plastics brought to these innocent chicks remind me of a torture technique employed for »thought prisoners« of the 1980s in Turkey. Starve the prisoners until their innards shrink. Feed them dry bread whose sharp edges would cut through the soft tissue as it entered. These prisoners would die a lingering death in absolute agony.

How could plastic remnants of our petty daily activities such as that Coca-Cola bottle cap, that BIC lighter, that Sharpie pen, end up in such a distant and pristine location? How could such an unimaginable amount of plastic accumulate in the most unlikely place on Earth, not our backyard, not cities, not even land, but out there in the open ocean, right under our noses, but unbeknownst to us? My mind went into overdrive.

»There is grandeur in this view of life, with its several powers, having been originally breathed into a few forms or into one; and that, whilst this planet has gone cycling on according to the fixed law of gravity,

86

FIG. 4
Pinar Yoldas, PACIFIC BALLOON TURTLE. *A plastivore turtle evolved to have an elastic back like the balloons its ancestors used to ingest shown in the 2014 exhibition* AN ECOSYSTEM OF EXCESS *in at the Ernst Schering Foundation, Berlin.*
© Pinar Yoldas, 2014, photo: Pinar Yoldas

from so simple a beginning endless forms most beautiful and most wonderful have been, and are being, evolved,« Charles Darwin says in a loving voice in awe and recognition of the intrinsic value of life.[2] I realized that just like everyone around me I was participating in slow-motion mass murder of the most beautiful and most wonderful beings. I thought about all the possible ways plastics could get entangled in living bodies. Epigenetics is the study of heritable phenotype changes that do not involve alterations in the DNA sequence. As American author and neuroendocrinologist Robert Sapolsky would put it, epigenetics is a result of gene–environment interactions. It takes 450 years for a plastic bottle to fully dissolve, and for some types of plastics it could be thousands of years.[3] In the purgatory of plastics between now and till 2421, they break down, they leach highly toxic substances while altering the chemical composition of the environment by way of microplastics. Therefore, plastic pollution could lead to a plethora of ecosystem changes directly and indirectly, through entering the food web as prey and predator, but also by instigating epigenetic changes in life forms. This could mean that we neither fully understand

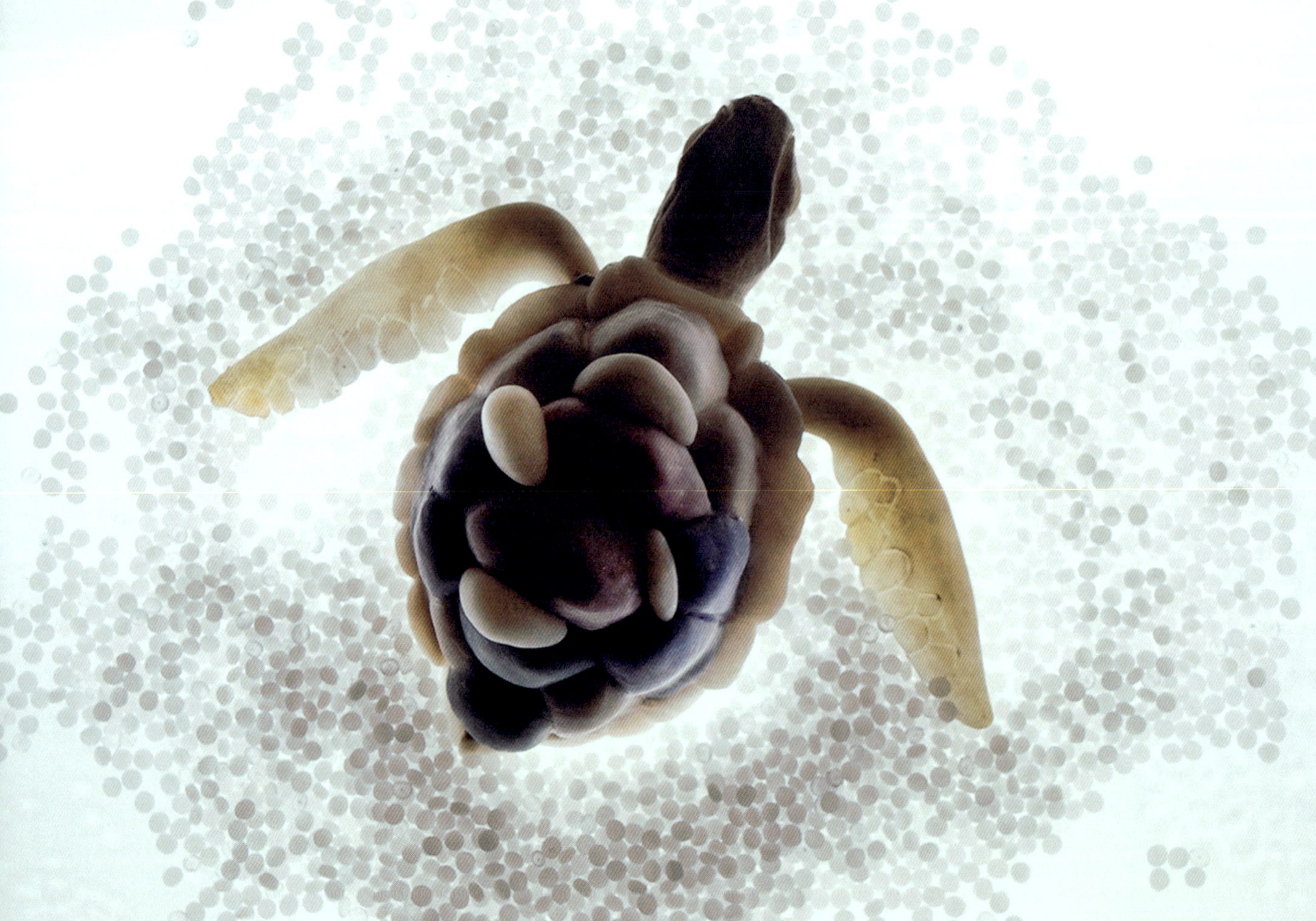

nor can effectively predict the accumulative impact of plastic pollution in marine ecosystems in the coming decades or coming centuries. At a time in which excessive consumerism and evolution coincide, an abnormal ecosystem comes to life.

AN ECOSYSTEM OF EXCESS, therefore, is the culmination of correspondence with scientists across the globe, research on key moments in the history of plastics and key figures around ocean conservation over several years.[4] One such protagonist is Captain Charles Moore who discovered the Pacific Trash Vortex. In a video interview Captain Moore is seen holding a jar full of saltwater mixed with debris from the PTV. He says »the ocean has turned into a plastic soup: this is the soup.« According to primordial soup theory, life started in the primitive oceans about 3.8 billion years ago. What if life had started in a primordial soup of plastics, what kind of life forms would have emerged?

Already in 2013 American xenobiologist Linda Amaral-Zettler of the Royal Netherlands Institute for Sea Research discovered new macrobiotic life emerging in plastic debris: »We unveiled a diverse microbial community of heterotrophs, autotrophs, predators, and symbionts, a community we refer to as the ›Plastisphere,‹« said Amaral-Zettler, »Plastisphere communities are distinct from surrounding surface water, implying that plastic serves as a novel ecological habitat in the open ocean.«[5] In addition to my correspondence with Dr. Amaral-Zettler, I chose to combine scientific data with speculative biology. The microorganisms were already evolving; all I had to do was spread the evolution over various taxa. In doing so, I also hoped to accomplish a curiosity-driven dispersion of scientific data about various forms of eco-destruction.

Following that principle, AN ECOSYSTEM OF EXCESS was an archipelago of different displays, each of which communicating a scientific publication. As a designer it was obvious that the world did not need yet another iconic chair design, or a kettle design, or a lemon squeezer design.[6] I challenged myself to design an entity far more complex than any industrial object imaginable: an ecosystem. In this ecosystem, each creature had adapted to a pollution-rich environment. This neo-Lamarckian approach becomes apparent in species like the Pacific Balloon Turtle (FIGURE 4) whose origin is a scientific publication of balloon pollution as a serious threat to Pacific sea turtles.[7] In the Plastisphere I suggested biological systems for organs that can sense and metabolize plastics (FIGURES 5—6).

According to legendary American marine biologist Sylvia Earle, »the Earth« is a misnomer. Our planet should be called »the Ocean.« She also calls the oceans our »blue heart.« Plastics are just one of many threats to our blue heart. In my next project, HOLLOW OCEAN, as opposed to isolating a particular problem, for example, plastics, I took a different approach (FIGURES 7—8).

88

An immersive space where the audience would become embroiled in a nexus of problems was the goal of the design. Is there a healthy recipe to integrate affective empathy with scientific data? How can curiosity be architecturally instigated? How can we design a memorable experience? When the content is highly negative (e.g. statistics on the death count among common dolphins in 2019 and its relation to overfishing and climate change), can we design experiences where the experience itself is regarded as positive? *Is there still space for beautiful narratives to surface, when the subject is the death of our planet as we know it?*

In an attempt to create meaningful categories for antropogenic eco-destruction of the oceans, I emphasized two movements: Extraction of resources (oil, fish, etc.) and introduction of pollutants (plastics, toxic chemicals, noise, heat, and carbon dioxide). Global warming became a shared component as it was connected to each problem, either by causing it or catalyzing it. Following in Dr. Earle's footsteps, I chose to use an ocean-centric map titled »Spillhaus projection.«[8] This map reverses the land-based bias of traditional cartographic projections and is laid out on the ground. Large-scale aquatic columns are distributed on this map. Each aquatic column represents a »chapter« in the eco-destruction of the oceans. The ocean has six chapters. Each chapter belongs to death.

FIG. 5
Pinar Yoldas, AN ECOSYSTEM OF EXCESS, 2014. *Installation view of the exhibition at the Ernst Schering Foundation, Berlin, Germany.*
© Pinar Yoldas, 2014, photo: Pinar Yoldas

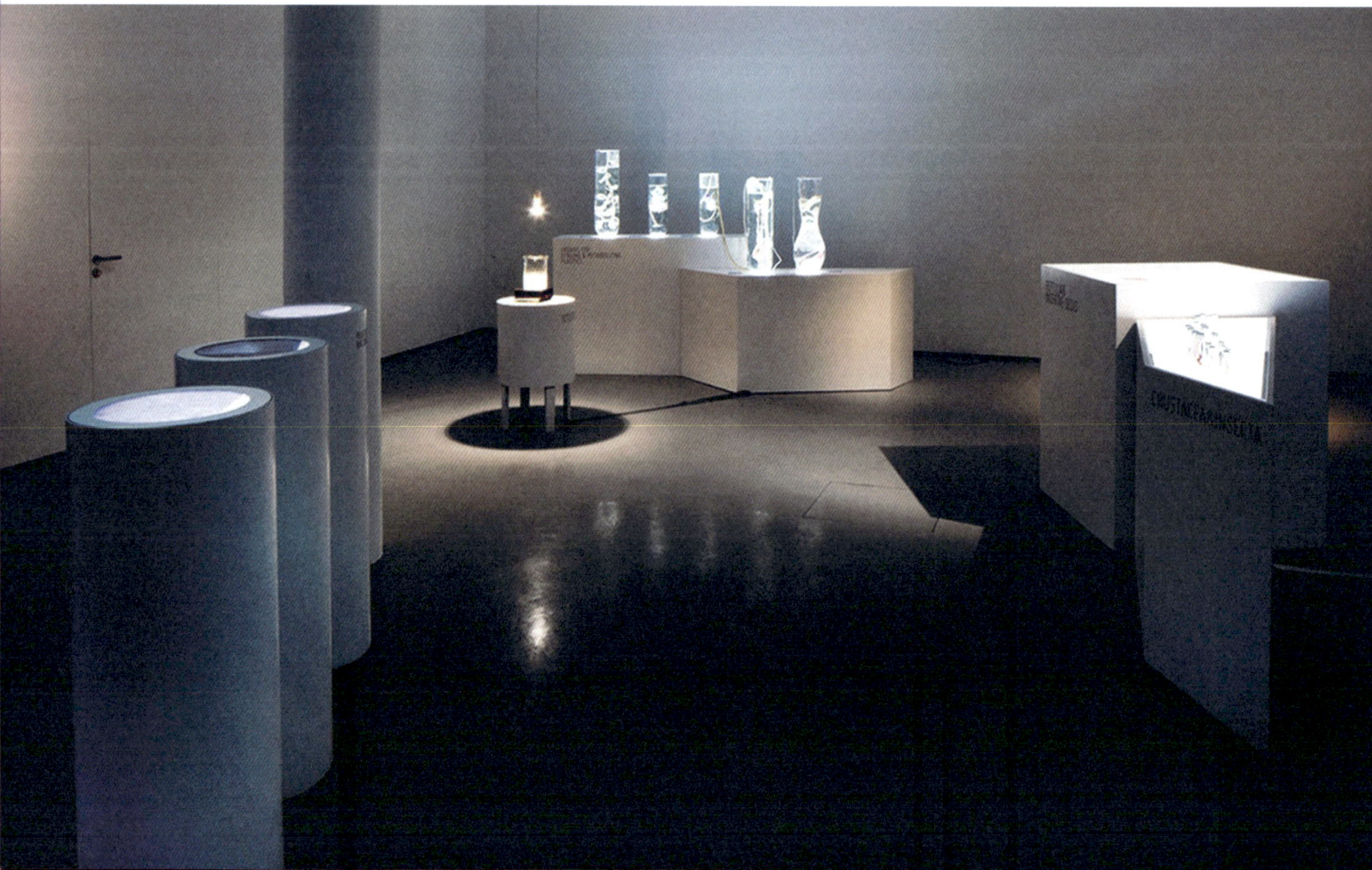

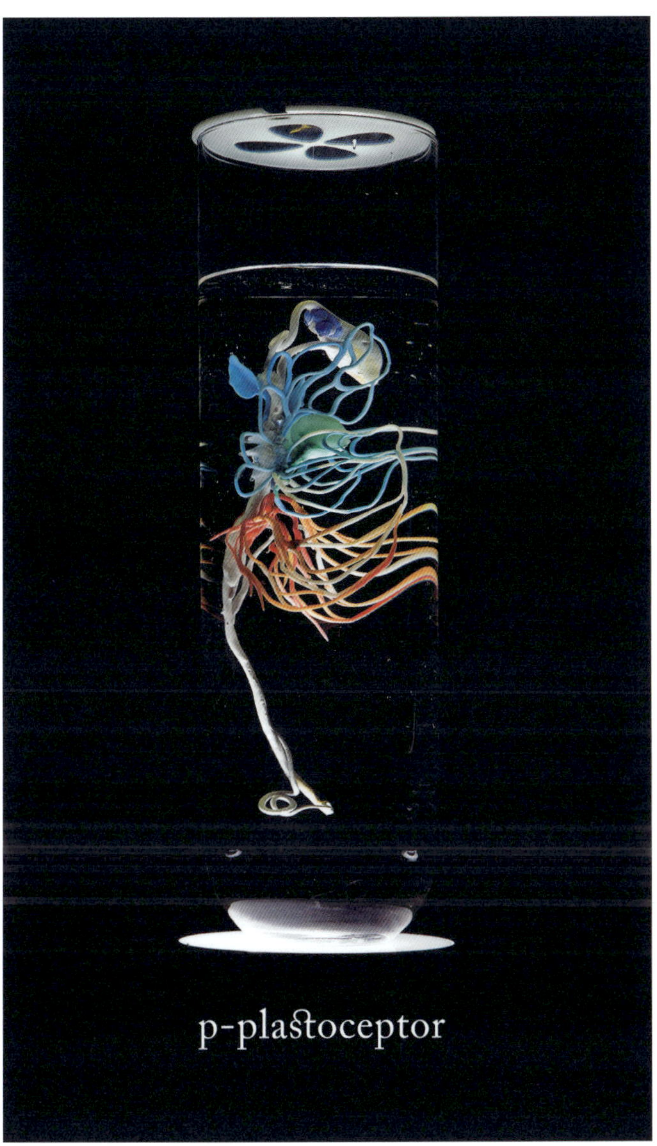

FIG. 6
Pinar Yoldas, P-PLASTOCEPTOR,
ORGAN FOR SENSING PLASTICS,
shown in the 2014 exhibition
AN ECOSYSTEM OF EXCESS at
the Ernst Schering Foundation,
Berlin, Germany.
© Pinar Yoldas, 2014, photo: Pinar Yoldas

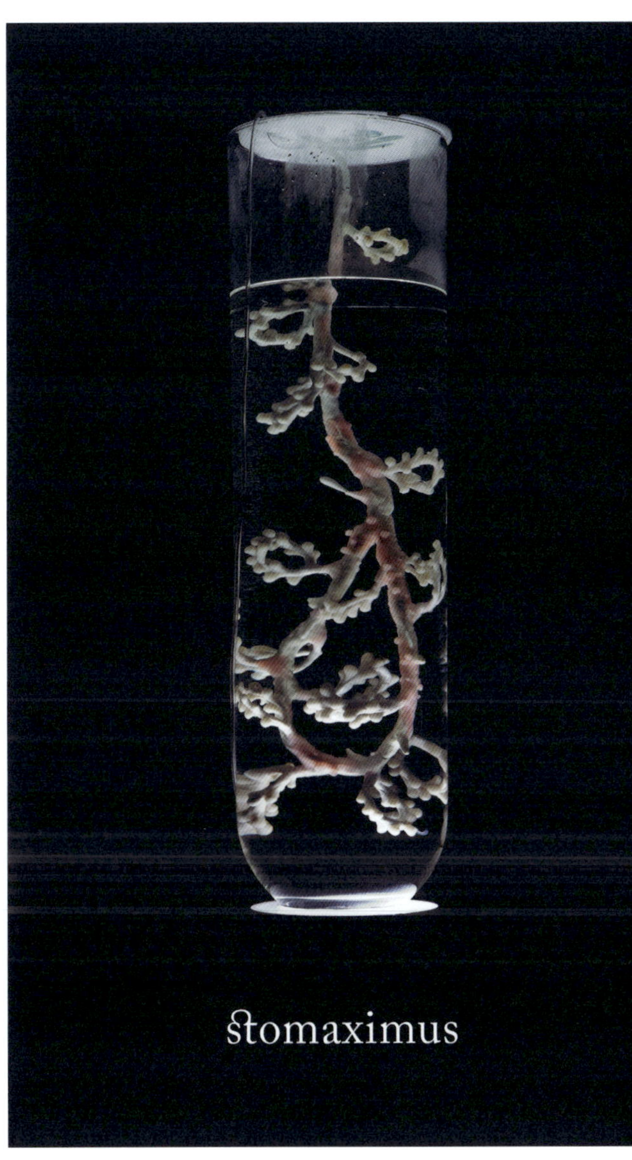

FIG. 7
Pinar Yoldas, STOMAXIMUS,
digestive organ for the Plastivore,
shown in the 2014 exhibition
AN ECOSYSTEM OF EXCESS at
the Ernst Schering Foundation,
Berlin, Germany.
© Pinar Yoldas, 2014, photo: Pinar Yoldas

90

Chapter 1: Plastic Ocean __ Over the last 50 years, humans have generated over six billion metric tons of plastic waste. Of this, only about 9% was recycled, 12% was incinerated, and 79% was left to accumulate in landfills or the natural environment. If current production and waste management trends continue, roughly 96 billion tons of plastic waste will be in landfills or the natural environment by 2050.[9] The adverse effects of petroleum — the real damage done by petro-consumerism is happening in our oceans. The most visible and disturbing impacts of marine plastics are the ingestion, suffocation, and entanglement of hundreds of marine species. Marine wildlife, such as seabirds, whales, fishes, and turtles, mistake plastic waste for prey, and most die of starvation as their stomachs are filled with plastic debris. They also suffer from lacerations, infections, reduced ability to swim, and internal injuries. Floating plastics also contribute to the spread of invasive marine organisms and bacteria, which disrupt ecosystems. According to a 2014 report by the International Union for Conservation of Nature, »Plastic debris has now become the most serious problem affecting the marine environment, not only for coastal areas of developing countries but also for the world's oceans as a whole.«[10] In this chapter titled »Plastic Ocean,« a column made of recycled plastics and composite materials is filled with water to create an aquarium for AN ECOSYSTEM OF EXCESS. Over time, new species emerge, setting an example for directed evolution, where the entire world has turned into a chemistry lab and environmental pressures that guide evolution are purely and severely anthropogenic.

Chapter 2: Phantom Ocean __ It takes about 1.03 seconds for the Google search engine to yield 1,380,000,000 results when you type »fish recipe.« However, when you search for »overfishing,« about 5,460,000 results are found in about a second. Thus for every three searches run on overfishing there are a thousand searches on how to cook and eat the said fish. Overfishing, and serious threats entangled in anything fishing-related, is perhaps the most overlooked and challenging topic as it pertains to our personal ethics and morality as much as to the corporations and governments involved.

The most shocking stats also come from the fishing industry, which I will dub the »salty slaughter industry.« A paper from 2005 claims that fishing has depleted large predatory fish communities worldwide by at least 90% over the past 50–100 years. Losing 90% of something — of any-thing — should be considered alarming if not terrifying. And yet we live on, oblivious to the detrimental effects of the global salty slaughter network.

In this chapter »Phantom Ocean,« oceans have become empty voids full of phantoms, spirits of sharks, marine turtles, whales, dolphins, and a myriad other species that form 40% of what's caught and killed from the oceans. However, most commercial fishing practices call this »surplus« of death »bycatch« and release the dead bodies of fishes, sharks,

and turtles back into the ocean. 60% efficiency by any standards
is a very low efficiency in catching commercially viable fish.

Another issue that is often hidden from the public eye is the
so-called »ghost nets,« which are abandoned or lost fishing nets (FIGURE 8).
Ghost nets are aptly titled because they linger in the vastness of the ocean
for eons: these discarded fishing nets will resist the test of time because
they are made of undying plastics. They continue to catch and kill
and they will continue to do so long after I am dead, and you are dead,
for they are nonbiodegradable and will remain where they are in one
form or another.

Chapter 3: Future Ocean __ In the following I want to imagine a cybernetic ocean
that is completely devoid of life as we know it. Oceans have been depleted
of all carbon-based life forms and are instead populated by synthetic
species that have been introduced into the oceans as a techno-fix of a
bygone era plagued by global warming, acidification, plastic pollution,
and overfishing. This chapter is one that is full of fascination with what
robotics and artificial intelligence (AI) can achieve if they follow in the
footsteps of carbon-based life. As part of a collaboration with the Soft
Robotics lab at the University of California San Diego, the chapter presents
an aquatic column showcasing a select number of »future« species
originally created with global missions such as:

1. to clean microplastics from the ocean;
2. to measure CO_2 levels for acidification-prone areas;
3. to locate and cut through ghost nets.

Alongside other rudimentary tasks, such as military intelligence, Internet-
cable protection, and prospecting and protecting offshore drilling sites,
these robots are an exaggerated form of our cultural tendency to create and
accept technological fixes above all else.

»Future Ocean« questions the negative, unintended consequences
of technology, while arguing that these consequences could be inherently
predictable and unavoidable. Does modern technology, in the presence
of continued economic growth, promote sustainability or hasten collapse?
Shall we put our last hope for protecting ocean life in the hands of a
technological elite? What are the perils of categorical thinking inherent
in engineering (Bay Area solutionism) our common future? Following
the aesthetic methodology of speculative biology present in AN ECOSYSTEM
OF EXCESS, this chapter offers a limiting case scenario, a glimpse into our
automated futurity.

Chapter 4: Dark Ocean __ Once you reach a depth of 200 m in the ocean, you
reach darkness. Only a small amount of light penetrates beyond this depth.
A depth of 200–1,000 meters is referred to as »the twilight zone« although

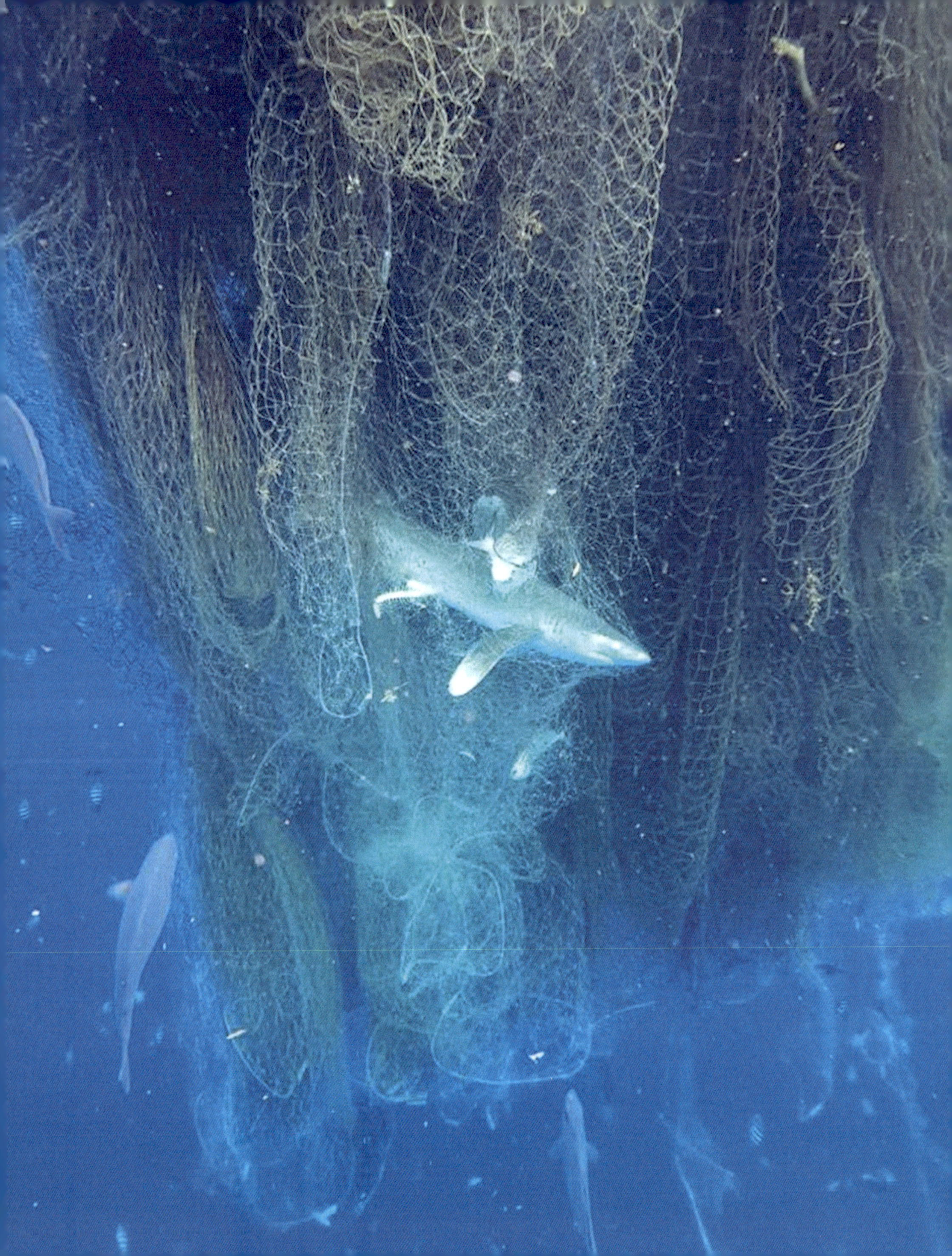

it is primarily dysphotic. It gets darker as you go down. Beyond 1,000 m there is no light. This is also the depth at which oil drilling typically takes place. The machinery operates in benthic dark, lurking in the abyss of the ocean like angular, mechanical, screechy leeches. As oil is sucked out of the ocean floor with a deafening noise, it creates what is very naively called »drilling muds.« Whereas ordinary mud can actually host life and minerals, drilling muds, for example, contain toxic substances like benzene, zinc, arsenic, radioactive materials, and other contaminants used to lubricate drill bits and maintain pressure. What's worse is that, depending on the level of toxicity, these muds can legally be released back into the marine environment. Think muddy poison sprinklers, giant ones, at the bottom of the ocean damaging the marine environment, which is so distant and alien to us that we literally make alien movies about it.

In 2018, in an essay written to former President Trump's revolutionary ecocide campaign, *National Geographic Magazine* declared that about 195 million gallons of gas are leaked into our oceans yearly from oil extraction, transportation, and consumption. When the Trump administration announced a five-year blueprint to expand offshore drilling and gas leasing in nearly all U.S. waters, environmentalists were the first to warn against the ecological shock that would be visited upon marine ecosystems.

In order to search for an offshore oil reserve, oil companies use a technique benignly named »Seismic Survey.« It should be called »Seismic Suffering and Death via High-Decibel Explosive Impulses.« As sonic bombs detonate in the name of »seafloor mapping and exploration,« one starts to question the efficacy of explorers who immediately kill what they explore. In 2014, the Bureau of Ocean Energy Management completed a Programmatic Environmental Impact Statement on proposed seismic surveying in the Atlantic, and estimated that 13.6 million marine animals would be disrupted.

Let us imagine that out of those 13.6 million animals, a fraction of ocean life will miraculously survive seismic attacks. What awaits them next is the increase in toxicity and drilling muds, which would continuously be released as long as the drill rig is operational. Let us imagine that a fraction of marine life could survive life-threatening levels of toxicity. What awaits the survivors next? Oil spills. A spill-free deepwater oil rig is very rare and thousands of oil spills occur each year in U.S. waters according to the National Oceanic and Atmospheric Administration (NOAA); most are small in size, spilling less than a barrel of oil. From the iconic 1969 oil well blowout in Santa Barbara, California, until 2017 when NOAA's report was published, there have been at least 44 oil spills that were all over 10,000 barrels (420,000 gallons), affecting U.S. waters. The largest was the 2010 Deepwater Horizon well blowout in the Gulf of Mexico, which spilled an estimated 200 million gallons of oil. A decade later many species are

This photograph was shot in 2018 close to the Cayman Islands by diver and fisherman Dominick Martin-Mayes and his companion Pierre Lesieur. It documents a discarded fishnet, one of many that are continuing to squander ocean wildlife, documenting the actual horror of overfishing.
Photo: Dominick Martin-Mayes and Pierre Lesieur

still struggling, their populations lower than before. Some species may never fully recover. Deepwater oil drilling is a systemic weapon against marine ecosystems that is deployed at will and ruthlessly in the name of creating more jobs or keeping the economy alive. It is perplexing to see these two words ecology and economy – both of which share a common origin *ekos* – at war with each other. Ecology threatens economy because sacrifices will have to be made to move away from business-as-usual to harmonious living with nonhuman life.

Economy threatens ecology because when executive decisions are made on a planetary level they affect all life forms that share a connected web of fragile ecosystems. Interconnectedness is key and as a society we are just scratching the surface of understanding what interconnectedness truly is. Money is a numerical abstraction of value, and there is peril in ignoring the impossibility of translating biological life and its manifestations in living things into number representations of financial wealth. Today, financial markets like Wall Street lose touch with reality, the very reality of living, breathing life. Ironically, the root word *ekos* means home. Unless the true value of home is understood, as well as shows that economy can work harmoniously with ecology, the fate of our oceans will remain dark. Offshore oil drilling is a direct act of violence against marine ecosystems. In the »Dark Ocean« chapter, an aquatic column will be dedicated to sonic, visual, and olfactory properties of offshore drilling, mimicking the terror that continues to take place in the eternal night of deep waters.

Chapter 5: Stifled Ocean __ »Stifled Ocean« focuses on the negative impact of global warming. This is a long chapter because the impact of global warming is, and will be, multifold and felt at so many different levels. Moreover, nobody really knows what exactly will happen for global warming may lead to a cascade of ecosystem collapse events. One area of focus is invasive species that prove to be more resilient to changing environments and anthropogenic pressures. One group under scrutiny is jellyfish. Contrary to the common belief that jellyfish will take over the world, as with any other animal group some jellyfish species struggle in disturbed systems, while others show more resilience and are able to survive in low oxygen and highly polluted environments. However, an estimated 2,000 species are expected to expand their range, appear earlier in the year, and steadily increase in population size each year. There will be many financial consequences of increasing volumes of jellyfish in the oceans, but nothing will compare to the ecological cost.

Chapter 6: Acid Ocean __ Prior to the Industrial Revolution, average ocean pH was about 8.2. Over the course of the last 200 years this value dropped to 8.1. Although the difference seems small at first glance, due to the fact that pH calculations are logarithmic, this in fact equals to a change of 25%. By the end of the century, seawater pH levels are estimated to drop to 7.8. Pure water has a pH of 7, which is considered neutral in the 14-step acidity/alkalinity chart. A womb carrying a fetus is considered healthy at around 7.1 to 7.3. Seawater is considered healthy at 8.2 to 8.1. To understand the significance of the right pH for vital functions we could think of it as air, as it is easier to imagine air with higher concentration of CO_2 and lower pH that make it less breathable.

Ocean acidification impacts shell builders directly by absorbing carbonate ions from water, which is necessary for building exoskeleton. Corals, for instance, build their hard stony skeletons over many years, much like a forest on land. Corals literally grow into an optimum habitat for many other species. Under severe acidity the coral's skeleton dissolves, which would be comparable to a forest fire or the disappearance of a forest, which would leave a myriad species homeless. For instance, pteropods like sea butterflies become subjected to shell dissolution, which kills them over time. Many larvae are affected via changing levels of protons in the ocean. Ocean flora is also negatively affected by excess CO_2 dissolved in seawater.

Acidification is one of the little-known threats to ocean life and life in general. In this chapter of HOLLOW OCEAN, shelled life at various scales will be displayed undergoing deformation and obliteration. The aquatic column will be treated as a volumetric data visualization projection, with acidity and global temperatures as two main axes. Using time as the third axis, morphological transformations in shelled life, such as barnacles, will be displayed.

Conclusion ＿＿ Starting with AN ECOSYSTEM OF EXCESS, I built art installations and exhibitions to communicate the findings of scientific research while creating a platform for education. I am convinced that art and artistic experience are valid learning environments and regardless of the intent of the artists behind them, artworks and artistic experiences posess an intrinsic quality to transform the way the audience is sensing, perceiving, and acting upon the world. Perhaps a thought may not be considered »an action,« but when thoughts are considered as connections between neurons, it can be claimed that artistic experiences »act« upon the nervous system. Art changes perception, elicits emotion, and leads to thought formation. In »Affective Learning: A Manifesto,« American scholar and inventor Rosalind Picard at the Massachusetts Institute of Technology (MIT), together with many other distinguished scientists including American scholar and social robotics expert Cynthia Breazeal, defined, in 2004, affective learning for the century to come. By pointing out the bias toward

95

cognition and the relative neglect of affect, they also show the developments and interest in affect: »On the most fundamental level, an accelerat-
ed flow of findings in neuroscience, psychology, and cognitive science
itself present affect as complexly intertwined with thinking, and performing important functions with respect to guiding rational behavior,
memory retrieval, decision-making, creativity, and more.«[11] Then what
about art and affective learning? And how does art offer an affective
learning experience?

Affective learning is defined as learning that involves our feelings,
emotions, and attitudes.[12] A spatial experience, such as walking around a
room, unlike a temporal experience, for example, motion pictures, cannot
offer a temporal continuum, or edit multiple experiences back to back.
However, a spatial experience is completely immersive and can stretch
through time. Spatial experiences when combined with emotional arousal
can lead to memorable experiences, so to speak highlights, in the viewers'
lives. Spatial experiences can also activate multiple modalities by using
sound, light, and temperature as agents of design. Being immersive and
multimodal are two major qualities of a successful learning environment.

A very essential form of activism is communication. Communicating planetary health topics to a wider audience is a worthwhile task
and presents many challenges. While some of the topics like plastics are
no longer new, there is still a lot to be learned and explored. Knowledge
is power, but then who holds the power?

With the goal of effectively communicating the plight of our
oceans, I set out to design affective experiences that offer an immersive
and multimodal learning environment. While my methods are not
empirical and I am not certain if the installations do serve their purpose,
there are many other subjects that have not been covered. I invite
every designer, artist, and cultural producer to take a stab at the
imperceptibility of our anthropogenic impact and ignorance about
eco-destruction so we can move EcoActivism forward.

1   In 2005, of all recycled material in the United States only 4% was plastics. Of all plastic trash generated, only 6% is recycled. Plastic is the least recycled material, but widely used in packaging. The other issue is composite materials that make it harder—if not impossible—to recycle them. Given the near impossibility of effective recycling, most product design ends up being disposable, although the disposability rate may differ.

2   Charles Darwin, *On the Origin of Species by Means of Natural Selection, or the Preservation of Favoured Races in the Struggle for Life*, London: John Murray (1859), 490.

3   Roland Geyer et al., »Production, Use, and Fate of All Plastics Ever Made,« *Science Advances* 3, no. 7 (2017), e1700782, American Association for the Advancement of Science (AAAS), doi: 10.1126/sciadv.1700782.

4   It took seven years to gain funding for the execution of a comprehensive artistic project on the detrimental effect of plastic on marine life. Particularly in the United States, I witnessed how most people around me chose to ignore the severity and urgency of plastic pollution. Curators, art institutions, and funding bodies were no exception. Today, as plastic pollution metastasized, it became harder to turn one's back on it. However, despite increasing curatorial interest in the topic, it is not enough to effect permanent change. The Transmediale festival in Berlin was the first institution to recognize the project's artistic merit, and it is thanks to Kristoffer Gansing and a generous grant from Schering Foundation Berlin that *An Ecosystem of Excess* was presented to the public at an exhibition in 2014.

5   In a paper aptly titled »Life in the ›Plastisphere‹: Microbial Communities on Plastic Marine Debris« by Erik R. Zettler, Tracy J. Mincer, and Linda A. Amaral-Zettler, *Environmental Science and Technology* 47, no. 13 (2013): 7137–7146. Online: https://doi.org/10.1021/es40128.

6   I do own the »Juicy Salif Citrus Squeezer« designed by Philippe Starck for Alessi, Italy. It is not very functional but it is the pinnacle.

7   Footage on Charles Moore can be accessed via Algalita. Online: https://algalita.org (last accessed January 23, 2021).

8   Named for its creator, the South African-born oceanographer, geophysicist, inventor, urban designer (having come up with the Minneapolis Skyway System), and comic artist Athelstan Spilhaus, the Spilhaus Projection »reverses the land-based bias of traditional cartographic projections.«

9   Roland Geyer et al., »Production, Use, and Fate of All Plastics Ever Made,« *Science Advances* 3, no. 7 (2017), e1700782, American Association for the Advancement of Science (AAAS), doi: 10.1126/sciadv.1700782.

10   Florian Thevenon, Chris Carroll, and João Sousa (eds.), *Plastic Debris in the Ocean: The Characterization of Marine Plastics and Their Environmental Impacts*, Situation Analysis Report (2014).

11   Rosalind W. Picard et al., »Affective Learning: A Manifesto,« *BT Technology Journal* 22, no. 4 (2004), 253.

12   David R. Krathwohl, Benjamin S. Bloom, and Bertram B. Masia, *Taxonomy of Educational Objectives: Handbook II: Affective Domain*, New York: David McKay Co. (1964).

Initiated in 2015 as a conceptual platform for
ideas, tactics, and collaborations, »Open Source
Estrogen« began with a simple question: What if
it were possible to synthesize estrogen in the
kitchen? From this seed came more questions as to
how gendered bodies are controlled and managed
through corporate and institutional science and
how endocrine-disrupting molecules exist already
around us as a state of environmental toxicity.
Is there hope for disobedient bodies amongst
capitalist and ecological ruins? To answer this
question, the project and its following mutations
position hormones and their molecular colonization as a constant entanglement with the self
—human, nonhuman, and planetary—with their
biopolitical omnipresence as the very source
from which we collectively hack, demystify, and
emancipate its molecular mystique.

All Washed Over by Hormones of Loving Grace

Mary Maggic

How is gender codified by hormones? How did we arrive at the black-boxed fact that a molecule like estrogen produces characteristics of femininity and testosterone of masculinity? It is widely accepted that gender is chemical, that »sex hormones« such as estrogen, progesterone and testosterone are responsible for the production of primary sex characteristics in the developing fetus, then of secondary sex characteristics at the onset of puberty.[1] However, the way that scientific institutions have defined female-ness and male-ness since the 1800s comes already entangled with preexisting gender biases. Canadian philosopher Ian Hacking writes in *Representing and Intervening*, »We did not just find sex hormones somewhere in a lost corner, like a desert island lost in the mist. We ourselves called sex hormones into existence.«[2] Despite the facts that hormones like estrogen and testosterone are responsible for the basic homeostasis of the body, as opposed to solely the reproductive system, and that they are produced in all bodies regardless of gender, scientists continued to source hormones from their codified gender assignments. For example, French-Mauritian scientist Charles-Édouard Brown-Séquard (1817–1894) who popularized the concept of organotherapy in the late 1800s, would inject himself with extracts of animal testicles in order to rejuvenate his own masculinity.[3] The hormones were literally »sexed«—given sexes of their own.

When the medical basis for organotherapy was later verified by the presence of hormones like estrogen and testosterone, these molecules would increasingly act as a kind of biosurveillance that controls and manages whole populations and their gender subjectivities. This state-scientific production of heterosexual bodies would eventually prove to be highly profitable for those institutions in charge of their gatekeeping. Operating under what Spanish philosopher Paul B. Preciado calls a »pharmacopornographic regime,« these molecules would reproduce notions of femininity and masculinity, where biotechnologies like Viagra, birth control pills and anti-AIDS medicine use sex, sexuality and sexual identity as the »somato-political centers for producing and governing subjectivity.«[4] The 1930s mark a period in the hormone historical timeline where several major European pharma-industrial companies set out to race and territorialize both human and nonhuman sources for hormones. From pregnant

women and horses to prisoners and psychiatric patients, there was no
end to the amount of consensual and nonconsensual bodies that could
be turned into biocommodities for a budding hormonal industry.
What emerged was a mutually beneficial triad: the clinic that diagnosed
and created more patients, the laboratory that researched and developed
hormonal therapies and finally, the pharmaceutical companies that
marketed these therapies to the masses.[5] Today, synthetic estrogen and
progesterone continue to be the most highly manufactured molecules
in the world.

While today's major petrochemical, agricultural and pharmaceutical corporations have compartmentalized into their distinct categories
of industry, their roles in the origins of chemical research are far more
intertwined. In the 1930s, British scientist Edward Charles Dodd pursued
an estrogenic molecule that would be referred to as the »mother substance,« one that would produce enough visibly feminizing effects to
be marketed as a hormone therapy to the masses. First synthesized in 1891
by Russian chemist Alexander Pavlovich Dianin (1851–1918), Bisphenol A
(BPA) was one of the molecules tested by Dodd and was determined to
be 1/37,000 as effective as estradiol, a naturally occurring estrogen. Finally,
Dodd came upon the far more estrogenic compound diethylstilbestrol
(DES), which eventually became a widely used hormone therapy for
»female problems« associated with menstruation and pregnancy, while
BPA's commercial potential would not be fully realized until after World
War II when it was industrially developed in the U.S. and Switzerland
as an epoxy resin and later as a polycarbonate plastic.[6] DES, which continued to be prescribed to women and factory animals for the next 30 years,
was finally banned in the 1970s for its carcinogenic effects and harmful
chemical legacy that spanned multiple generations of women. In the
same decade, as petrochemical companies like General Electric,
Shell and Dow Chemicals ramped up production on plastics, BPA had
already reached half a billion pounds in the U.S.[7]

We're All Living in an Estroworld ＿＿ Since the continued rise of industrial
capitalism, every corner of the planet is left with some residue of synthetic
molecules, many of which are hormonal. These synthetic molecules are
synonymously known as persistent organic pollutants (POPs), endocrine
disrupting compounds (EDCs) and xenoestrogens because of their estro-
gen-mimicking and estrogen-displacing properties. From the discovery
of polycarbonates in the Marianas Trench, to whole populations of birds,
frogs and fish failing to produce viable offspring due to pesticide contamination, to the transgenerational cancers inherited from grandmothers
who were prescribed DES, this microscopic moment on the scale of geologic time is already (and continues to be) marked by unprecedented levels
of environmental toxicity, irreversible planetary changes and collective

FIG. 1

*PHALLOMETRICS. Digital illustration
for the project GENITAL(＊)PANIC, 2019.*
© Mary Maggic

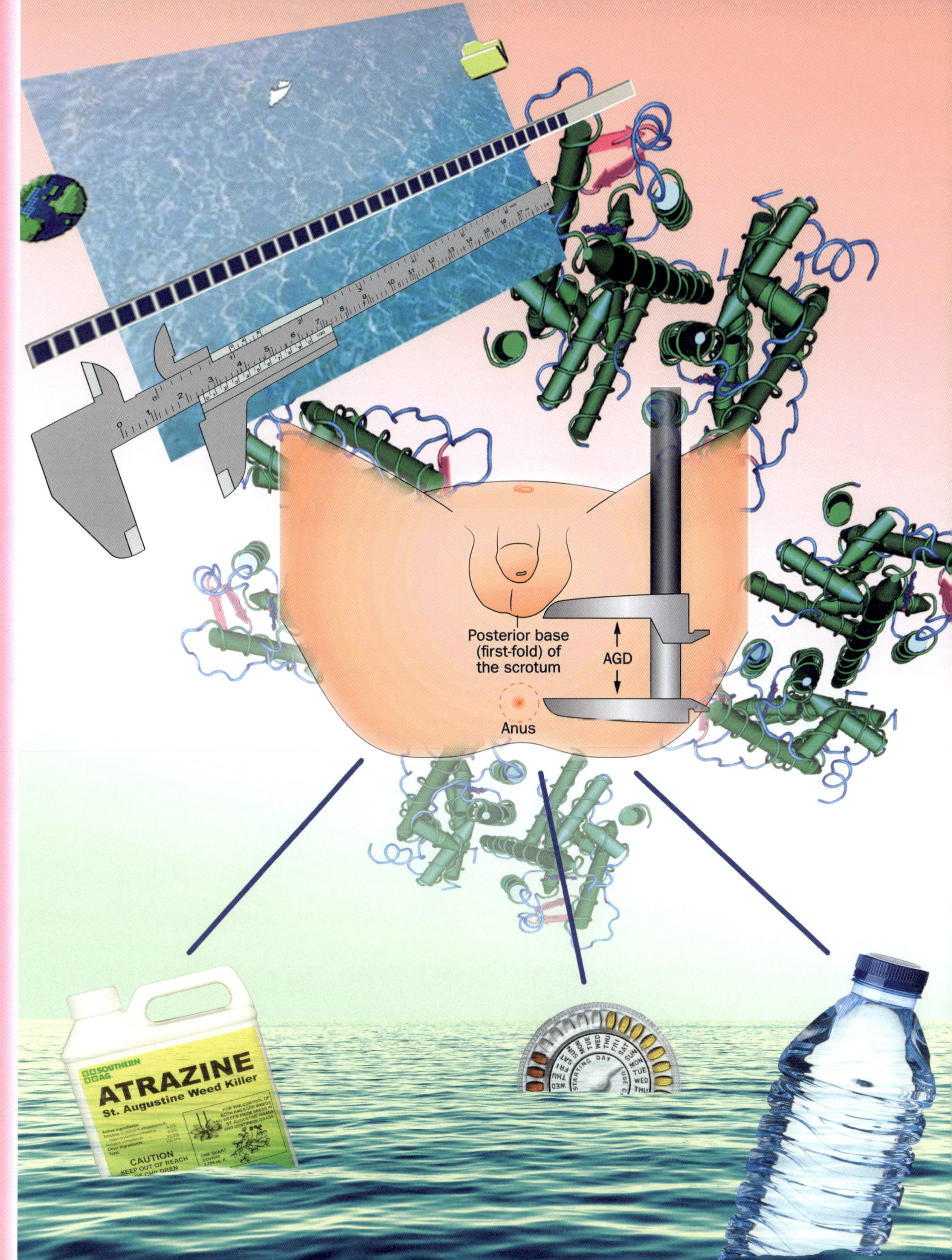

Posterior base
(first-fold) of
the scrotum
AGD
Anus
SOUTHERN AG
ATRAZINE
St. Augustine Weed Killer
CAUTION
KEEP OUT OF REACH
OF CHILDREN

species mutations. Canadian researcher and writer Heather Davies describes the accumulation of plastics as a kind of »geologic indigestion,« a marker of the Anthropocene where the natural can no longer be disentangled from the synthetic.[8] The effects of these synthetic molecules on the human body have been linked to neurological (autism, lower IQ, mood disorders) and physiological effects (diabetes, obesity, early-onset puberty, worldwide sperm count drop) as well as various reproductive cancers. These molecules drift, seep, wander, flow, invade wherever they please, carried by both air and water in invisible and unimaginable ways (FIGURE 1). This type of »slow violence« is in direct contrast to blatant catastrophic events such as the Chernobyl nuclear disaster; more akin to climate change, the effects of environmental toxicity are gradual and therefore difficult to perceive and mirror the preexisting lines of inequality where toxicities are more likely to be distributed among nonwhite, indigenous and marginalized communities.[9]

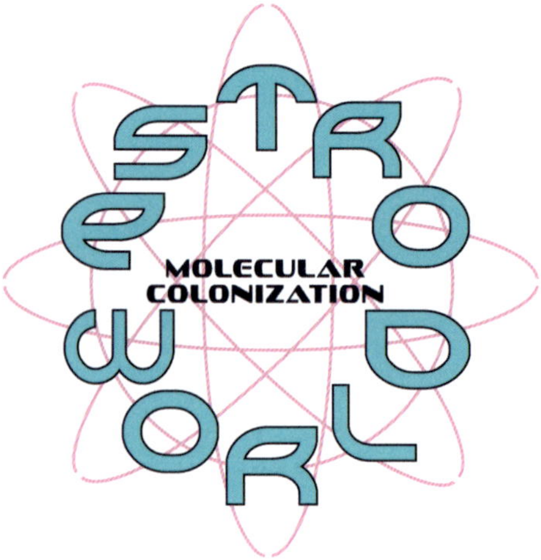

None of us consent to live in a permanently polluted world, where harmful molecules continue to leave their chemical traces despite their classifications as endocrine disruptors as well as the many environmental lobbying and activist efforts to pass legislation that would further regulate the unrelenting capitalist production of toxic molecules. While bodies are being unknowingly polluted, non-normative bodies continue to be policed and pathologized on the basis of our oppressive and exclusionary gender binarism, from violent surgeries of intersex infants and the disqualification of intersex Olympic athletes from their gendered categories to the denial of hormonal therapies to transindividuals. Dominant political discourses continue to feed our collective panic with apocalyptic visions of the future while framing reproductive futurism as the ultimate privilege of cis-, abled heteronormative bodies, suggesting that queer, trans and intersex bodies have no place in the survival of the human species.[10] Marked by an all-pervasive environmental toxicity deeply tied to patriarchy, neoliberal capitalism and colonial hegemony, and a resulting epistemic crisis to our preexisting taxonomies of ecological heteronormativity and gender binarism, this is the ESTROWORLD (FIGURE 2) that we all live in, and this is the molecular colonization that we never signed up for.

There Is Hope for All Disobedient Bodies ＿＿ So what does it mean if our bodies are industrially modulated, that our sex, gender and reproduction are not as fixed and recalcitrant as we always believed? How do we situate our bodies, identities and fears in the midst of toxic and alienating environments? Are we able to reformulate old notions of the normative body in order to build more inclusive futures? A collaborative research project initiated in 2015, *Open Source Estrogen* began with a simple speculative question: What if it were possible to synthesize estrogen in the kitchen? From this seed came many more questions as to whose bodies are affected,

FIG. 2
ESTROWORLD INC.
Digital illustration, 2020.
© Mary Maggic

who is producing and distributing these hormones, who gets to have access, and what are the ethics of self-administering self-synthesized hormones? The project's scope eventually expanded to the multitude of hormonal molecules that now pervade the planet as a state of environmental toxicity, revealing a strange tension between *active* queering (through gaining access to hormonal therapies) and the *passive* queering that happens unbeknownst to almost every organism on the planet. Emerging from this tension is the urgency to refigure new feminist care strategies for living and coping in a permanently polluted world, where body sovereignty is utmost at stake, and where notions of purity are not and should not be a viable option (FIGURES 3—4).

The term »open source« in the project's name takes on a double meaning. »Open source« signifies one of the project's main methodologies, the collaborative prototyping and sharing of DIY/DIWO (do-it-yourself/ do-it-with-others) protocols for hacking hormones. If we frame these molecules as black-boxes wherein knowledge has been institutionalized from the realm of amateur exploration, then the social rupture of these black-boxes usurps the institutional process of knowledge production. The ethos of hacker culture has always questioned power and power relations when asked who gets to produce knowledge because whoever gets to produce knowledge also gets to define and enforce subjectivities.

105

FIG. 3
*Video still from the speculative
fiction film* HOUSEWIVES MAKING
DRUGS, *2017* (10:12 min.).
© Mary Maggic

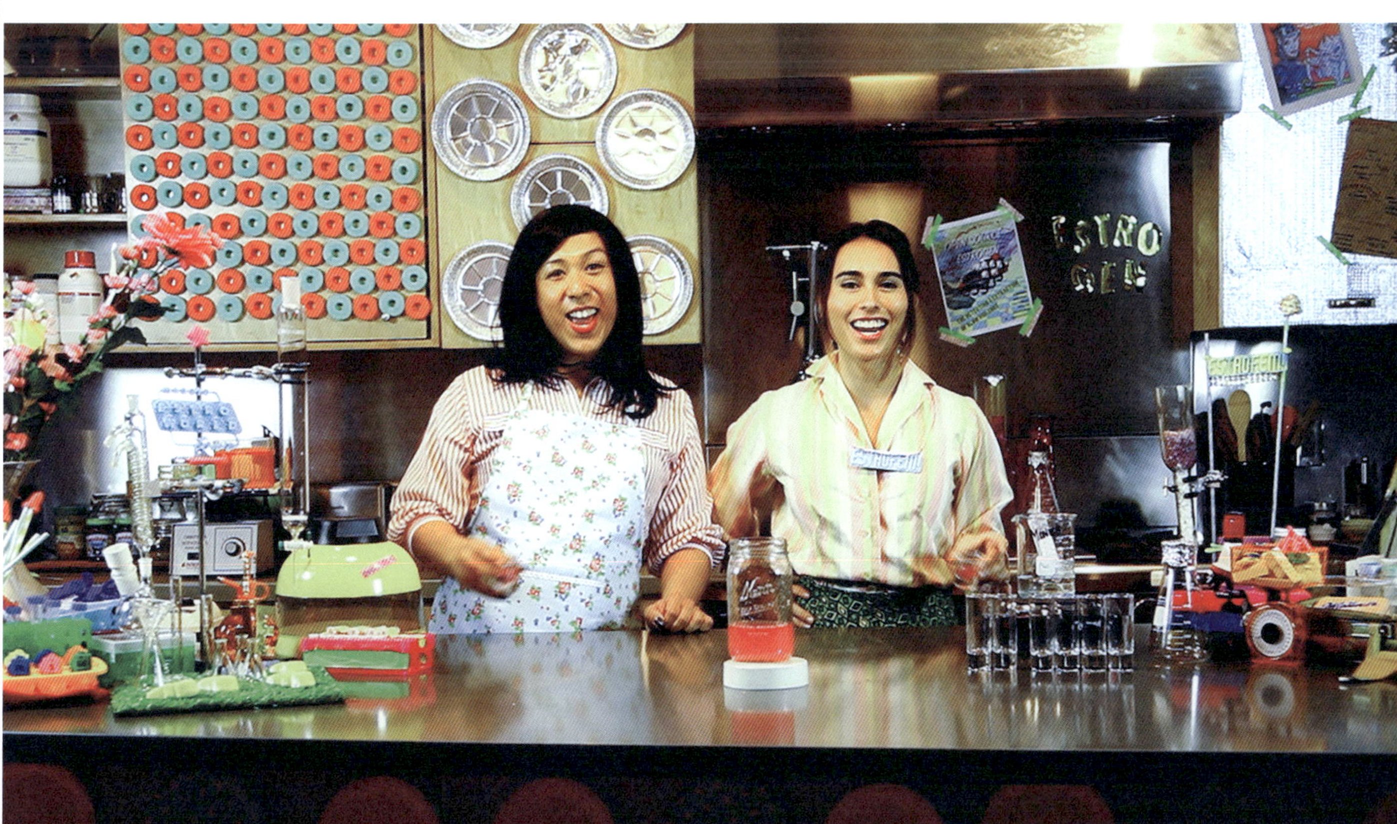

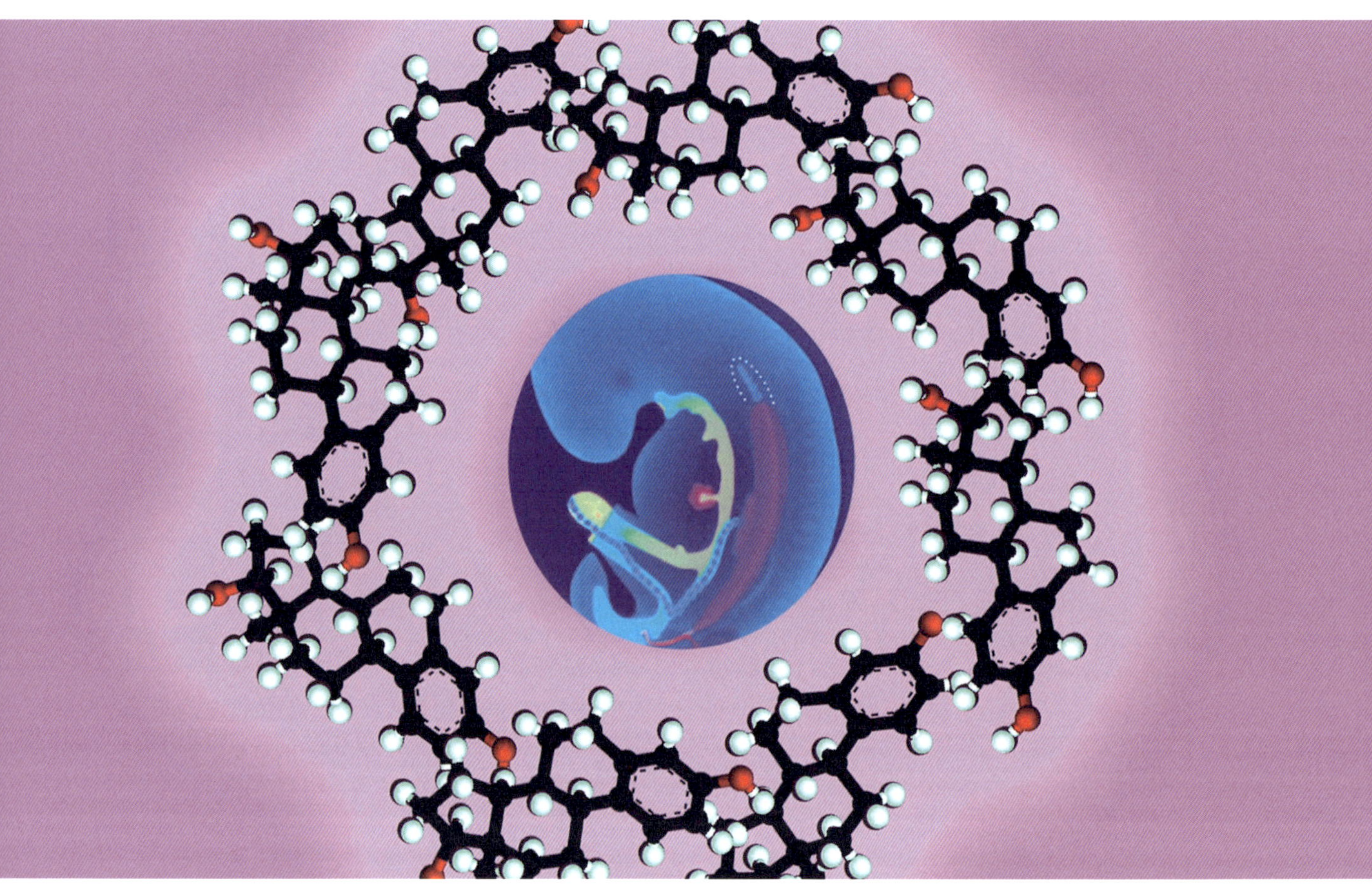

FIG. 4
*Projection still from the parti-
cipatory performance*
MOLECULAR QUEERING AGENCY, 2017.
© Mary Maggic

Hacking and demystifying hormones undoubtedly creates critical spaces
for redefining those power relations and fostering new collaborative
queerings of the status quo. If hormones are the tools of biotechnological
surveillance, then hacking them readily dismantles their institutional
biopower. Due to the biopolitical omnipresence of these molecules,
the project's name takes on its second meaning. »Open source« is also
the potential state of these molecules—readily available to hack, mutate,
collaborate with, become with, and build new possibilities for our
collective future. In the famous words of *The Xenofeminist Manifesto*,
»Alienation is the labor of freedom's construction.«[11] While the biopolitical
ubiquity of these molecules renders our planet more alienated than
ever before, it is with this very alienation that we can explore the limits
of the possible. There are visions of a just and inclusive future outside the
ones prescribed to us by dominant patriarchal discourses that allow us
to move beyond apocalyptic panic. There is hope for all disobedient bodies.

Hormone Queering Resistance ＿＿ The intersection of art and biotechnology
has always ruptured the institutional boundaries that have defined
power and knowledge, opening the doors for greater public accessibility.
One such intersection is the practice of »biotechnical civil disobedience,«
which originates from Critical Art Ensemble, a renowned U.S. art and
tactical media collective that pioneered works around bioterrorism,
germ warfare, transgenic foods, and the unconsenting commodification
of marginalized bodies. Their projects often employed processes of public
demystification in order to address critical concerns around corporate
biotechnology where the lay public becomes increasingly divested of its
political and socioeconomic implications.

Hacking therefore bridges these gaps in power and understanding
through a process that U.S. writer and artist Claire Pentecost regards as
»public amateurism,« where people consent to learn and fail together
in public, removing the hierarchy of the expert.[12] In a similar spirit, *Open
Source Estrogen* also devises its own cultural strategy for civil disobedience,
taking influence from the chapter »Transgenic Production and Cultural
Resistance: A Seven-Point Plan« in *Molecular Invasion* by Critical Art
Ensemble, 2002[13]:

Six Point Plan for Hormone Queering Resistance
1. Unearth the dominant patriarchal agents of hormonal production
   and pollution; build public understanding of the xeno forces at play.
2. Demystify the institutionalized »black-boxed« knowledge of
   biochemistry, endocrinology and ecotoxicology; pave the way
   for hormone hacking, freak science and amateur exploration.
3. Resist neoliberal pharmaco-capitalist profiteering of (un)consent-
   ing bodies.
4. Reject glorifications of »the natural,« condemnations of
   »the unnatural« and above all, rhetorics of technosolutionism
   that promise to elucidate both.
5. Undermine deeply entrenched notions of (eco)heteronormalcy
   and purity; use »queering« as a reclaimed potential for resistance.
6. Consider the microperformativity of hormones as an agential
   power of not only molecular colonization but also of molecular
   collaboration.

While *Open Source Estrogen* makes up the conceptual foundation for
why we demystify hormones, ESTROFEM! LAB provides the *how*. Through
a nomadic workshopology and collaborative prototyping practice
often referred to as »estrogen geeking« or »freak science,« ESTROFEM! LAB
focuses on the tools and protocols that make visible our shared molecular
colonization (FIGURE 5). A term that emerged through the online and
in-person gatherings of the global Hackteria network for open source

FIG. 5
*YES-HER Yeast Biosensors
Mobile Lab, Linz, 2016.*
© Mary Maggic

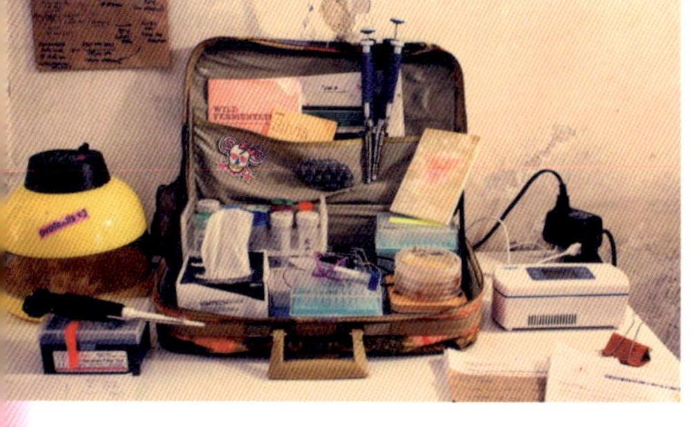

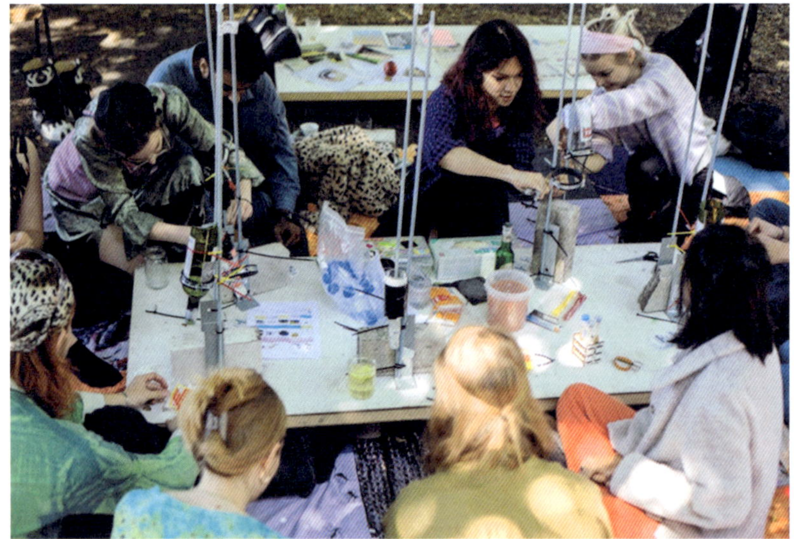

FIG. 6
ESTROFEM! LAB *workshop hosted
by hhintersection, Hamburg, 2020.*
© Mary Maggic, photo: Maik Gräf

biological art, workshopology is the constant feedback loop of work-
shopping in and of itself—it represents the concept of *homo ludens*
or communal tinkering, play and geeking as an iterative process that
continuously develops new accessible ways for amateur knowledge-
production. Because of the Hackteria network and collaborative spirit
of workshopology, the hormone-hacking protocols of ESTROFEM! LAB
owe themselves to collaborations with Paula Pin from Spain and Gaia
Leandra from Italy of Transhackfeminism and Gynepunk Lab, Slovenian
new media artist and collaborator with *Aliens in Green* Špela Petrič,
as well as many others in the larger art-science scene, such as U.S.
interdisciplinary artist Rian Hammond of Open Source Gendercodes
and Canadian artist and professor Byron Rich, who first conceptualized
the idea of an open source birth control pill for *Open Source Estrogen*.
A series of mobile labs outfitted into suitcases, these protocols have included (1) YES-HER yeast biosensors, a low-cost detection method using transgenic yeast containing a human estrogen receptor; (2) Urine-Hormone
Extraction-Action, a DIY column chromatography method for the isolation
of urinary hormones; (3) DIY Solid Phase Extraction, a method of concentrating hormones and EDCs from dilute environmental water samples;
and lastly, (4) fungal bioremediation with various species of white-rot
mushrooms. The protocols have been largely disseminated through public
workshopology (FIGURE 6), acting as discursive and reflexive exercises
in building technical knowledge as well as greater body sovereignty. Only
from this social excavation of harmful toxicities can we begin devising
strategies of collective care for a more equitable future.

The Fluid Commons ___ One such strategy was explored during a 10-month research residency called »River Gynecology« which took place in Yogyakarta, Indonesia, in 2019. Collaborating with Lifepatch, a citizen initiative in art, science, and technology on the Jogja River Project, the goal was to engage communities living along the river to monitor its health as though it were their own bodies, exploring strategies founded in decolonial feminism and care. Kali Code is the most heavily polluted river in Yogyakarta and at first glance it is a surreal landscape that is colonized by plastic waste dumped from residences, hotels, hospitals, and general shops and eateries in the area (FIGURES 7–8).

109

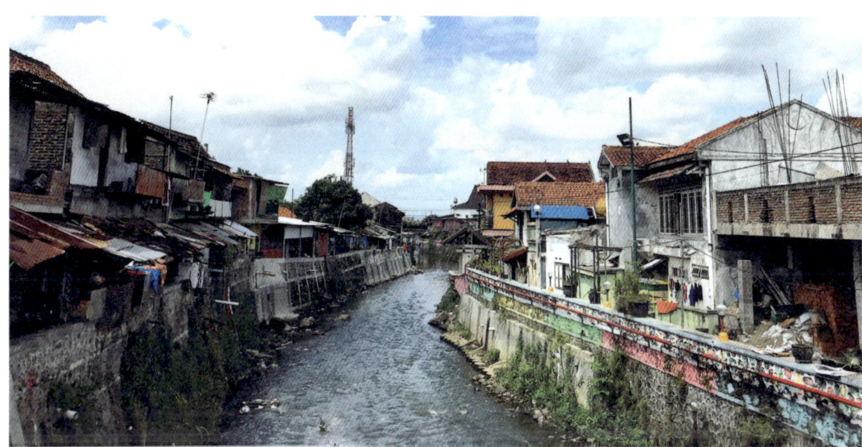

FIG. 7
*View from a bridge across the
Kali Code river in the city center
of Yogyakarta, Indonesia, 2019.*
© Mary Maggic

FIG. 8
*Children playing at the Kali Code
river during the summer dry season,
Yogyakarta, Indonesia, 2019.*
© Mary Maggic

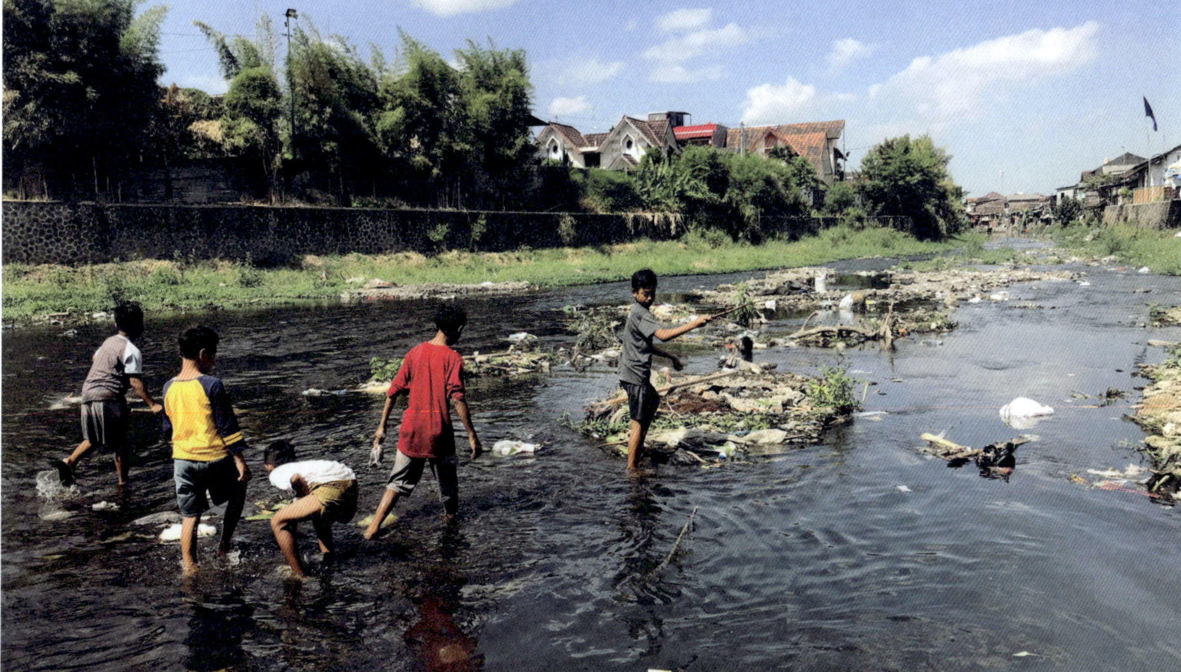

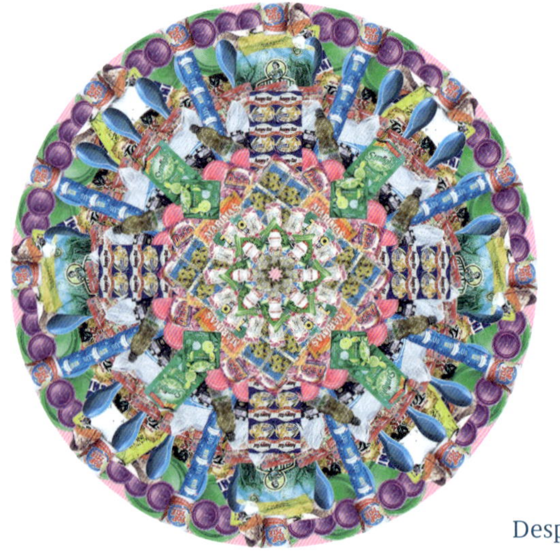

FIG. 9
*Projection still of a rotating
mandala composed of trash found
in the Kali Code river, Indonesia,
from the installation* MILIK BERSAMA
REKOMBINAN, *2019.*
© Mary Maggic

Despite the polluted conditions of this river, the citizens still believe the
water is safe for daily use, such as cooking, cleaning, washing, drinking,
playing, and fishing. While the root of the problem is complex and
multifaceted (income level, education and awareness, lack of government
infrastructure), it can also be seen as a sociocultural issue. Because
Javanese mysticism, Kejawèn, is especially embodied in the Indonesian
river communities living along the river, the project tried to incorporate
these elements into the presentations, workshops, and artworks stemming
from this research residency. Through interviews with local citizens
it was discovered that while several spirits have been sighted to travel
along the river, the river itself is not a spiritual entity but rather a highway
that connects two spiritual kingdoms: the volcano Merapi in the north,
and the South Sea.

Because the river was observed as »other« from the bodies of the
people, the project took inspiration from the theoretical framework
of Hydrofeminism as a way of creating solidarity with and across watery
bodies.[14] Although water is the stuff of life, it is also the primary carrier
of all these harmful toxicities that connects us all in a collective multispe-
cies struggle. While the citizens of Kali Code are undoubtedly those who
never consent to live in an intoxicated landscape, they themselves are
perhaps incognizant of their own porosity with the environment and the
plastic waste that is the source for their becoming-with. In the trilogy
of works titled *MILIK BERSAMA REKOMBINAN* (2019), this very plastic waste is
heavily represented in all three works of the installation. First, long latex
sculptures, embedded with trash, flank the left and right side of the
installation and represent aerial views of the river on the east and west side
only as if the water was replaced with skin. Next, a sinuous bamboo sculpture resembling the shape of Kali Code is marked with petri dishes of
bio-remediating fungi (inspired from one of *ESTROFEM! LAB* protocols) as they
sit along blue-stained agar that invites microbial contamination throughout the duration of the exhibition. And lastly, a rotating mandala projec-

tion composed of the same trash in the latex sculpture rotates constantly while producing new kaleidoscope-like combinations, and includes audio of a child's voice reciting a story about a river who is in pain because it cannot digest all the plastic so it must return it all to humanity (FIGURE 9).

The original impetus of *Open Source Estrogen* asked: What does it mean to truly take back control over our bodies from patriarchal hegemonic forces, and refigure strategies for living, acting, and caring in a permanently polluted world that is the Estroworld? From the hands-on and collaborative experimentations of ESTROFEM! LAB that generated protocols from detection to extraction to remediation, we are able to seize the means of hormonal knowledge production and reveal molecules that were previously invisible or heavily obscured. This then allows a reflexive way to confront the policing of our bodies and at the same time the existence of our bodies in a toxic Anthropocene. The practice of hacking functions not only at the level of scientific technicality, but as sites of mutations and queering that are necessary for social resistance, consciousness-raising, and deconstructions of the very definitions of »normal« and »natural« that have produced inequalities for so many thus far. And like the hormonal molecules it investigates and emancipates, the project is akin to a responsive body that never stops queering, a gender with no final destination. *Open Source Estrogen* urges us all to undo the trap of eco-heteronormativity and ultimately rewrite a future that undoubtedly embodies queerness. It's all about what kind of world we really want to build here. We are all living in an Estroworld, but it can also be *our* Estroworld.

1 Anne Fausto-Sterling, *Sexing the Body: Gender Politics and the Construction of Sexuality*, New York: Basic Books (2000).

2 Ian Hacking, *Representing and Intervening: Introductory Topics in the Philosophy of Natural Science*, Cambridge: Cambridge University Press (2010).

3 Charles-Édouard Brown-Séquard, »The Effects Produced on Man by Subcutaneous Injection of a Liquid Obtained from the Testicles of Animals,« *The Lancet* 137, 3438 (1889): 105–107.

4 Paul B. Preciado, *Testo Junkie: Sex, Drugs, and Biopolitics in the Pharmacopornographic Era*, New York: The Feminist Press at the City University of New York (2013).

5 Nelly Oudshoorn, *Beyond the Natural Body: An Archaeology of Sex Hormones*, London: Routledge (1994).

6 Sarah A. Vogel, »The Politics of Plastics: The Making and Unmaking of Bisphenol A ›Safety,‹« *American Journal of Public Health*, November, 99 (Suppl. 3) (2009): S559–S566.

7 Elvira Greiner, Thomas Kaelin and Goro Toki, »Bisphenol A,« in *Chemical Economics Handbook*, Menlo Park, CA: SRI Consulting (2004).

8 Heather Davis, »Toxic Progeny: The Plastisphere and Other Queer Futures,« *PhiloSOPHIA: A Journal of Continental Feminism* 5, no. 2 (2015): 231–250.

9 Rob Nixon, *Slow Violence and the Environmentalism of the Poor*, Cambridge, MA: Harvard University Press (2013).

10 Lee Edelman, *No Future: Queer Theory and the Death Drive*, Durham, NC: Duke University Press (2004).

11 Laboria Cuboniks, *The Xenofeminist Manifesto*, London: Verso (2015).

12 Claire Pentecost, »Outfitting the Laboratory of the Symbolic: Toward a Critical Inventory of Bioart, ibid., 107–123.

13 Critical Art Ensemble, *The Molecular Invasion*, New York: Autonomedia (2002).

14 Astrida Neimanis, »Hydrofeminism: Or, on Becoming a Body of Water,« in *Undutiful Daughters: Mobilizing Future Concepts, Bodies and Subjectivities in Feminist Thought and Practice*, eds. Henriette Gunkel, Chrysanthi Nigianni, and Fanny Söderbäck, New York: Palgrave Macmillan (2012).

Life is the most exceptional form of poetry,
albeit complicated, messy, fragile, and quickly
dwindling. Biodiversity is nature's art.
What will become of this art as we continue to
extinguish life in the name of monetary growth?
Today's environmental problems are global
in scale and complex. To face this milieu of issues,
we need the creativity of artists, scientists,
and those focused on other disciplines combined
to creatively address such challenges we
and other species currently face. My continued
project SEARCHING FOR THE GHOSTS OF THE
GULF responds to missing Gulf of Mexico species
through visual artworks and collaborative
actions with coastal communities that are
themselves culturally endangered.

Some Working Notes on *Searching for the Ghosts of the Gulf*

Brandon Ballengée

Ten Years after Deepwater Horizon __ The Gulf of Mexico is a special place for many of us—and over ten thousand other species—she is our sanctuary, our home, our mother, provider and *sometimes* destroyer. As an artist, I find the Gulf of Mexico to be an inspirational source of color, form, intrigue, tranquility, and even fear. From the science side, the Gulf is among the most important and biologically diverse marine environments in the world. She is resilient, powerful, seductive, but also dangerous, damaged and suffocating in her own *sang noir*, a regional term describing crude oil.

2020 marks the ten-year anniversary of the Deepwater Horizon (DWH) oil spill, which is considered the largest environmental disaster in the history of the United States and the largest known petrochemical spill by volume in human history.[1] The tremendous amount of oil spilled during DWH was estimated at greater than 200 million gallons, which resulted in an immediate contamination area of 149,000 km^2 and continued to spread through currents widely in the Gulf and beyond for years.[2]

From an ecological and economic standpoint, the DWH spill could not have occurred in a more disastrous location, for the Gulf of Mexico is one of the most important and biologically diverse environments in the world. The Gulf is a nursery for thousands of marine species and has numerous endemic[3] organisms, including 77 fishes found nowhere else in the world.[4] Gulf seafood is an important source of food for millions of people in North America, and, since marine species migrate by following the Gulf Stream, people throughout Europe also rely on these fish for protein.

Cleanup efforts exacerbated the spill's toxicity and reach by utilizing Corexit 9500 and other chemical dispersants. Such dispersants molecularly bind and break down oil into smaller droplets that more readily mix with the water and sink. According to the Material Safety Data Sheets (MSDSs) for Corexit 9500, produced by the chemical manufacturer Nalco,[5] no toxicity studies were conducted prior to its use in the Gulf.[6] However, numerous earlier toxicology studies found such dispersants to be teratological[7] to marine wildlife and possibly carcinogenic to humans.[8] Regardless, an estimated two million gallons of chemical dispersants were used for DWH in deep-sea as well as in surface water. Findings have

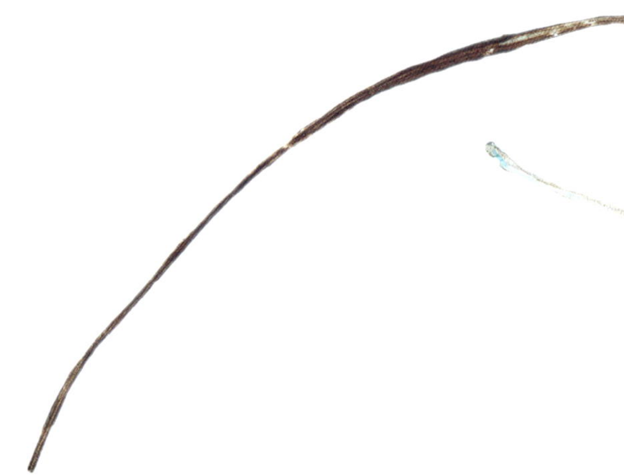

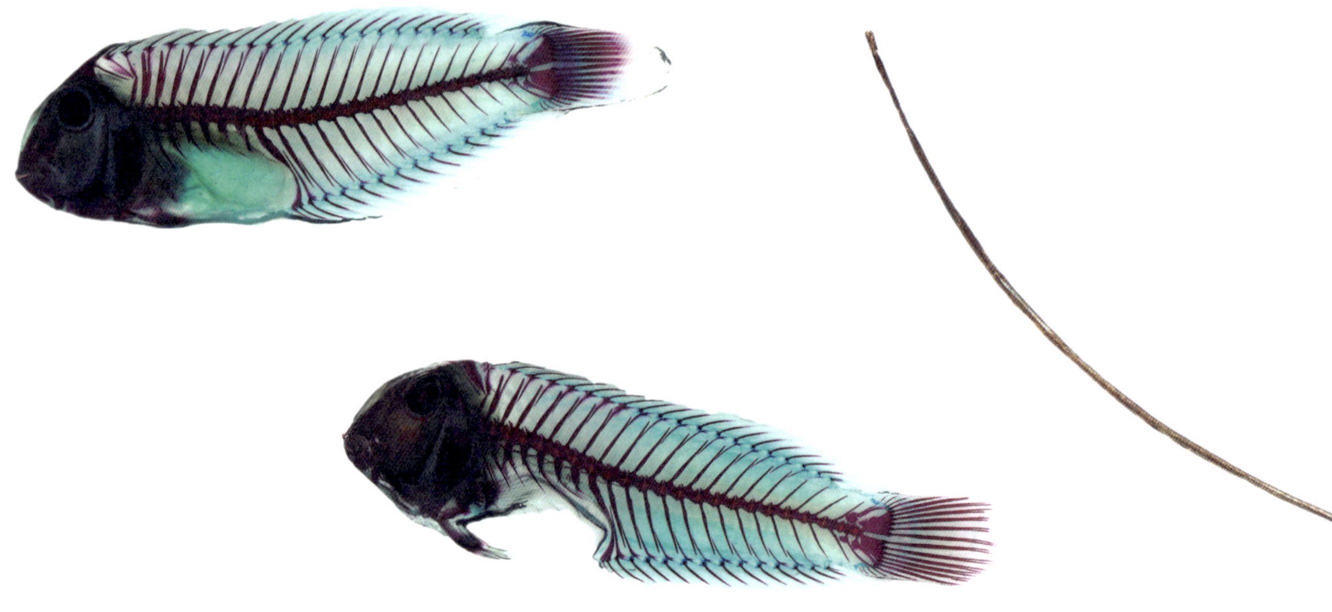

FIG. 1
RIP PARROT FISH from the GHOSTS OF THE GULF, 2014. Giclée print on handmade Japanese rice paper.
© Ballengée Studio

FIG. 2
RIP AFRICAN POMPANO from the GHOSTS OF THE GULF, 2014. Giclée print on handmade Japanese rice paper.
© Ballengée Studio

Some Working Notes on *Searching for the Ghosts of the Gulf*

Brandon Ballengée

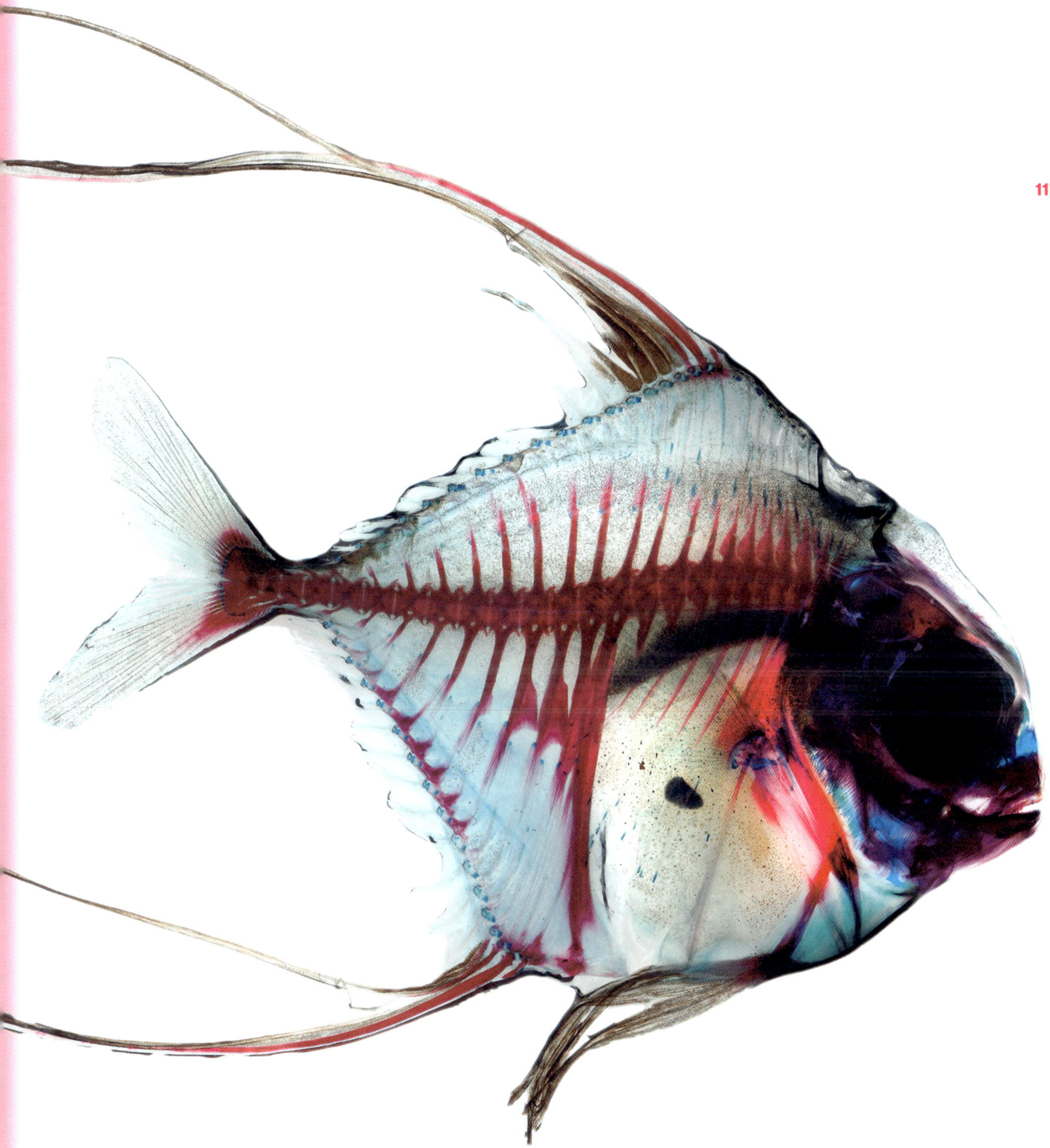

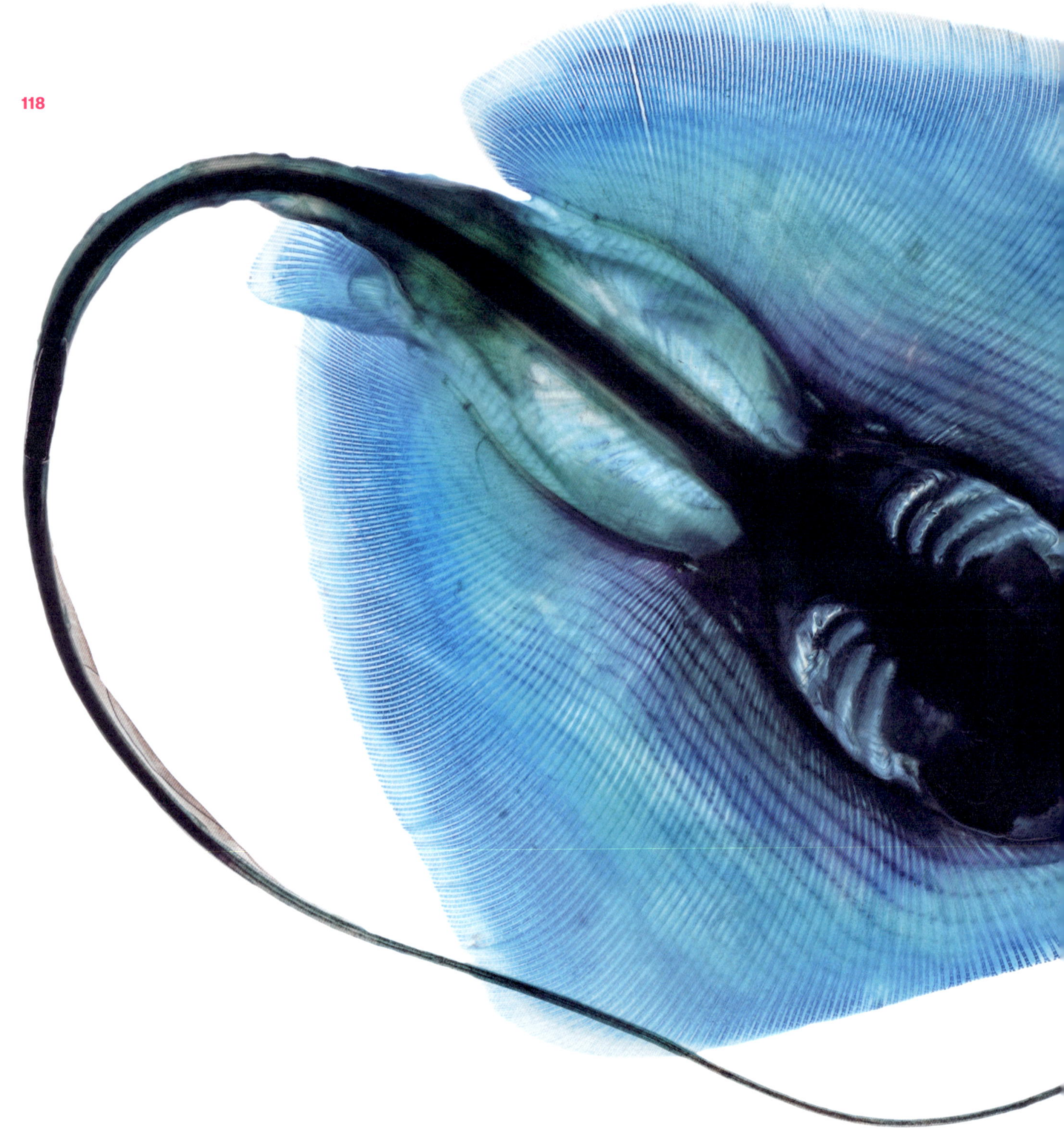

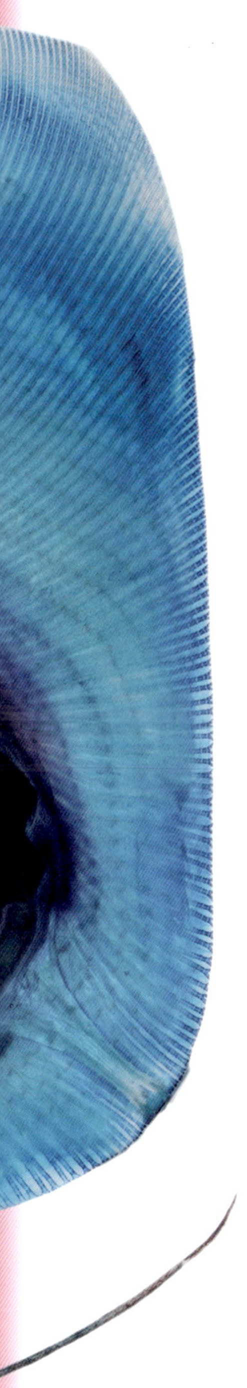

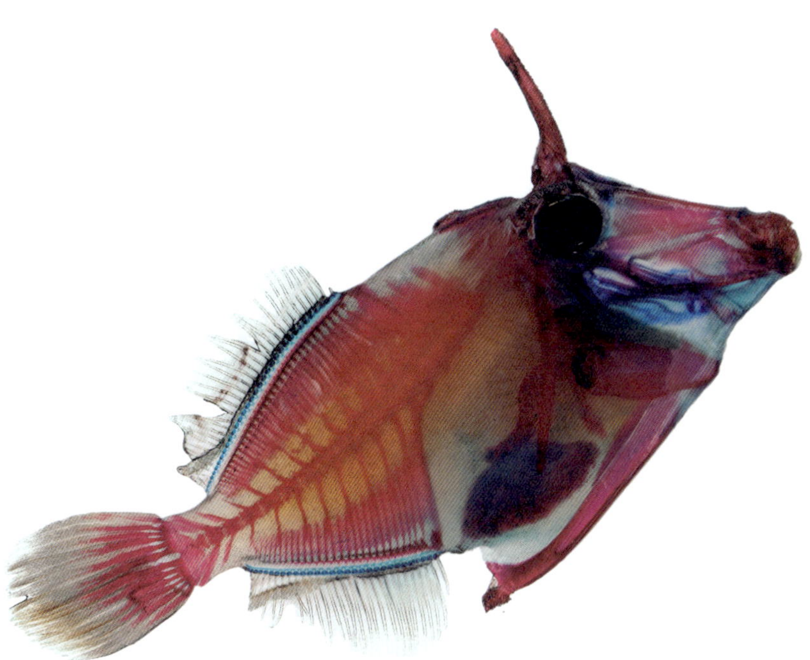

FIG. 3
RIP BLUNTNOSE STINGRAY from the GHOSTS OF THE GULF, 2014. Giclée print on handmade Japanese rice paper.
© Ballengée Studio

FIG. 4
RIP TRIGGERFISH from the GHOSTS OF THE GULF, 2014. Giclée print on handmade Japanese rice paper.
© Ballengée Studio

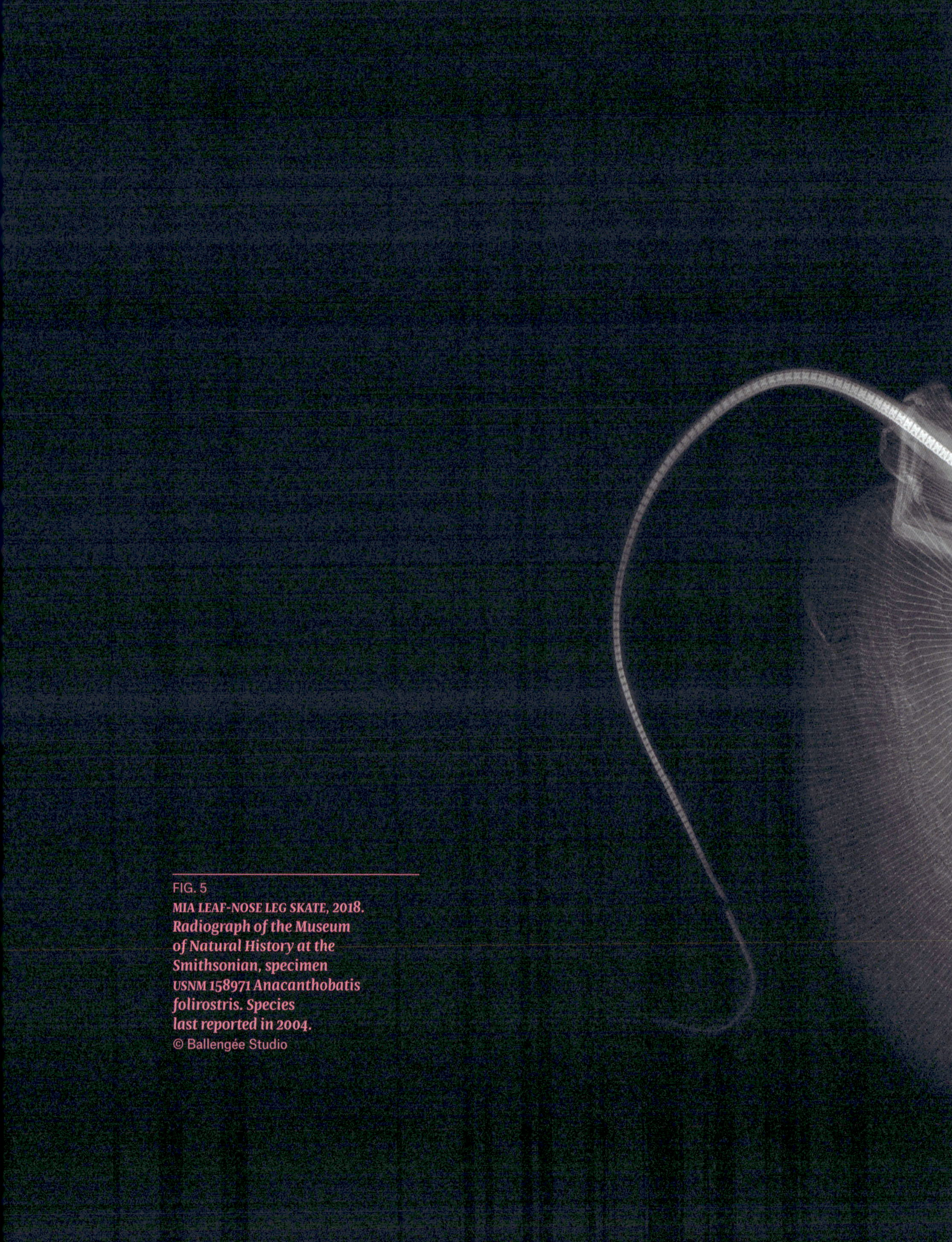

FIG. 5
MIA LEAF-NOSE LEG SKATE, 2018.
Radiograph of the Museum
of Natural History at the
Smithsonian, specimen
USNM 158971 Anacanthobatis
folirostris. Species
last reported in 2004.
© Ballengée Studio

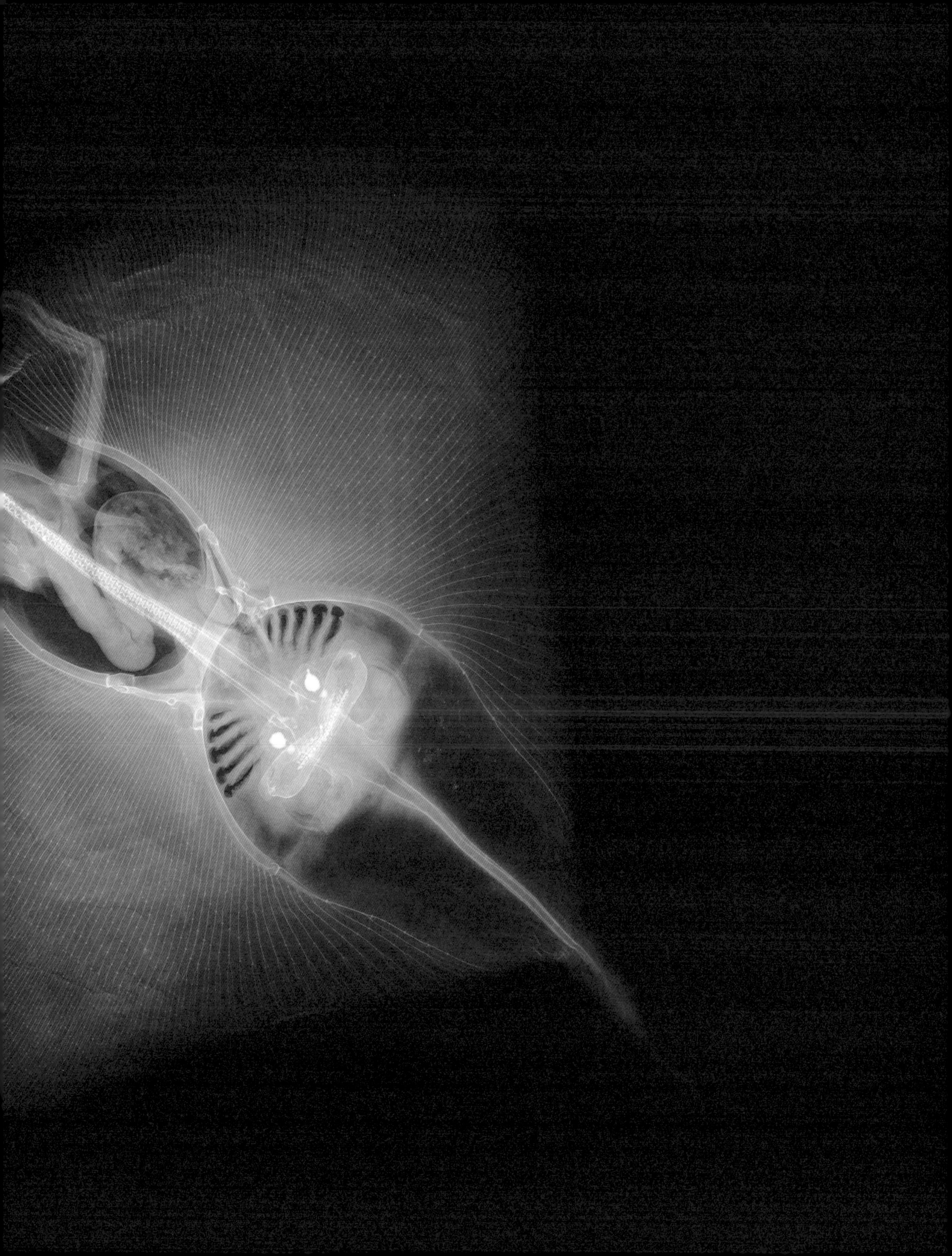

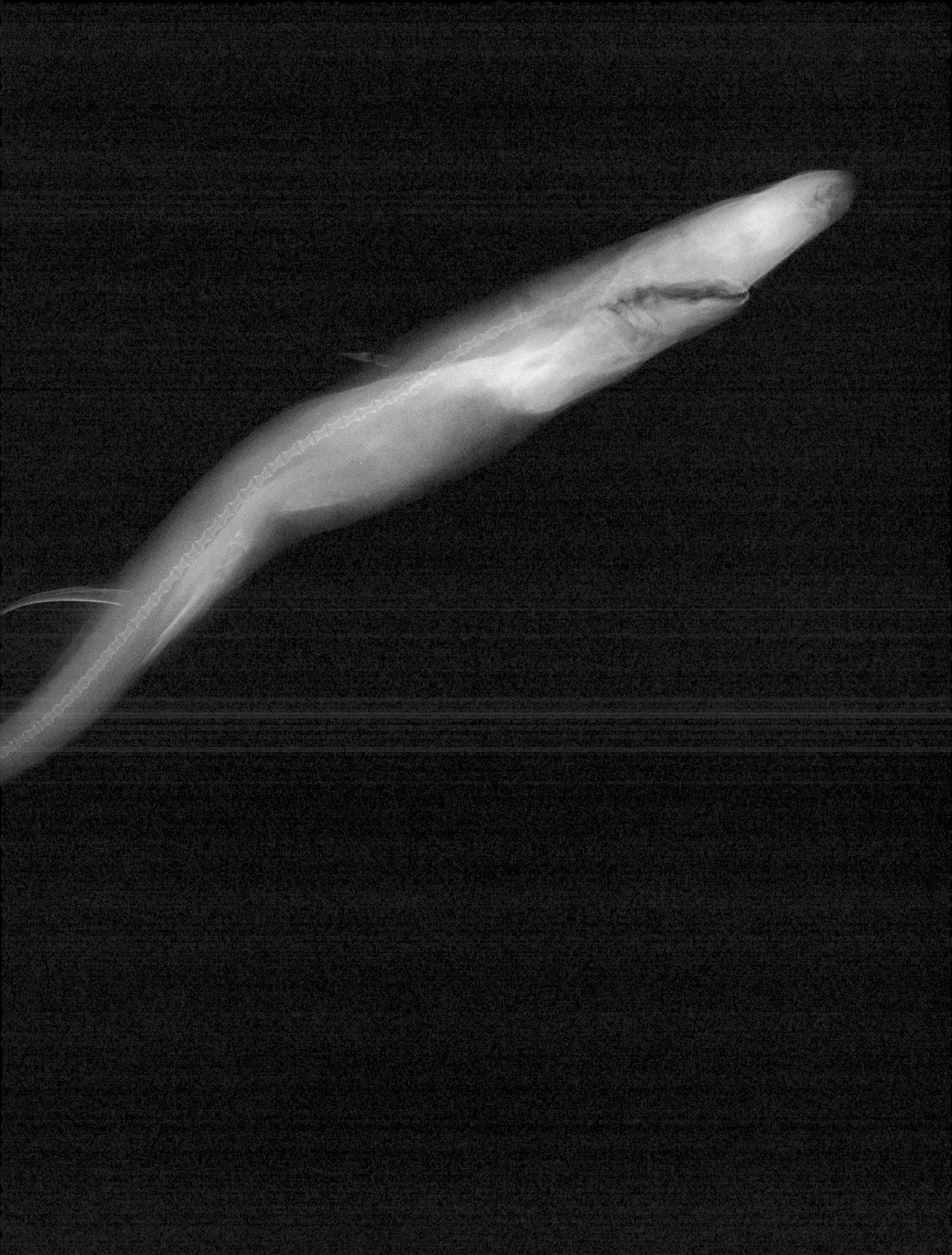

suggested this made DWH oil as much as 52% more toxic and more difficult
to clean up and increased the negative impact on wildlife.[9] To date,
it is suggested that almost 100 million gallons of DWH oil combined with
dispersants remain in the Gulf.[10]

The DWH impacts to Gulf of Mexico species and ecosystems are still
not fully understood, however; some persistent ecological effects include
damage to deep ocean coral communities, harm to oyster fisheries over
several years, loss of marshlands as well as population declines of marine
mammals, sea turtles and seabirds.[11] There is considerable evidence that
some species, especially fishes, continue to be physically and developmen-
tally challenged or even have become absent following DWH.[12] Further,
risks from oil-production-related polycyclic aromatic hydrocarbons (PAH)
exposure and concentrations in fishes are widespread in the Gulf of Mexico
and will likely continue as extraction of petrochemical intensifies.[13]

Responding through Art and Actions __ Since the DWH oil spill, much of my
art and research science has focused on the perilous environmental state
of the Gulf, post-spill. Even ten years after DWH, the long-term impact
on Gulf communities, fishes, other biota, and Gulf ecosystems is still
not well understood. Additionally, there have been over 2,000 smaller
spills since DWH and another spill, the Taylor or MC20, began even earlier,
in 2004, and continues uninterrupted today. Through my installations,
photographs, crude paintings, and programs, I want to give meaningful
visual form to these environmental insults and inspire individual actions
towards systemic change and foster the change of individual attitudes
and behaviors among coastal residents, which will (hopefully) result in
increased community resilience to climate change. These socioecological
actions may include: increased practices of sustainable fishing and aqua-
culture; development of ecotourism; support and transfer to renewable
energy production from wind and solar instead of gas and oil; more access
to quality science and art education; return to local farming; others to
be determined.

During the DWH spill (April through September 2010), I collected
thousands of dead specimens from Louisiana beaches. These were preserved and over 20,000 of them were later utilized to create the monumental installation *COLLAPSE* (2010–2012) (FIGURE 11). A subset of these specimens were chemically cleared and stained,[14] then made into the initial
GHOSTS OF THE GULF photographic series in 2014. From 2016, I exhibited
these preserved specimens and prints in the collaborative *CRUDE LIFE*
project, which was displayed at Gulf community events such as Mardi
Gras or Blessing of the Shrimp Fleets as well as at schools, in parks
and even at the Louisiana State Senate—all in all, the project reached
more than 5,200 individuals (FIGURE 7).[15]

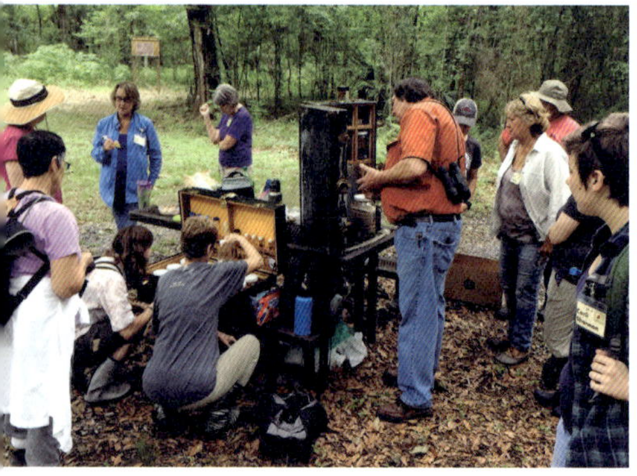

FIG. 7
*CRUDE LIFE Portable Museum
installation and ECO-ACTION
outreach event in 2017 at
Indian Bayou, Saint Martin
Parish, Louisiana.*
Photo: Brandon Ballengée

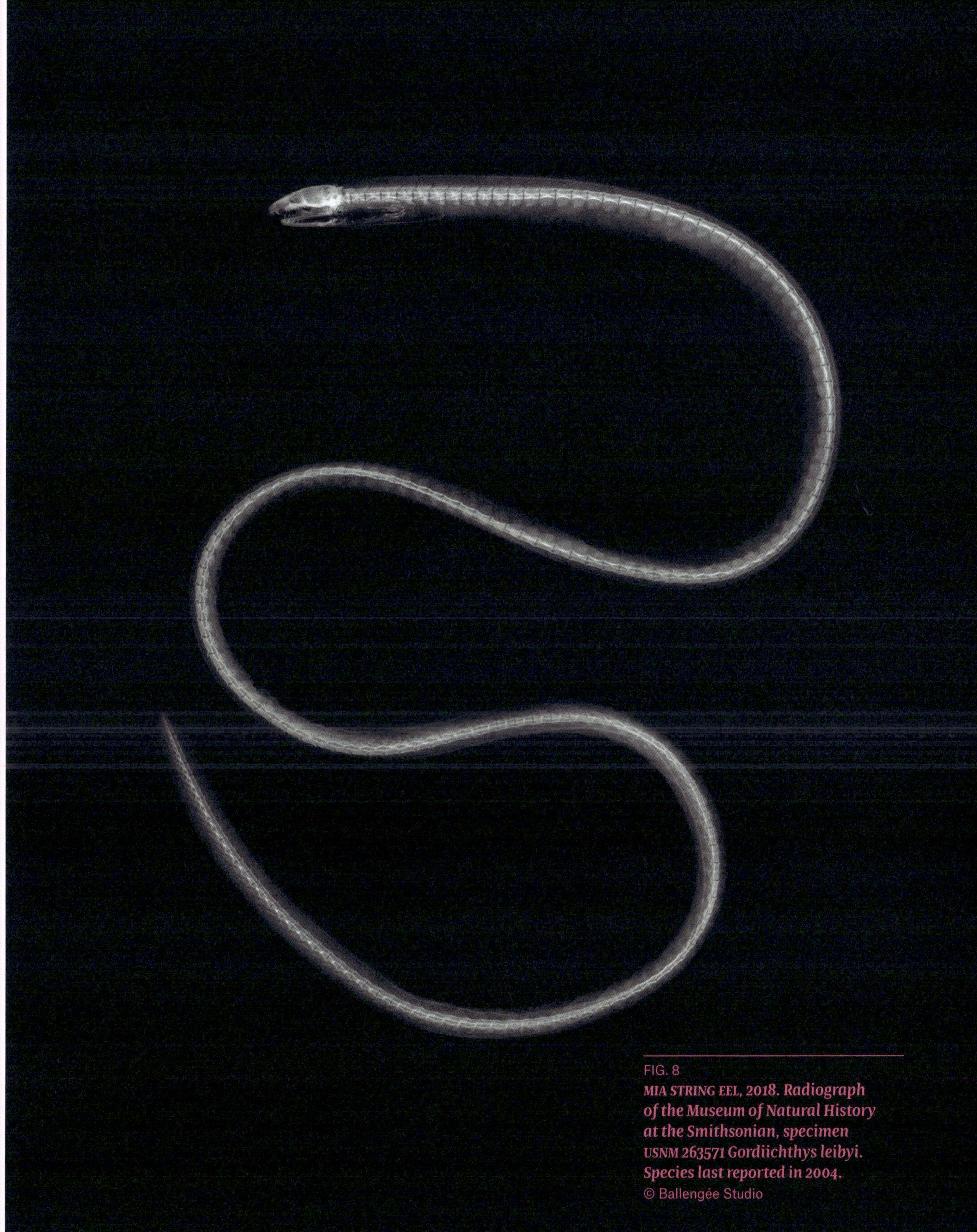

FIG. 8
MIA STRING EEL, 2018. Radiograph
of the Museum of Natural History
at the Smithsonian, specimen
USNM 263571 Gordiichthys leibyi.
Species last reported in 2004.
© Ballengée Studio

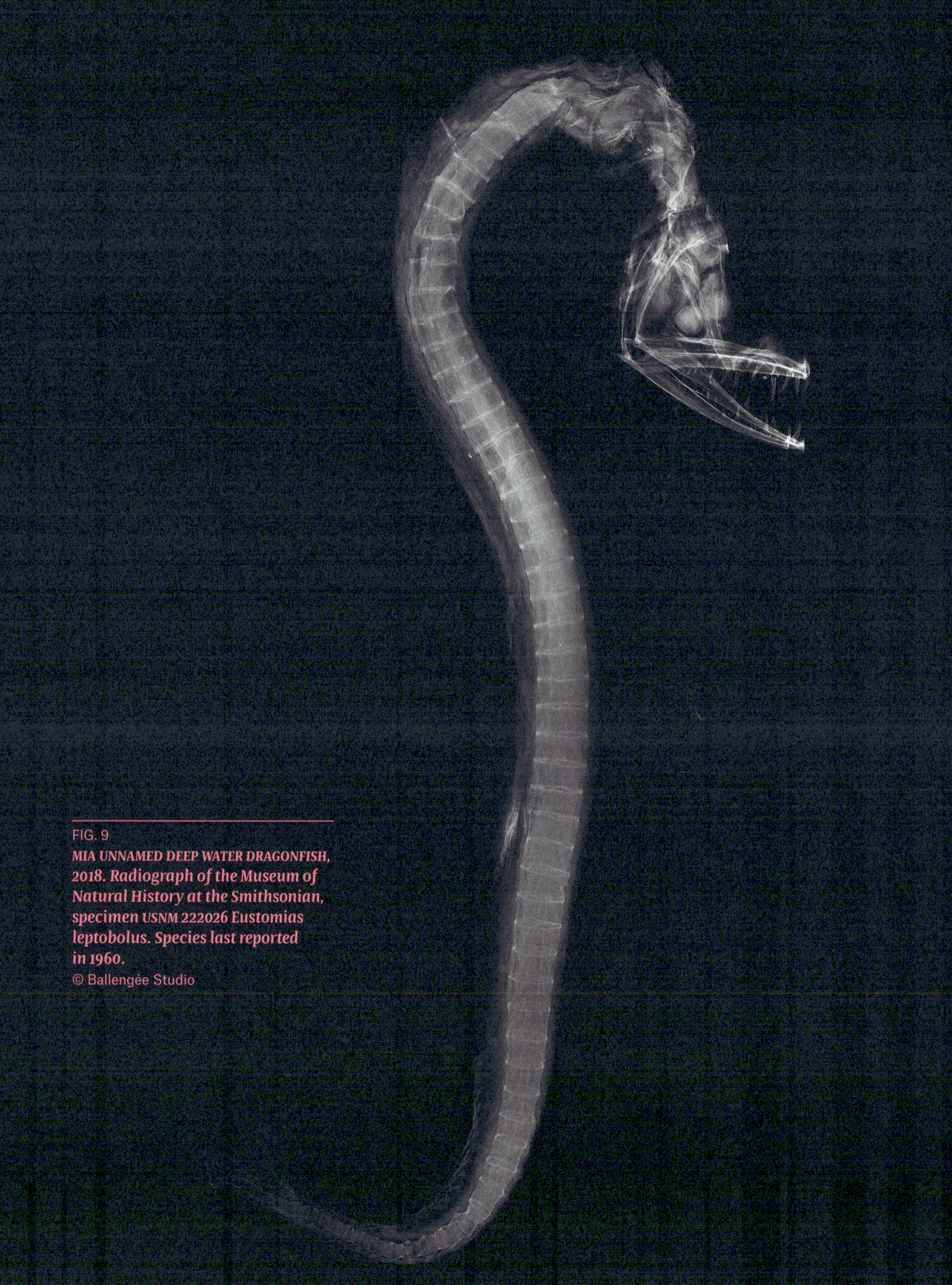

FIG. 9
MIA UNNAMED DEEP WATER DRAGONFISH,
2018. Radiograph of the Museum of
Natural History at the Smithsonian,
specimen USNM 222026 Eustomias
leptobolus. Species last reported
in 1960.
© Ballengée Studio

Inspired by my ongoing work with Louisiana coastal communities, *SEARCHING FOR THE GHOSTS OF THE GULF* is an ongoing interdisciplinary art and environmental advocacy project seeking to portray absent biodiversity and activate coastal residents through three intertwined components: *portraying* (drawing, photographing and radiographing) missing species from natural history collections to create prints and drawings made from dried crude oil material; *activating* coastal communities through participatory ecological-art field trips and programs; *exhibiting* works in pop-up exhibitions in unconventional venues (FIGURE 10).

127

Portraying ___ In 2016, I became part of an interdisciplinary Louisiana State University research team (where I am currently a research associate in the Museum of Natural Science), which published that 14 fish species, endemic to the Gulf of Mexico, have not been reported following the DWH spill.[16] Even prior to the spill, several Gulf fishes remained elusive and had not been found in decades (1950 through 2005). Little is known about these species and the only records we have of their existence is a handful of preserved specimens scattered among natural history collections.[17] As an artist and biologist, I am inspired to portray and to tirelessly search for these GHOSTS.

In response, I create portraits of these missing species, which I refer to as GHOSTS OF THE GULF, as a way to give form to each of the lost species (FIGURES 1–4). The GHOSTS are drawn from historic specimens in the Tulane University Biodiversity Research Institute's (TUBRI) Suttkus Fish Collection, the second largest preserved fish collection in the world, located in Belle Chase, Louisiana, USA, and others I photographed and radiographed as a 2017 artist-in-residence at the Smithsonian National Museum of Natural History in Washington D.C., USA, which is housing the largest fish collection in the world. Some portraits are printed radiographs while others are drawn using solidified DWH »tar balls« collected from Gulf beaches or from »fresh« crude oil from the Taylor spill (FIGURES 5–6, 8–9). These GHOSTS intend to convey mystery as well as melancholy as a means to engage audiences towards introspective contemplation, asking what is lost from our collective treatment of the Gulf. I am also interested in what creating portraits of missing animals means at a point in history where we find ourselves in Earth's sixth mass extinction event under way, when species are disappearing so fast that we cannot even scientifically record them.

Activating ___ Coastal Louisiana's economy strongly relies on its fisheries and the oil industry, both of which have become increasingly unstable in recent decades. Rising seas, coastal erosion, sediment diversions, and multiple oil spills have depleted fish populations and devastated oyster farming. What is happening to coastal ecosystems is happening to us—we are intertwined, and it takes art to make this clear to us.

To activate coastal communities, I lead ECO-ACTIONS — artistic and
environmental inquiries into socioecological stressors facing our coast.
For these, residents join me in conducting ecological field sampling
in disappearing marshland habitats, while making their own art envision-
ing future scenarios where their communities will hopefully adapt and
survive. Likewise, I learn from these residents about their knowledge
of the Gulf and her species. The ECO-ACTIONS focus on missing Gulf species,
land and habitats, while also exploring loss of coastal culture.

Additional ECO-ACTIONS are planned (post-COVID-19) for the Suttkus
Fish Collection. There, programs will include introductory fish drawing
and photography workshops for fisherfolk, oil workers and others —
a strategy to connect stakeholders with Gulf biodiversity in a novel way.
My goal is to generate discourse around the complexity of the Gulf of
Mexico food web, species and habitat disappearances, and create situations
where fisherfolk and biologists will strategize means to help Louisiana
fisheries to once again become sustainable.

Fundamentally, these programs are intended to build resilience
as they connect human to nonhuman communities, improve knowledge
of local ecosystems and provide strategies for expressing these findings
with others through art. My underlying idea is that by examining complicated socioecological issues, grounded in scientific facts, through the lens
of art, participants may brainstorm solutions and begin to take creative
actions towards positive changes.

Exhibiting __ To share these works, pop-up exhibitions are set up in unconventional
venues such as marinas, seafood markets, parks, libraries, festivals,
and other places where fisherfolk, oil field workers and their families
gather. My intention is to reach new audiences, and also to enlist residents
in ECO-ACTIONS and other activities. My hope is that, collectively, we
may yet develop novel strategies to adapt and sustain our communities,
human and nonhuman alike.

Working with coastal Louisiana communities over the past
decade has taught me that art can be an important icebreaker for meeting
residents and a means to discuss complex socioecological challenges.
Art may also act as an olive branch with fisherfolk and oil workers — many
of whom remain resistant to the concept of human-caused environmental
impact like climate change, while at the same time being the part of
the population faced with the greatest threats to their culture and livelihoods. Through these pop-up exhibitions, I am able to meet and recruit
potential project participants, communicate my environmental concerns
and learn about others' perspectives, while brainstorming creative ideas
towards survival.

Speculative Futures ___ Land in coastal Louisiana is being lost at the fastest rate on Earth and, in recent decades as discussed above, several Gulf species have gone missing. As habitats and biodiversity disappear, so do the cultures that rely on them. The fate of the Gulf's children remains precarious. I believe art grounded in science will play an important part in building regional socioecological resilience. As an artist, I continue to develop an aesthetic of »loss,« giving a visual form to the growing absence of life on our rapidly degrading planet. As a scientist, I find it increasingly important to share research findings about such losses with the public. Through art, I am able to speculate future outcomes, question our current behaviors, express my concerns as well as mourn. As a biologist, I must remain analytical and report unbiased information on species found within or missing from ecosystems. Combined, art and science are complementary ways of trying to understand our world and ourselves, as well as a means to address the complex socioecological challenges we and other species currently face. In a joint effort, we may find some of the Gulf's lost children and save ourselves in the process.

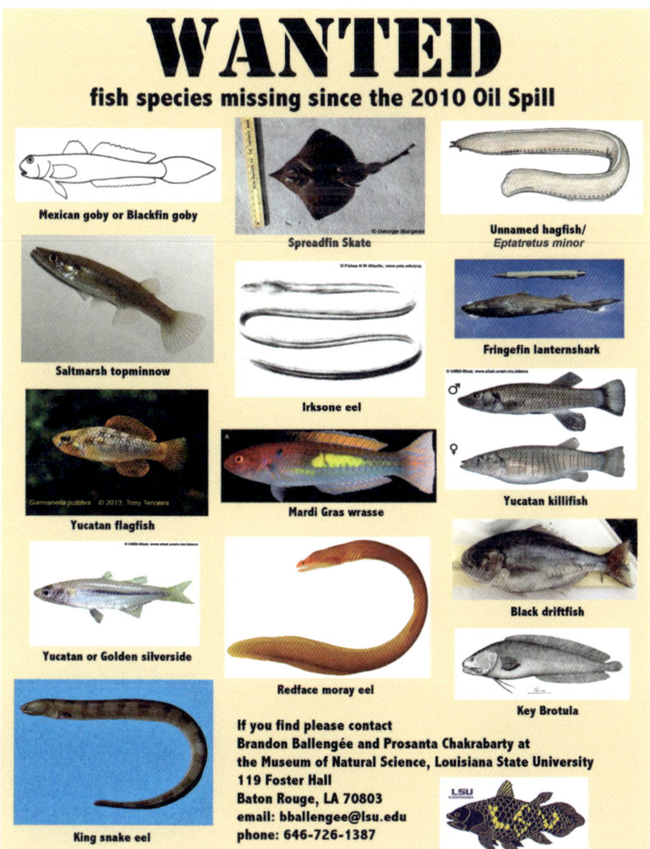

FIG. 10
Gulf of Mexico endemic fishes »missing« since the 2010 oil spill. This poster has been given away during outreach events and posted in marinas, schools, groceries, the Louisiana State Senate, and other venues.
© Ballengée Studio

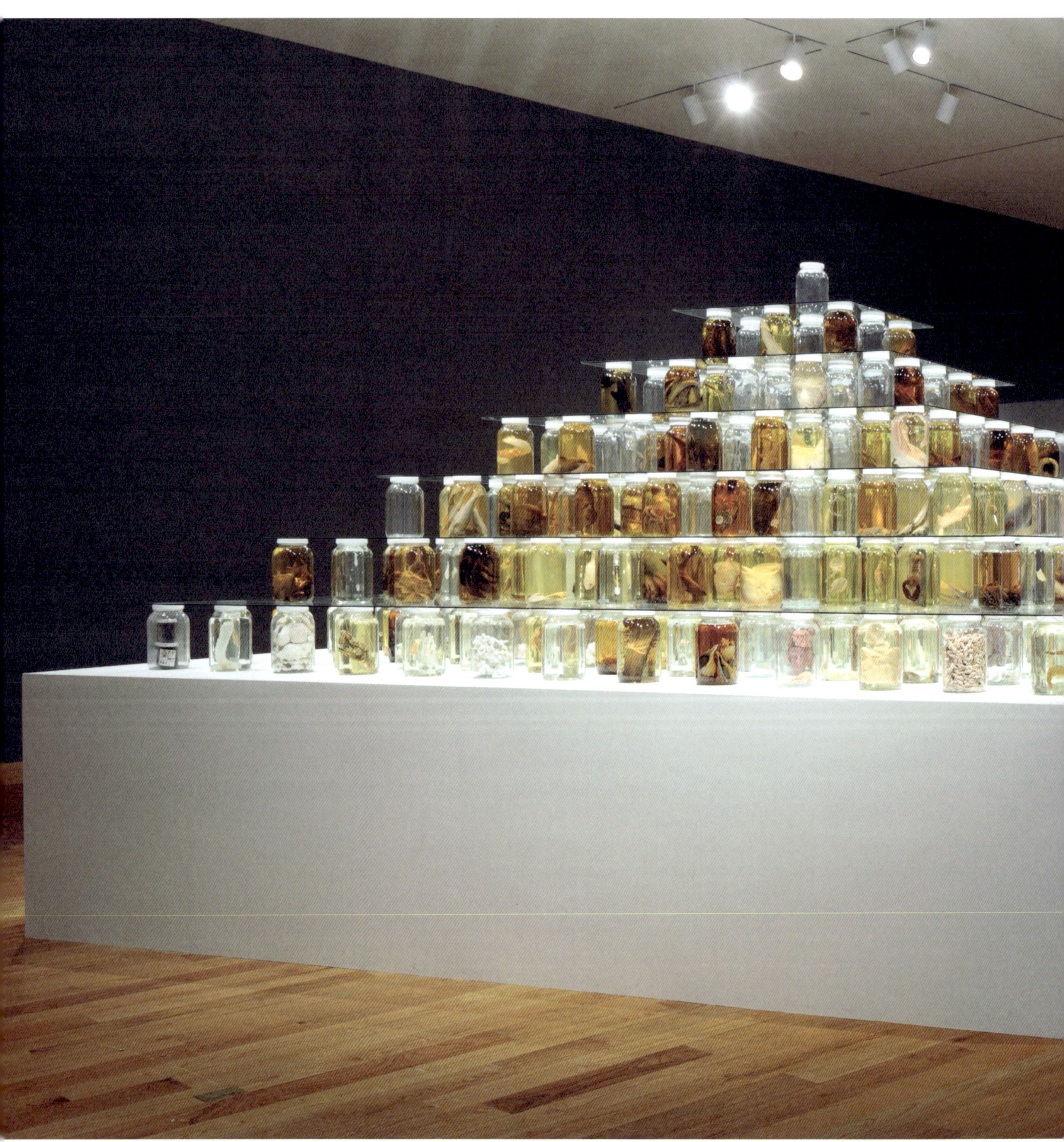

Some Working Notes on *Searching for the Ghosts of the Gulf*

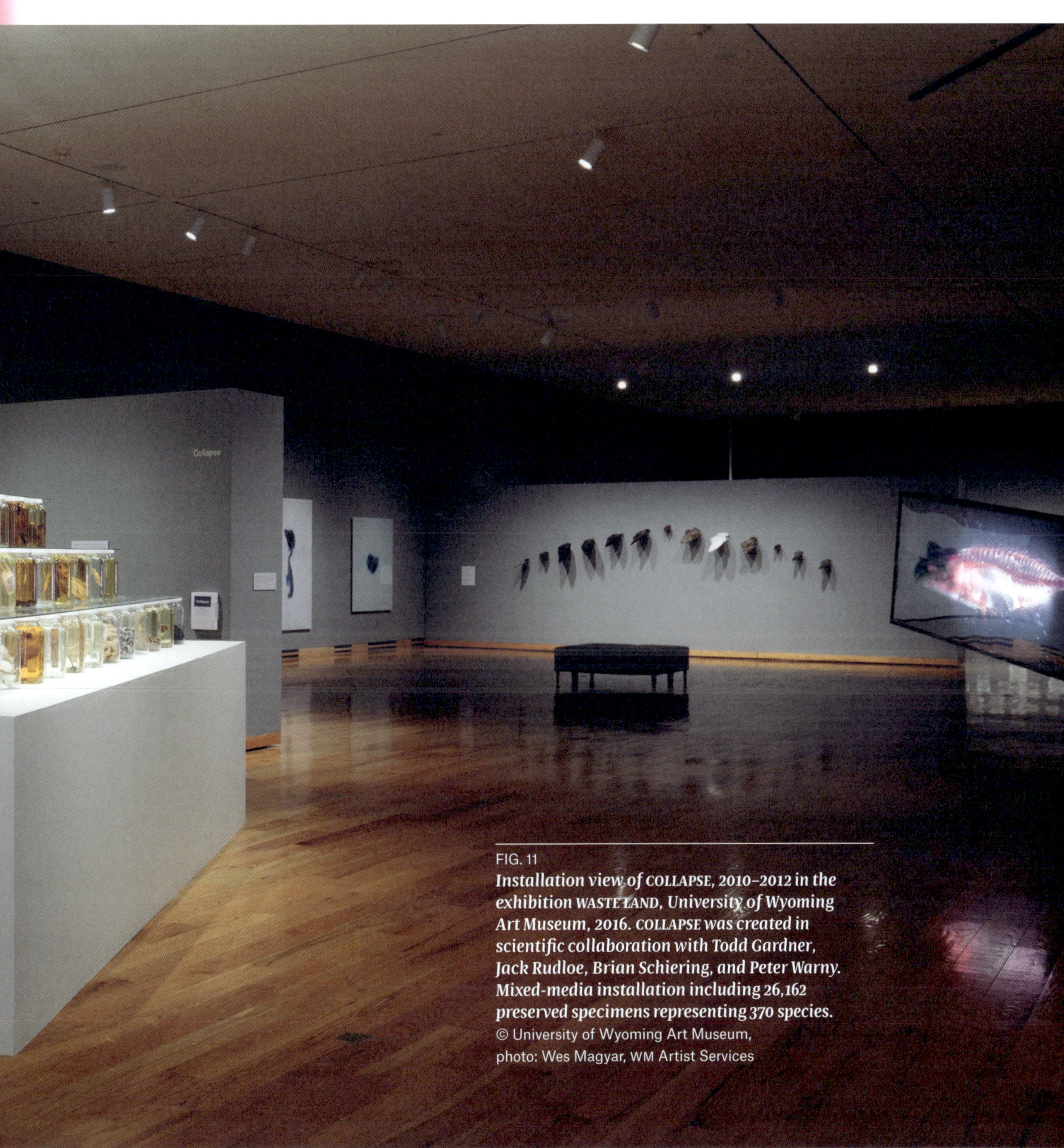

FIG. 11

Installation view of COLLAPSE, 2010–2012 in the exhibition WASTE LAND, University of Wyoming Art Museum, 2016. COLLAPSE was created in scientific collaboration with Todd Gardner, Jack Rudloe, Brian Schiering, and Peter Warny. Mixed-media installation including 26,162 preserved specimens representing 370 species.
© University of Wyoming Art Museum, photo: Wes Magyar, WM Artist Services

1 Michael Belanger, Luke Tan, Nesime Askin, and Carin Wittnich, »Chronological Effects of the Deepwater Horizon Gulf of Mexico Oil Spill on Regional Seabird Casualties,« *Journal of Marine Animals and Their Ecology* 3, no. 2 (2010): 10–14; Nancy N. Rabalais, »Assessing Early Looks at Biological Responses to the Macondo Event,« *BioScience* 64, no. 9 (2014): 757–759.

2 Nancy N. Rabalais, »Troubled Waters of the Gulf of Mexico,« *Oceanography* 24, no. 2 (2011), 200; Jonathan L. Ramseur and Curry L. Hagerty, »Deepwater Horizon Oil Spill: Recent Activities and Ongoing Developments,« *Congressional Research Service*, January 31 (2013), 2013; Ian R. MacDonald, O. Garcia-Pineda, A. Beet, S. Daneshgar Asl, L. Feng, G. Graettinger, D. French-McCay et al., »Natural and Unnatural Oil Slicks in the Gulf of Mexico,« *Journal of Geophysical Research: Oceans* 120, no. 12 (2015): 8364–8380; Igal Berenshtein, Claire B. Paris, Natalie Perlin, Matthew M. Alloy, Samantha B. Joye, and Steve Murawski, »Invisible Oil Beyond the Deepwater Horizon Satellite Footprint,« *Science Advances* 6, no. 7 (2020), eaaw8863.

3 »Endemic« here means species only found in the Gulf of Mexico.

4 Prosanta Chakrabarty, Glynn A. O'Neill, Brannon Hardy, and Brandon Ballengée, »Five Years Later: An Update on the Status of Collections of Endemic Gulf of Mexico Fishes Put at Risk by the 2010 Oil Spill,« *Biodiversity Data Journal* 4 (2016), e8728.

5 Nalco Water or Nalco Holding Company is an Illinois-based water and chemical treatment company that manufactured the oil dispersants widely used on the Deepwater Horizon oil spill. An estimated five million liters of these oil dispersants were used—the largest use to date. David Biello, »Is using dispersants on the BP Gulf oil spill fighting pollution with pollution,« *Scientific American*, June 18 (2010).

6 Elizabeth B. Kujawinski, Melissa C. Kido Soule, David L. Valentine, Angela K. Boysen, Krista Longnecker, and Molly C. Redmond, »Fate of Dispersants Associated with the Deepwater Horizon Oil Spill,« *Environmental Science & Technology* 45, no. 4 (2011): 1298–1306.

7 Teratology is the scientific study of physical abnormalities in developing organisms such as birth defects or other bodily deviations during growth. See Pamela Taylor, *Practical Teratology*, Cambridge, MA: Academic Press (1986).

8 Andrew Rogerson and Jacques Berger, »The Toxicity of the Dispersant Corexit 9527 and Oil-dispersant mixtures to Ciliate Protozoa,« *Chemosphere* 10, no. 1 (1981): 33–39; Michael M. Singer, Saji George, Diana Benner, Susan Jacobson, Ronald S. Tjeerdema, and Michael L. Sowby, »Comparative Toxicity of Two Oil Dispersants to the Early Life Stages of Two Marine Species,« *Environmental Toxicology and Chemistry* 12, no. 10 (1993): 1855–1863; Alan Scarlett, Tamara S. Galloway, Martin Canty, Emma L. Smith, Johanna Nilsson, and Steven J. Rowland, »Comparative Toxicity of Two Oil Dispersants, Superdispersant-25 and Corexit 9527, to a Range of Coastal Species,« *Environmental Toxicology and Chemistry* 24, no. 5 (2005): 1219–1227; Elizabeth T.H. Fontham and Edward Trapido, »Oil and Water,« *Environmental Health Perspectives* 118, no. 10 (2010), A422.

9 Roberto Rico-Martínez, Terry W. Snell, and Tonya L. Shearer, »Synergistic Toxicity of Macondo Crude Oil and Dispersant Corexit 9500A® to the *Brachionus plicatilis* Species Complex (Rotifera),« *Environmental Pollution* 173 (2013): 5–10.

10 Jonathan L. Ramseur and Curry L. Hagerty, »Deepwater Horizon Oil Spill: Recent Activities and Ongoing Developments,« *Congressional Research Service*, January 31 (2013).

11 Mace G. Barron, Deborah N. Vivian, Ron A. Heintz, and Un Hyuk Yim, »Long-Term Ecological Impacts from Oil Spills: Comparison of Exxon Valdez, Hebei Spirit, and Deepwater Horizon,« *Environmental Science & Technology* 54, no. 11 (2020): 6456–6467.

12 Andrew Whitehead, Benjamin Dubansky, Charlotte Bodinier, Tzintzuni I. Garcia, Scott Miles, Chet Pilley, Vandana Raghunathan et al., »Genomic and Physiological Footprint of the Deepwater Horizon Oil Spill on Resident Marsh Fishes,« *Proceedings of the National Academy of Sciences* 109, no. 50 (2012): 20298–20302; John P. Incardona, Luke D. Gardner, Tiffany L. Linbo, Tanya L. Brown, Andrew J. Esbaugh, Edward M. Mager, John D. Stieglitz et al., »Deepwater Horizon Crude Oil Impacts the Developing Hearts of Large Predatory Pelagic Fish,« *Proceedings of the National Academy of Sciences* 111, no. 15 (2014): E1510–E1518; Benjamin Dubansky, Andrew Whitehead, Jeffrey T. Miller, Charles D. Rice, and Fernando Galvez, »Multitissue Molecular, Genomic, and Developmental Effects of the Deepwater Horizon Oil Spill on Resident Gulf Killifish (*Fundulus grandis*),« *Environmental Science & Technology* 47, no. 10 (2013): 5074–5082; Brette Fabien, Ben Machado, Caroline Cros, John P. Incardona, Nathaniel L. Scholz, and Barbara A. Block, »Crude Oil Impairs Cardiac Excitation-Contraction Coupling in Fish,« *Science* 343, no. 6172 (2014): 772–776; Edward M. Mager, Andrew J. Esbaugh, John D. Stieglitz, Ronald Hoenig, Charlotte Bodinier, John P. Incardona, Nathaniel L. Scholz, Daniel D. Benetti, and Martin Grosell, »Acute Embryonic or Juvenile Exposure to Deepwater Horizon Crude Oil Impairs the Swimming Performance of Mahi-mahi (*Coryphaena hippurus*),« *Environmental Science & Technology* 48, no. 12 (2014): 7053–7061; Matthew Alloy, David Baxter, John Stieglitz, Edward Mager, Ronald Hoenig, Daniel Benetti, Martin Grosell, James Oris, and Aaron Roberts, »Ultraviolet Radiation Enhances the Toxicity of Deepwater Horizon Oil to Mahi-mahi (*Coryphaena hippurus*) Embryos,« *Environmental Science & Technology* 50, no. 4 (2016): 2011–2017; Prosanta Chakrabarty, Glynn A. O'Neill, Brannon Hardy, and Brandon Ballengée, »Five Years Later: An Update on the Status of Collections of Endemic Gulf of Mexico Fishes put at risk by the 2010 Oil Spill,« *Biodiversity Data Journal* 4 (2016), e8728.

13 Erin L. Pulster, Adolfo Gracia, Maickel Armenteros, Gerardo Toro-Farmer, Susan M. Snyder, Brigid E. Carr, Madison R. Schwaab, Tiffany J. Nicholson, Justin Mrowicki, and Steven A. Murawski, »A First Comprehensive Baseline of Hydrocarbon Pollution in Gulf of Mexico Fishes,« *Scientific Reports* 10, no. 1 (2020): 1–14.

14 The clearing and staining process is a post-preservation chemical process whereby varied hard and soft tissues are »stained« utilizing biological dyes and other tissues are »cleared« using enzymes resulting in a specimen that resembles a brightly colored x-ray; for full description, please see Brandon Ballengée, »Ghosts of the Gulf,« *Cultural Politics* 11, no. 3 (2015): 346–360.

15 *Crude Life* (2016–18) was an interdisciplinary art–science project created in collaboration with Prosanta Chakrabarty (scientist), Suzanne Fredericq (scientist), Owen Miller (artist), Rachel Mayeri (artist), Monique Verdin (artist), and consisted of »citizen science« surveys of Gulf wildlife and the creation of a Portable Museum of Gulf Biodiversity utilized for public outreach. For further information, please see Brandon Ballengée, »Crude Life,« *Issues in Science and Technology* 34, no. 2 (2018): 66–71.

16 Prosanta Chakrabarty, Glynn A. O'Neill, Brannon Hardy, and Brandon Ballengée, »Five Years Later: An Update on the Status of Collections of Endemic Gulf of Mexico Fishes Put at Risk by the 2010 Oil Spill,« *Biodiversity Data Journal* 4 (2016), e8728.

17 Up to 44 of the 77 known Gulf endemic fishes are rare, have not been reported, or recent data is insufficient to understand their population status. See Prosanta Chakrabarty, Glynn A. O'Neill, Brannon Hardy, and Brandon Ballengée, »Five Years Later: An Update on the Status of Collections of Endemic Gulf of Mexico Fishes Put at Risk by the 2010 Oil Spill,« *Biodiversity Data Journal* 4 (2016), e8728.

AQUATOCENE: THE SUBAQUATIC QUEST FOR SERENITY
is an ongoing sound art project that investigates
the phenomenon of human-generated underwater noise pollution in the world's oceans, seas,
lakes, and rivers. This project utilizes audio
compositions of subaquatic soundscapes
to reflect on the anthropogenic sonic impact
on underwater habitats and marine wildlife
in order to raise awareness and underscore the
importance of preserving safe sound environments for animals inhabiting the world's waters.

Aquatocene: A Subaquatic Quest for Serenity

Robertina Šebjanič

Over 70% of the Earth's surface is covered with water habitats. 97% of the world's water is salt water, 2% is fresh water in the form of ice, and only the remaining 1% is drinking water, which is distributed around the planet very unevenly. The oceans are some of the most complex, challenging, and harsh environments for humans to access. To enter this world requires specially designed tools and technology. But as hidden as it is for us, its inhabitants, diversity is breathtaking, for it is assumed to contain somewhere between 500,000 and 10 million marine species, bacteria, other microbes, and viruses. Most ocean species remain undiscovered and up to 2,000 new species are described each year—with more exploration the numbers rise.

Advances in technology have reached a level that enables us to examine and explore the oceans in a systematic, scientific, and—at least partly—noninvasive way. Our ability to observe ocean environments and their inhabitants has caught up with our imagination and helps us to understand in ways we would never have imagined just a few decades ago. The ongoing development of remote sensing systems technology means we can use the data collected to understand more about the deep sea and its seafloor. Yet after many decades of exploration we still know more about space than about the deep sea.[1] In 1966, Nobel Laureate John Steinbeck (1902–1968) published an open letter in the September 1966 issue of *Popular Science* magazine in which he called for increased efforts to explore the world's oceans and seas. Steinbeck argued that investigation of the Earth's oceans was critical to the success of humanity and deserved the same funding and organization as space exploration.[2]

To enter and explore the otherworldly, unknown, and inhospitable deep aquatic environments where life on Earth once emerged, extensive training and appropriate equipment are required. However, our own physical limitations as well as the limitations of the technology we use for our investigations are likely the reasons why we are still not familiar with the vivid soundscapes of underwater worlds. Sound is the principal and most important way of communicating for marine creatures and an orientation sensor: the underwater acoustic environment is as rich and fascinating as any other known to humankind. The majority of us are unaware of these vibrant underwater acoustics, the sounds of the ocean's

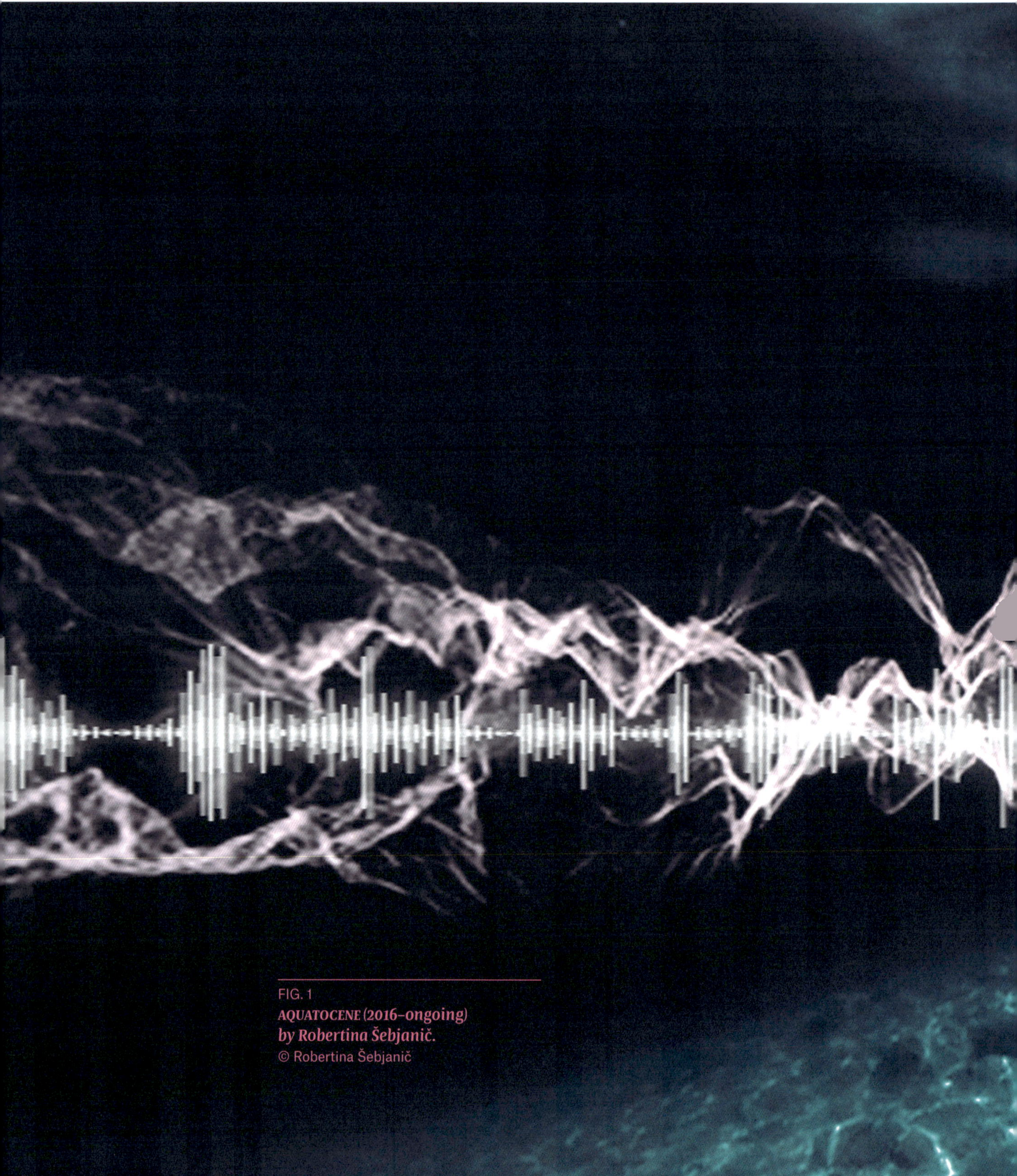

FIG. 1
AQUATOCENE (2016–ongoing)
by Robertina Šebjanič.
© Robertina Šebjanič

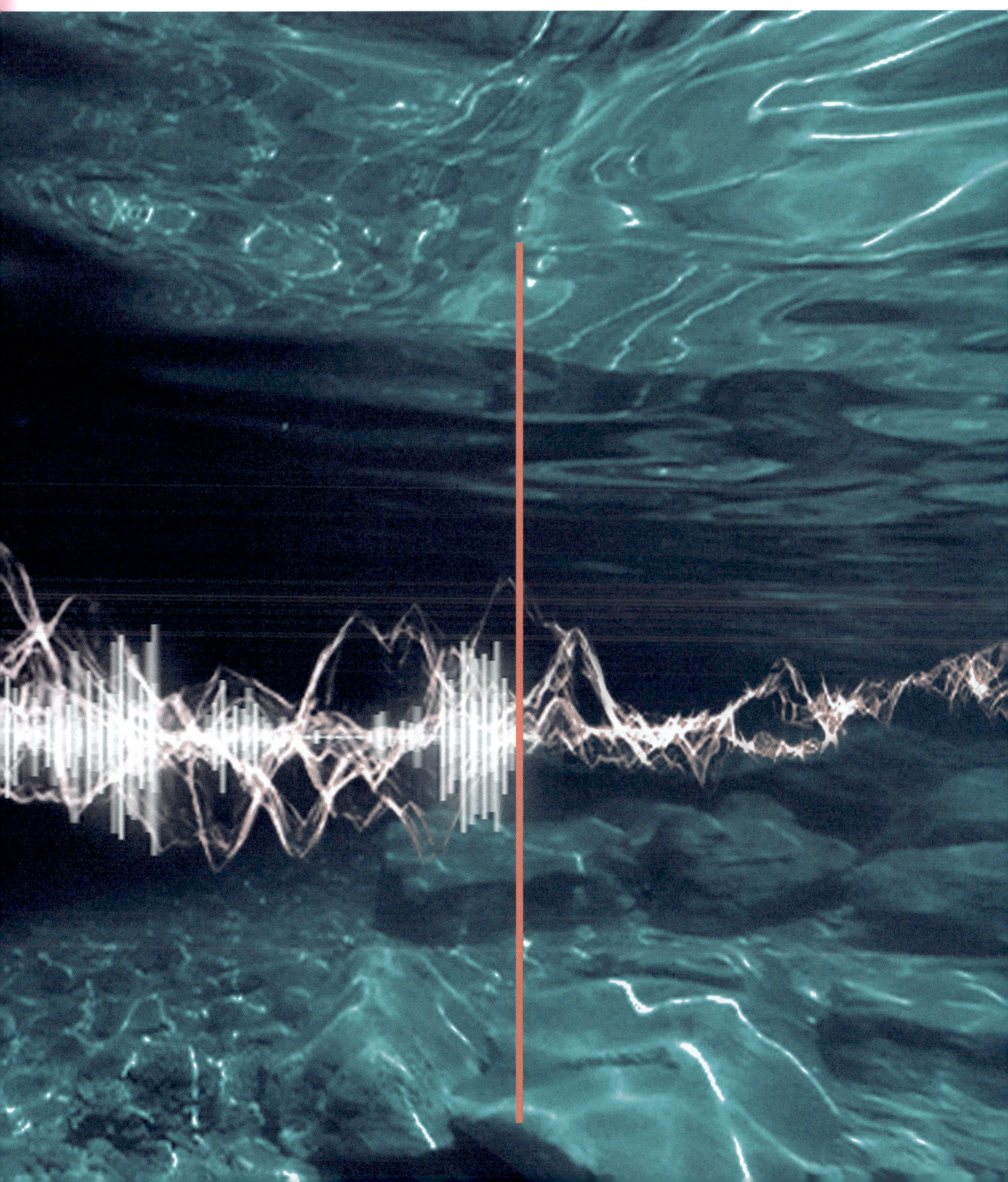

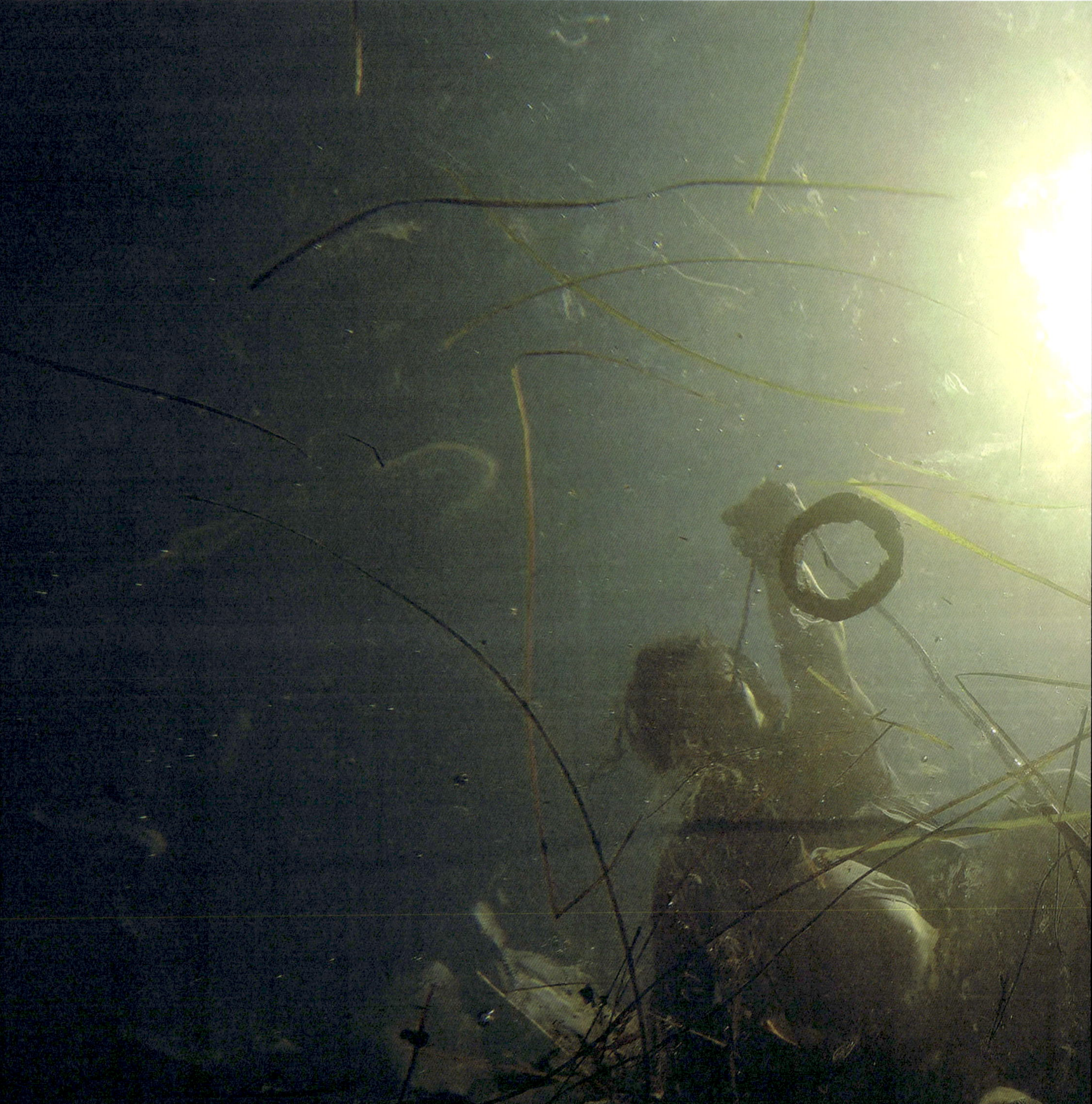

Aquatocene

Robertina Šebjanič

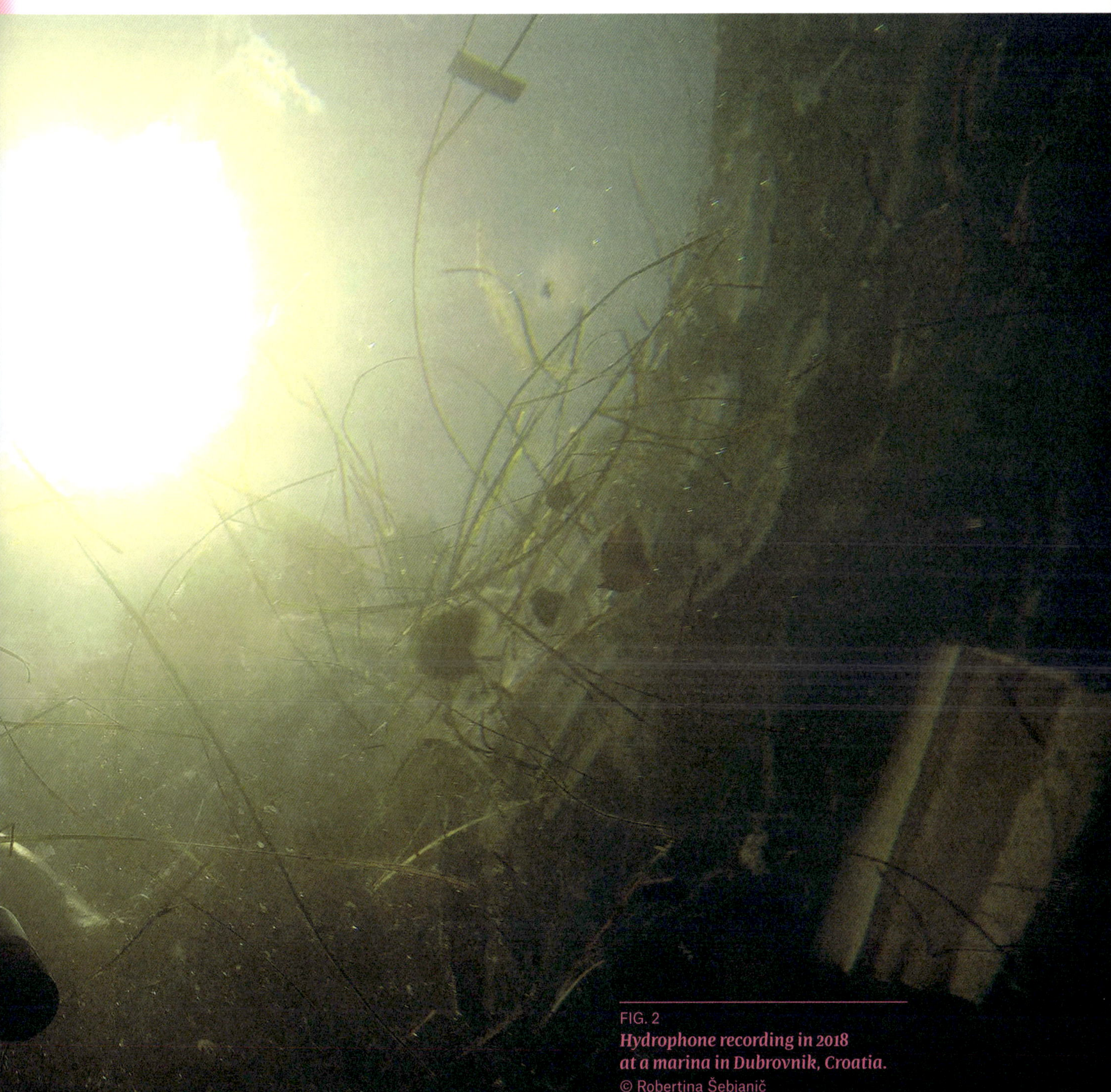

FIG. 2
Hydrophone recording in 2018
at a marina in Dubrovnik, Croatia.
© Robertina Šebjanič

depths. Our acquaintance with these sound worlds is often limited to the sound of waves crashing on our shores and beaches. However, whales, shrimps, seals, clown fish, dolphins, and a myriad of other creatures of the deep sea live in a liquid sonic environment. On top of our lack of awareness of the diversity of marine sounds, we are often ignorant of the fact that human-induced sonic pollution has already profoundly changed the soundscape of the waters of our planet. It has changed them to the terrifying degree that natural communication between marine animals is severely impeded in nearly all oceans and seas. Today, human-made noise caused by ships, oil rigs, submarines, boat engines, sonar, drilling, and mining, amongst others, lead to severe disruption of subaquatic habitats with huge consequences for marine life.[3] This anthropogenic technological sonic interference is the core of my ongoing art and science research project *AQUATOCENE* (FIGURES 1—8).

»Aquaforming« in the Age of the »Aquatocene« __ In the context of my artistic work, I use two terms to define our relationship with terrestrial waterbodies: »aquaforming,« which relates to the status of waters in the age of the Anthropocene and the human impact on the basic structure of aquatic habitats; and »Aquatocene,« which primarily refers to underwater noise pollution. The epoch of the Anthropocene in which we live today marks the end of the climate of the Holocene epoch of geologic time and is characterized by human »terraforming« and »aquaforming.« The incredible environmental challenges we are facing are the direct result of a complex amalgamation of sociopolitical and economic factors. It would be impossible to describe all of these numerous worrying issues within the scope of this article, but I will mention a few.

The steady rise in temperature of the world's oceans over the past decades has serious effects on marine life and marine ecosystems, like coral bleaching and the loss of breeding grounds for marine fishes and mammals. Similarly, overfishing of big predators and other marine species has seriously endangered numerous species or even led to their extinction and thus to a great loss of biodiversity.

The human impact on water habitats and its inhabitants is enormous, all the way from the nano and micro to the macro level. We find plastic and microplastic particles everywhere on this planet, even in the deepest and darkest corners of our oceans—a very telling and alarming fact.[4] The enormous degree of pollution is, of course, a direct result of the huge volume of output of global industrial production and mass consumerism, especially in the areas of fisheries, marine farming, marine transport, oil extraction, and drilling. At the same time, these problems themselves point to possible solutions to avert the further brutal exploitation of natural resources. I believe that now it is high time for profound introspection that would (hopefully) eventually lead, among other things,

FIG. 3
A view of the Atlantic Ocean in 2019 from the ship RV CELTIC EXPLORER *operated by the Marine Institute in Galway, Eire, during an art and research residency by the Aerial/Sparks project.*
© Robertina Šebjanič

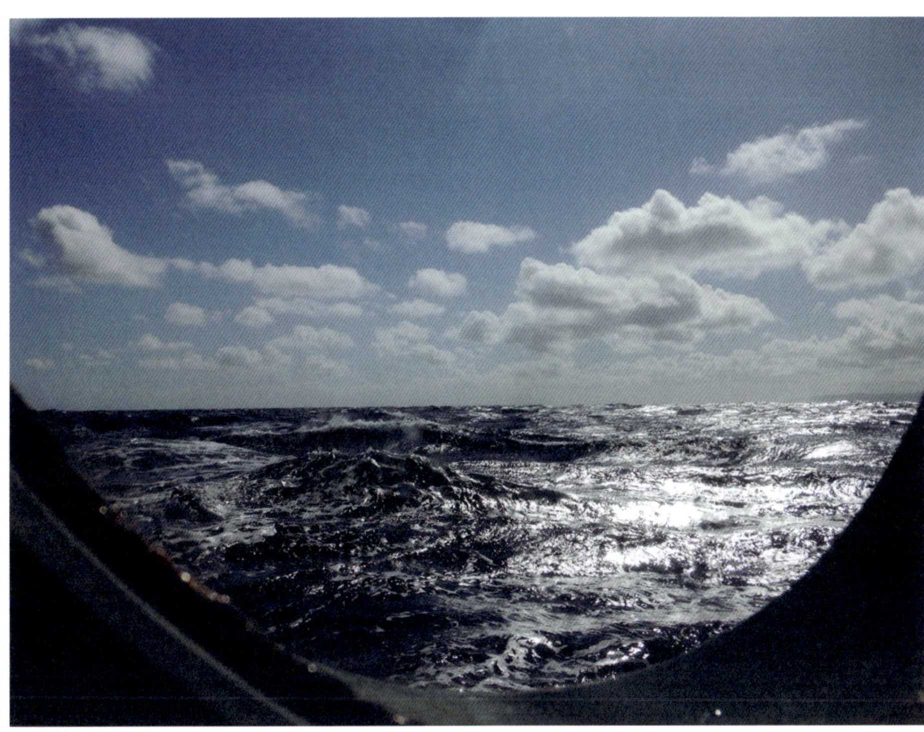

143

FIG. 4
Hydrophone recording in 2020 near Izola on the Adriatic coast during an art and research residency at PiNA (Association for Culture and Education), Slovenia.
© Robertina Šebjanič

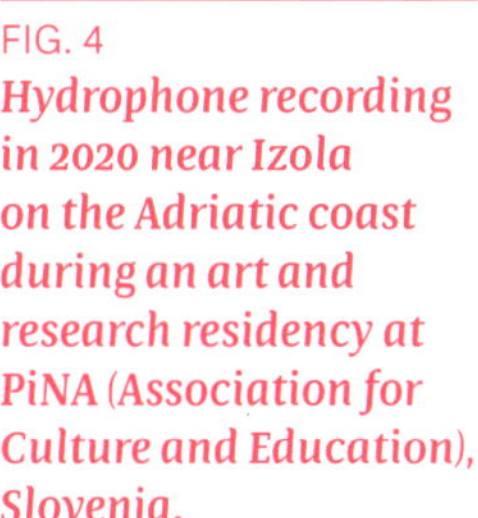

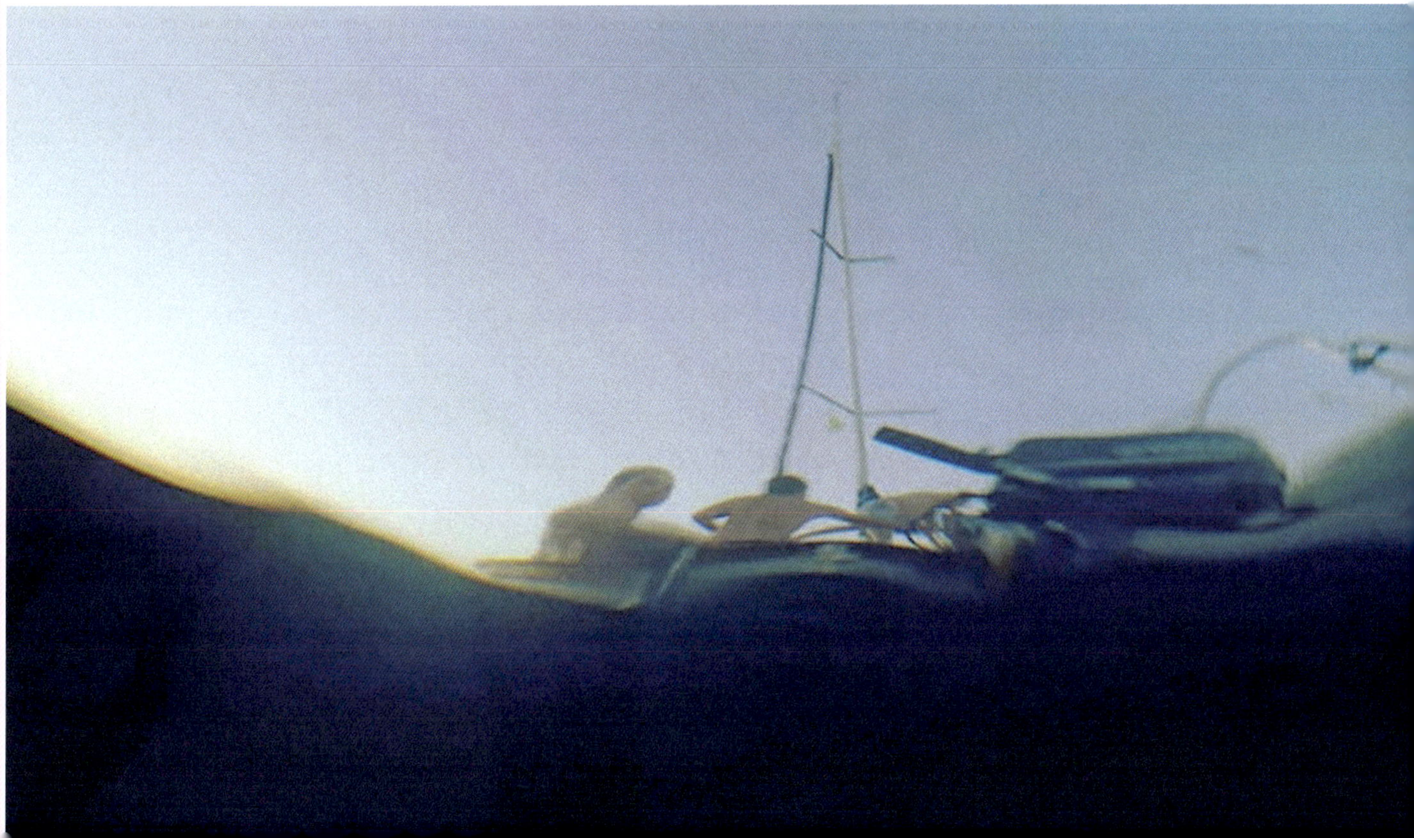

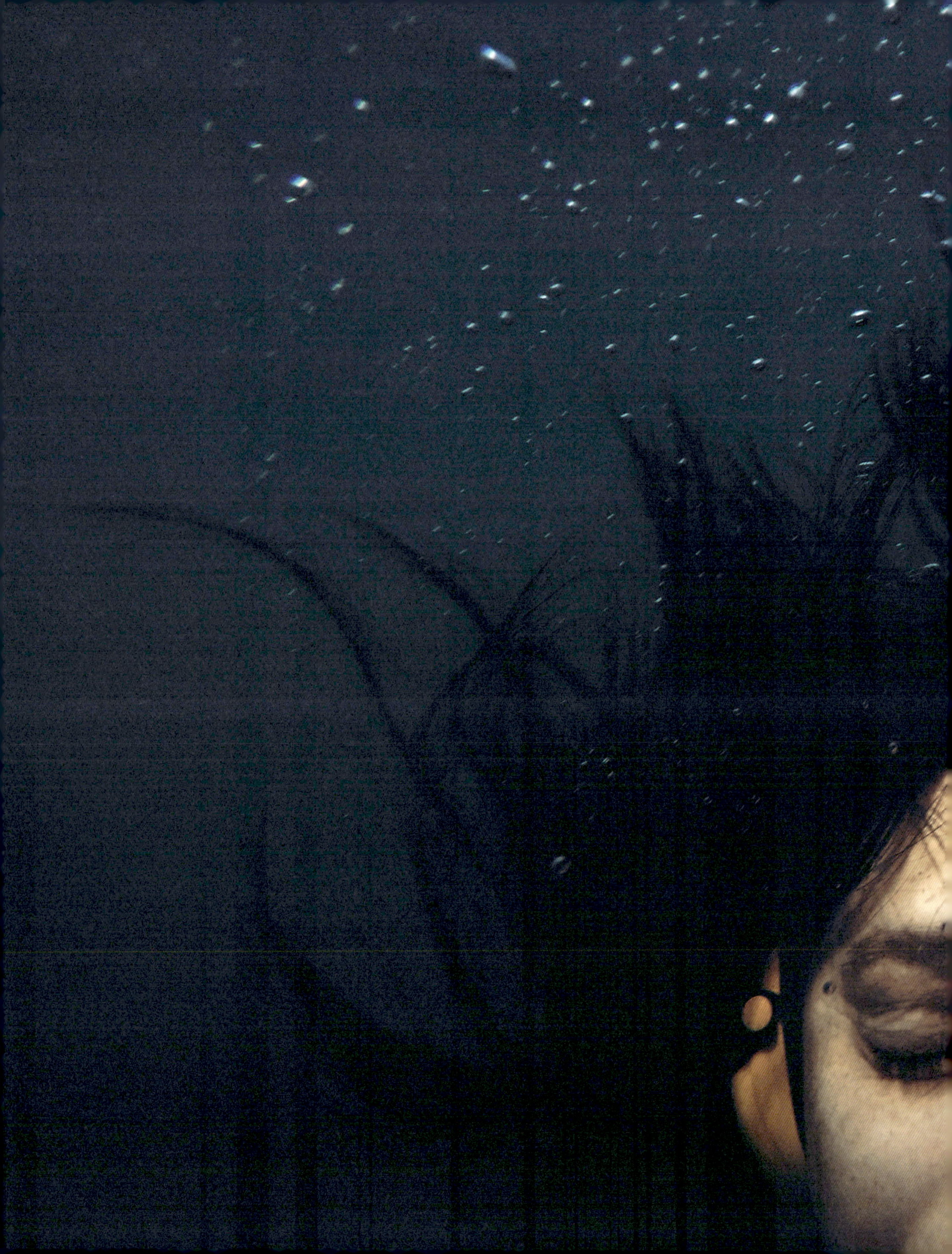

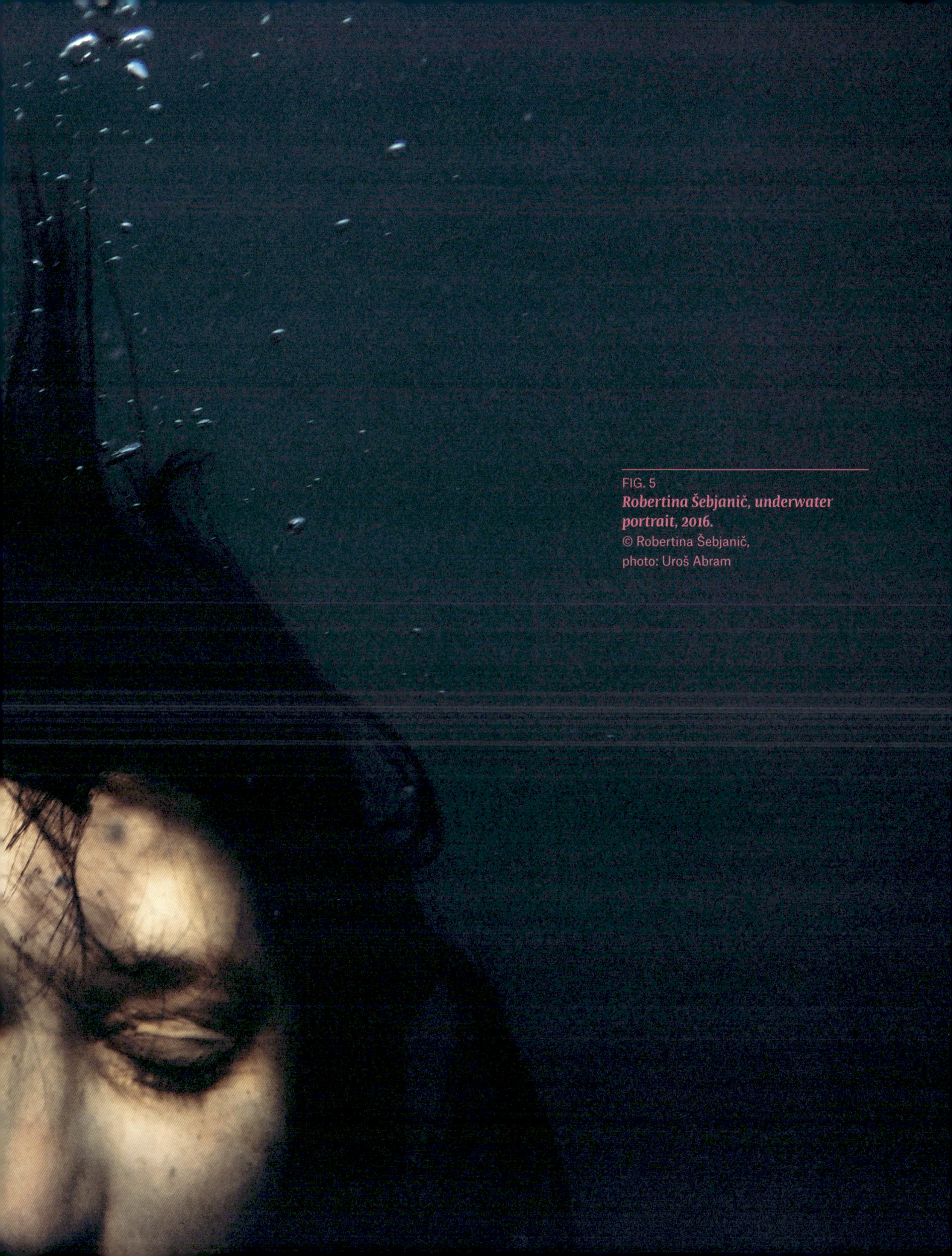

FIG. 5
Robertina Šebjanič, underwater portrait, 2016.
© Robertina Šebjanič,
photo: Uroš Abram

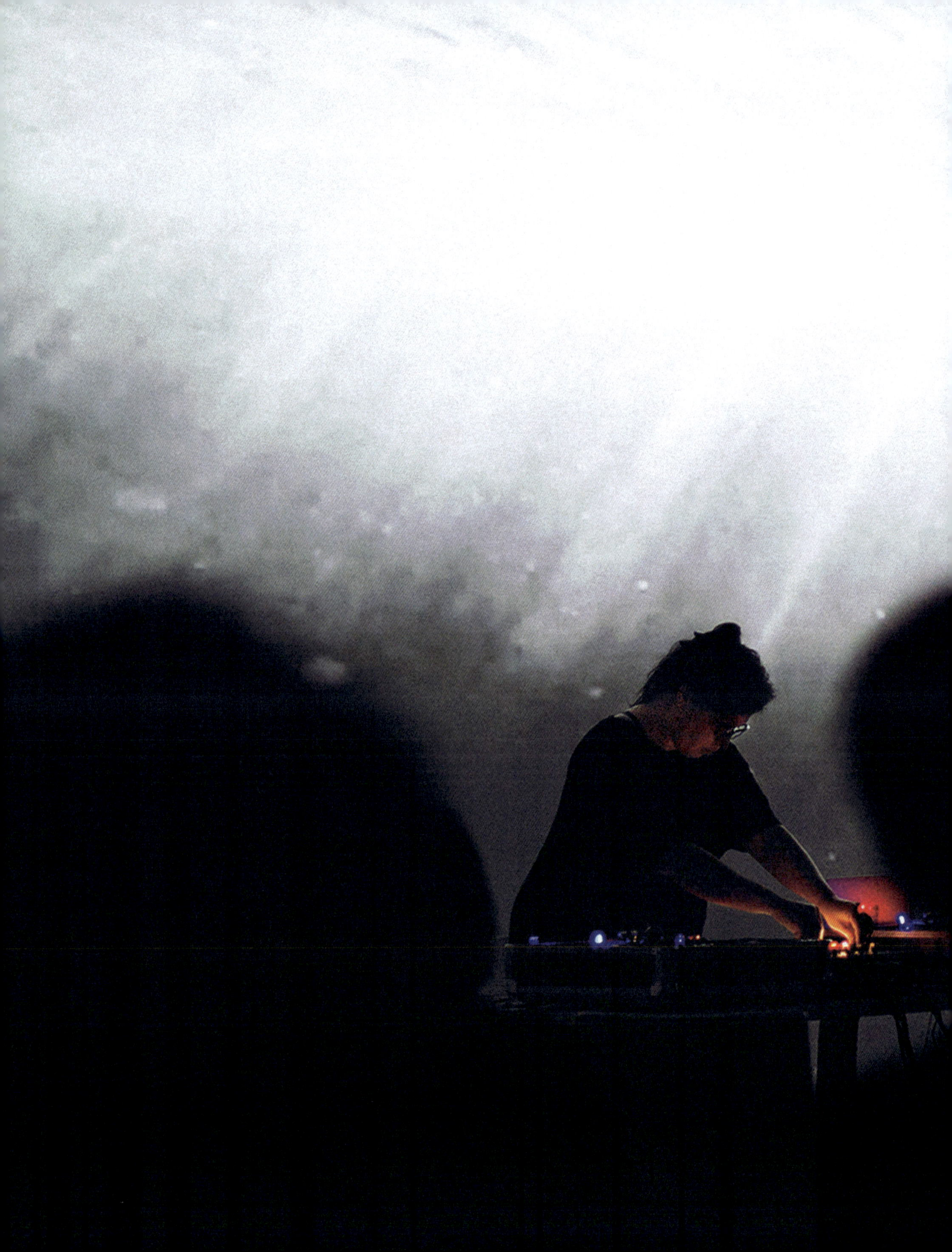

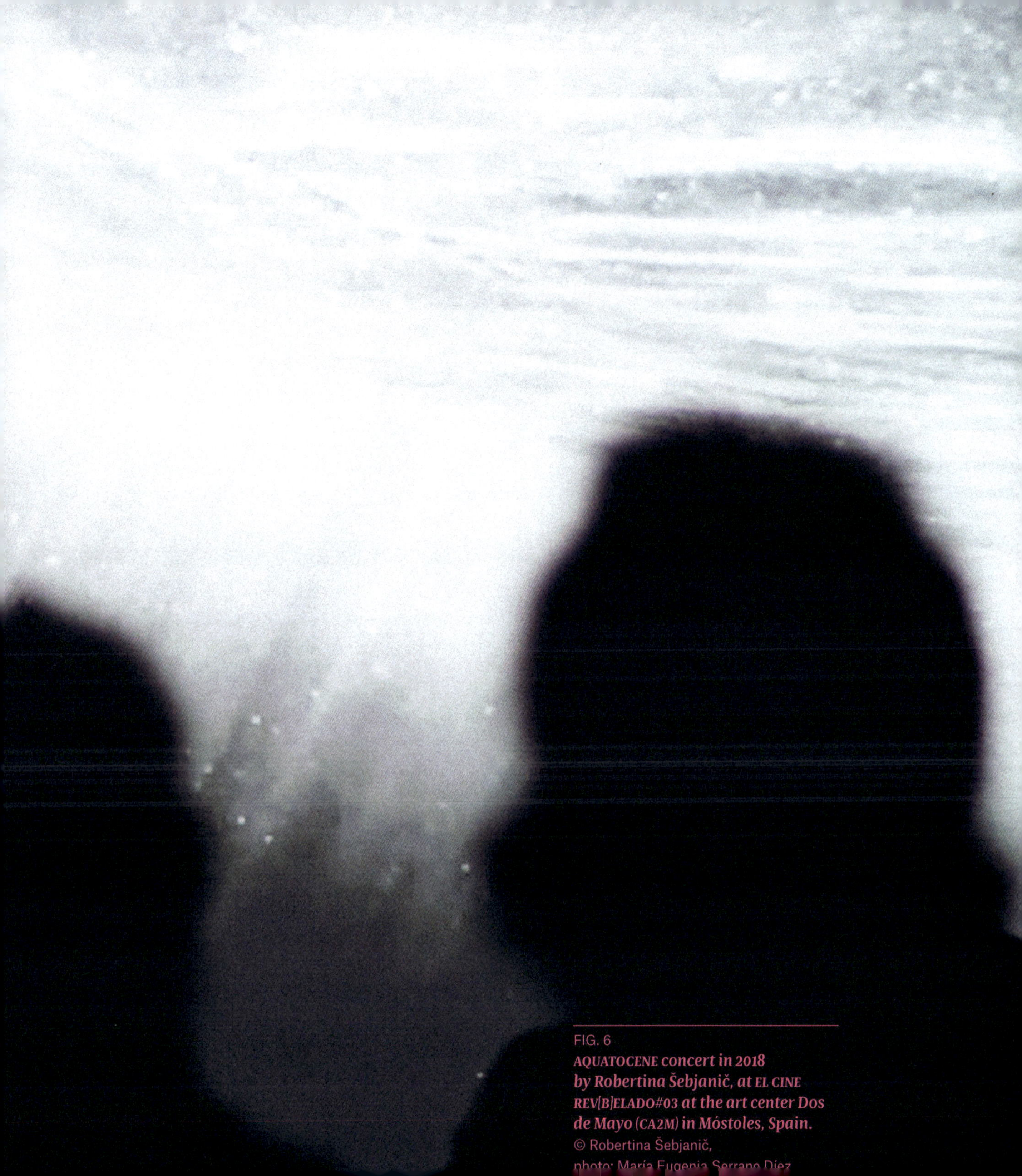

FIG. 6
AQUATOCENE concert in 2018 by Robertina Šebjanič, at EL CINE REV[B]ELADO#03 at the art center Dos de Mayo (CA2M) in Móstoles, Spain.
© Robertina Šebjanič,
photo: María Eugenia Serrano Díez

to the development of ecological awareness. The use of strategies of empathy with the awareness that care needs to be a conceptual part of development in the agricultural industry and transport industry, when looking for new and more sustainable options of food production and distribution. There is great interest in moving farming to (under)water habitats, but this could quickly lead to negative effects and turn out to be just as problematic for the environment if approached carelessly. The same goes for the transportation industry as most of the goods are transported on ships along routes in the oceans, seas, and rivers. The broad spectrum of environmental challenges is very concerning and therefore it is absolutely necessary to reconcile the interests of humanity, economy, science, and (geo)politics in order to save and preserve our water habitats. Furthermore, I think it is important to state that, along with a thorough and in-depth understanding of the challenges we are facing, it is imperative that we endeavor to meet them with empathy and solidarity and our main guiding principle should be in balance with the natural world that surrounds us.

Sounds of a Troubled World: Songs of the »Aquatocene« ___ »Aquatocene« is a term I coined to describe the processes of aquaforming in the Anthropocene, focusing on anthropogenic noise pollution that is widespread in our underwater worlds. Within my ongoing art-sci research, I choose particularly to focus on the biological, chemical, and geopolitical perspectives of the issues pertaining to the ecology of water habitats.

In 2014, during a residency in Izmir, Turkey, I found myself in the Gulf of Izmir with a hydrophone trying to record the sounds of swarms of jellyfish and other marine animals living in this part of the Mediterranean Sea. Listening to the material I recorded, I was astonished to find that my hydrophone had picked up so many sounds of boats, ships, and other human-made noise sources. I tried hard to filter out these sounds from my recordings to get »true« and »authentic« sounds of marine life, but did not succeed. The recordings of this »reality« affected me deeply and left a lasting impression. The more I listened to the human tech presence—noise—the more I realized it would make no sense to try and »photoshop« the sonic reality of the sea. I was interested in the real soundscape of the underwater world, the biodiversity, and also our imprint on it. Thus, I started to investigate the phenomenon of human-made underwater noise pollution, which resulted in the first stage of my ongoing sound artwork AQUATOCENE: SUBAQUATIC QUEST FOR SERENITY.

Since my stay on the shores of the Aegean at the Gulf of Izmir in 2014, I have been fortunate to spend a significant part of my life by or on the sea. Whether it was the Mediterranean Sea, the Adriatic, the North Sea, or the Atlantic Ocean, Lake Xochimilco or the Danube, Manzanares, Ljubljanica rivers and more, I always made sure to take my hydrophone with me and did some recording. With every new field recording added to

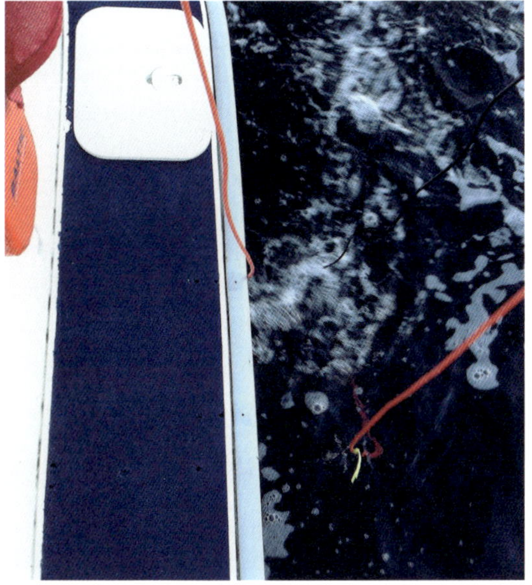

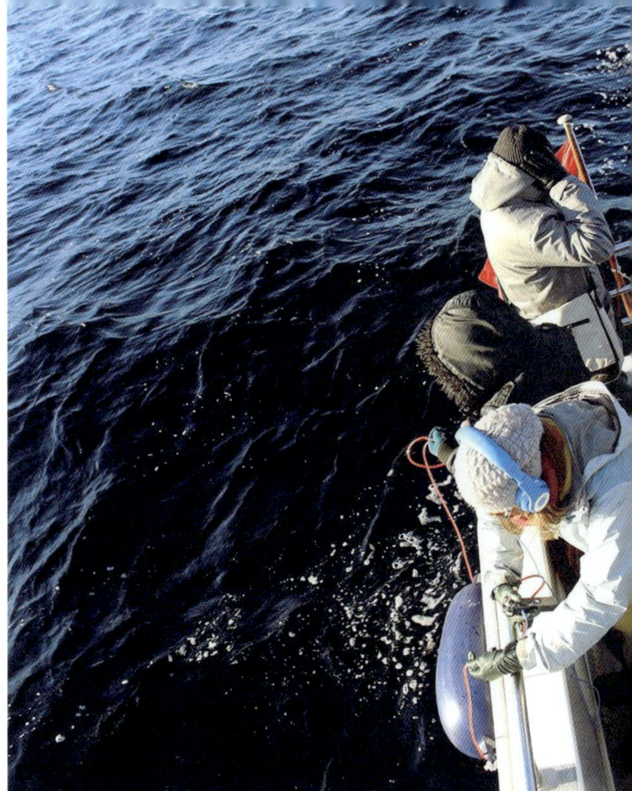

FIG. 7—8
Hydrophone recording in 2015
at the Pikslo deep diving workshop
at the Pixel festival in Bergen,
Norway.
© Robertina Šebjanič

my collection, I was more and more surprised to realize that human-made noise is so omnipresent in the underwater worlds. Most of us associate a sense of infinite tranquility and relaxation with the sea. This had also applied to me, but my perspective changed dramatically over the years with my recordings from various (underwater) locations. My expectations about underwater tranquility were fundamentally challenged. Sound waves travel through water much faster and much further than through air, for example, which is the reason why sound works so well as the communication and orientation of so many marine species. Light in the sea is only present near the surface (in the upper 200 meters) while in deeper layers light disappears and the ocean gradually becomes a world of utter darkness. That is why, down there, animals rely much more on sound than on vision. One of recent changes to the sonic environment of the waters is a rise in temperature and acidity: the subsequent change in pH directly influences acoustic transmission under water and thus global warming is also impacting the ocean soundscape.[5]

The sound art project AQUATOCENE seeks to acquaint listeners with the reality of what we think are underwater soundscapes of deep and shallow waters all over the globe. The compositions are the results of my field recording at specific locations. In my compositions listeners are confronted with the overwhelming human impact on the bioacoustics of the underwater habitats and they can also experience sounds of marine life and natural aquatic acoustics. During the concerts the audience is immersed in my personal interpretation of recorded soundscapes—a mixture of all the sounds I recorded in specific locations at various seas and oceans.

The heavily altered underwater sound environment caused by the large amount of invasive sounds scares and confuses marine animals while they navigate across the ocean. It affects their communication abilities, and the animals' attempts to avoid the unfamiliar sound environment often has drastic consequences, such as the beaching of whales and dolphins, to mention only the most well-known phenomenon.[6] As the premise of AQUA-TOCENE is simple – making »invisible« underwater sound pollution »visible« – the issues raised by it are present in all the world's waters. With its audio compositions of subaquatic soundscapes, the project reflects upon the human-made, technological sonic impact on marine life and the ecological changes, and therefore challenges. With »Aquatocene« I would like to raise awareness and underscore how important it is to preserve safe sound environments for animals living in the world's oceans, seas, lakes, and rivers. With their immersion in the sound compositions at my concerts or at home when listening to my sounds on digital media or vinyl,[7] I hope to open the listeners to the experience of these unknown sonic worlds.

When composing with material from field recordings I also try my best to communicate authentic experiences to the listeners. My hydrophone can sink to max. 30 m (the length of my cable), but as the sounds that I record are from afar, on finalized compositions it is possible to hear the diversity of the bioacoustics with a strong presence of the highly industrial sounds of our technological imprint. I think that being with a different species in *eye-to-eye* or rather *ear-to-ear* contact can lead to a deeper relationship and experience that is very enriching and can help to establish empathy with the unknown – with the Other.

As the nature of sound is different in water than it is in the air (in water, sound travels five to seven times further than it does in the air), you can imagine how present these underwater sounds are. One very interesting example of this was done in an experiment conducted in the early 1990s, which showed that sound emitted from Heard Island (near Australia) was picked up at sites in the Northern and Southern Atlantic and the Pacific Ocean as well as the Indian and Southern Oceans.[8] The sound does not stay within the borders or areas of a country, and as presented in the 1990s experiment, the sound travels very far. The trade routes and transport infrastructure on the oceans and seas in response to globalization play a huge role in driving the world's economy and it is necessary to talk more about this in the context of common geopolitical strategies for its constant increase is threatening the world's biodiversity.[9]

Investigating habitats, which are hostile to humans, is challenging, yet at the same time extremely rewarding. Over the last decade, I have been grateful for the opportunity to collaborate with experts from a vast number of scientific institutions, ranging from marine institutes to private aquariums. Despite these experts' huge efforts, there is still a vast amount of unexplored terrain and much is unknown about Earth's waters. But one thing is certain: technological interventions in the ocean soundscape by ships' engines and sonar can create huge disturbances in fragile marine habitats and have been connected to a number of effects ranging from the beaching of whales mentioned above as well as to the Lombard effect where certain species get louder and louder to overcome the background noise, thereby further increasing the acoustic intensity of the entire habitat. When recording with hydrophones, I try to interact with marine life in a noninvasive way. I also give workshops on how to build DIY hydrophones to teach others to listen more closely to the underwater world. In the course of these workshops, I introduce participants and the wider public to the pressing issues of aquatic ecology, as well as encourage them to become proactive in their communities. Dropping a hydrophone beneath the waves and experiencing the scale of underwater sounds first hand is always an *ear-opening* moment for everyone involved.

...Me, Us, Them Together in the »Aquatocene« ___ I locate my artistic practice at the intersection of cultural, (bio)political, and biological realities of aquatic environments. The methods of my arts-based research as well as the final presentations of my projects rely strongly on the cross-pollination of knowledge from diverse fields and ways of knowing. From the point of view of my expression in art, it is vital to discuss and critically frame the dominant scientific narratives and the way they shape the ways we perceive our realities. To find a good balance, it is essential to compare science to artistic and philosophical narratives rather than solely analyzing reality through statistics. For me, it is meaningful that I integrate and merge different disciplines and perspectives such as art, science, and technology. Combining the knowledge of different disciplines can be very challenging, but rewarding too.

This approach appeals so much to me because transdisciplinary thinking takes a unique approach to solving problems. Both science and art are human endeavors to understand and describe the world around us. The subjects and methods have different traditions and different audiences are addressed, but I think the motivations and goals of both arts and science are fundamentally the same.

As sound is one of the main sensors for a great deal of marine life, the following questions arise: How do we understand the use of our senses? How do marine species use their senses? While we are taking our time to ponder on or completely ignore the ramifications that our economy and ways of production have on nature, marine life has already started to adapt. As French writer Pascal Quignard states in his book *La haine de la musique*, we as humans are able to close our eyes, but it is impossible to do the same with our ears.[10] The latter also holds true for marine life, which relies so heavily on hearing organs to navigate and, ultimately, to survive. Animals are unable to »shut their ears to human-made sounds,«

and the acoustic changes underwater are most palpable along the main marine traffic routes.[11] During the first months of 2020, we experienced how silence took over the world due to the COVID-19 pandemic, and this silence affected the land as well as the sea. Might this be the opportunity scientists have been waiting for to record the sounds of the ocean as they had been during the early days, before propeller-driven ships started traversing the globe? The global pandemic made it evident: it is possible to silence the world again and scientists took that opportunity to do their work. The first results have already been published. If we follow the explanations of British scientist Peter Tyack at the University of St. Andrews, Scotland, who is well known in the field of marine mammal biology and concerned about the effects of anthropogenic sound on wildlife, it becomes clear that the recordings give scientists a glimpse of the ocean with little human interference, a situation which had never occurred before.[12] Due to this incident, we may ask: How big will the impact of this unique situation be on current marine life, and what will it mean for the future? Will be this a glimpse into the past or will it also provide some indication of possible future developments?

I have a hope that the public that is immersed in the soundscapes of AQUATOSCENE sound compositions could partly have this kind of experience when hearing the beauty of the sounds of inhabitants of the waters, and at the same time also feel the anxiety that is caused because of our technological presence. As mentioned at the beginning of my contribution, there is still so much that is unexplored in our water habitats. This surely signifies great potential for times ahead, but our biggest concern should be to raise public awareness about the pressing circumstance that water habitats have to be handled with care. As economic interests in exploiting our waters will remain as high as ever or even increase, we must not delude ourselves into thinking that we are not prone to repeating or continuing the mistakes of the past. Developments in the field of technology, which have contributed greatly to the prosperity that comes with economic growth, have shown there are many ethical questions we need to address. And while all this might sound rather dystopian, research and the zeitgeist might warrant a slightly darker interpretation of what is to come. What humankind—or rather, the world's waters— would really need is a good coordinated development program of empathy and solidarity as a strategy to give water habitats the attention they deserve and ensure not only the survival of marine life but also of our civilization which depends strongly on the latter. As marine scientist Sylvia Earle recently stated: »*Why is it that scuba divers and surfers are some of the strongest advocates of ocean conservation? Because they've spent time in and around the ocean, and they've personally seen the beauty, the fragility, and even the degradation of our planet's blue heart.*«[13]

In spring 2021 3 × vinyl AQUATOCENE was published by Inexhaustible
Editions, sublabel SonoLiminal:

Aquatocene 60.3913° N 5.3221° E _North Sea

Aquatocene 48°43'38.2"N 3°59'33.4"W _Atlantic Ocean

Aquatocene 42°57'31.62 N 17°8'12.08 E _Adriatic Sea

Aquatocene 38°26'24.7"N 27°07'12.9"E _Aegean Sea

Aquatocene 53°19'37.3"N 5°07'18.5"E _Wadden Sea

Aquatocene 42°38'26.5"N 18°06'20.4"E _Adriatic Sea

1   Jon Copley, »Just How Little Do We Know about the Ocean Floor?« *Scientific American*, October 9 (2014). Online: www.scientificamerican.com/article/just-how-little-do-we-know-about-the-ocean-floor/ (last accessed January 19, 2020).

2   Rose Pastore, »John Steinbeck's 1966 Plea to Create a NASA for the Oceans,« *Popular Science*, May 20 (2014). Online: www.popsci.com/article/technology/john-steinbecks-1966-plea-create-nasa-oceans/ (last accessed January 19, 2020); John Steinbeck, »Let's Go after the Neglected Treasures beneath the Seas,« *Popular Science*, September (1966): 84–87.

3   Jim Robbins, »Oceans Are Getting Louder, Posing Potential Threats to Marine Life,« *The New York Times*, January 22 (2019). Online: www.nytimes.com/2019/01/22/science/oceans-whales-noise-offshore-drilling.html (last accessed January 19, 2020).

4   Alan J. Jamieson, Lauren S.R. Brooks, William D.K. Reid, Stuart B. Piertney, Bhavani E. Narayanaswamy, and Thomas D. Linley, »Microplastics and Synthetic Particles Ingested by Deep-Sea Amphipods in Six of the Deepest Marine Ecosystems on Earth,« *Royal Society Open Science*, February 27 (2019). Online: https://royalsocietypublishing.org/doi/10.1098/rsos.180667 (last accessed January 19, 2020).

5   »How Does Sound in Air Differ from Sound in Water?« The Discovery of Sound in the Sea (DOSITS), Online platform developed by the University of Rhode Island (URI)'s Graduate School of Oceanography (GSO) in partnership with Inspire Environmental of Newport, RI. Online: https://dosits.org/science/sounds-in-the-sea/how-does-sound-in-air-differ-from-sound-in-water/ (last accessed January 19, 2020).

6   U.S. National Research Council, *Ocean Noise and Marine Mammals: Effects of Noise on Marine Mammals*, Washington, D.C.: National Academies Press (2003). Online: www.nap.edu/read/10564/chapter/5, (last accessed January 19, 2020).

7   Robertina Šebjanič, »Aquatocene: Subaquatic Quest for Serenity,« https://robertina.net/aquatocene/ (last accessed January 19, 2020).

8   »Underwater Noise and Its Impact on Marine Life,« SAFETY4SEA, National Oceanic and Atmospheric Administration, June 23 (2015). Online: https://safety4sea.com/noaa-underwater-noise-and-its-impact-on-marine-life/ (last accessed January 19, 2020).

9   Vanessa Pirotta, Alana Grech, Ian D. Jonsen, William F. Laurance, and Robert G. Harcourt, »Consequences of Global Shipping Traffic for Marine Giants,« *Frontiers in Ecology and the Environment*, December 5 (2018). Online: https://esajournals.onlinelibrary.wiley.com/doi/full/10.1002/fee.1987 (last accessed January 19, 2020).

10 Pascal Quignard, *The Hatred of Music*, New Haven, CT: Yale University Press (2016); title of the French original: *La haine de la musique*, Paris: Calmann-Lévy (1996).

11 Rob Williams, Andrew J. Wright, Erin Ashe, Louise K. Blight, and Matthew A. Wale, »Impacts of Anthropogenic Noise on Marine Life: Publication Patterns, New Discoveries, and Future Directions in Research and Management,« *Ocean & Coastal Management* 115 (2015): 17–24. Online: www.sciencedirect.com/science/article/pii/S096456911500160X (last accessed January 19, 2020).

12 Maurice Tamman, »Pandemic Offers Scientists Unprecedented Chance to ›Hear‹ Oceans as They Once Were,« Reuters, June 8 (2020). Online: www.reuters.com/article/us-health-coronavirus-climate-research-i/pandemic-offers-scientists-unprecedented-chance-to-hear-oceans-as-they-once-were-idUSKBN23F1M3 (last accessed January 19, 2020).

13 Sylvia A. Earle, published December 24 (2017). Online: https://twitter.com/sylviaearle/status/944716255105355777 (last accessed January 19, 2020).

NOISE AQUARIUM is a participatory art project that
comes after a series of works which explored how
to immerse the audience using scale and sound
to bring forward the importance of the invisible
and inaudible. This essay follows the trajectory
of a number of collaborative artworks that have
inspired and led to the concept and framework
of NOISE AQUARIUM: NANO (2003), NANOMANDALA
(2003), WATER BOWLS (2006), and BLUE MORPH (2008).
All of these works involve scale, sound, and
working with ways to slow down the audience
by using technology to amplify the experience
that requires balancing, stillness, and deep listening to give access to the invisible and inaudible
realms. I aim to describe the coemergent process
that was developed over the past two decades
in order to create works that are ecocentric
and transdisciplinary.

NOISE AQUARIUM: Iterations, Variations, and Responsive Ecotistical Work

Victoria Vesna

The artwork NOISE AQUARIUM used existing visualizations as an anchor to create an experience that amplifies the importance of plankton for life on Earth by submerging the participants in sounds of underwater noise pollution and prompting them to center and balance themselves. Many of us are still not aware that plankton is a central source of the oxygen we breathe and that noise pollution underwater is destroying so much of the marine life that has a direct impact on all living beings. The aim is to show that even though we do not see the plankton with our naked eye or hear the noise deep down in the oceans and seas, we are all implicated due to the destructive effects of our excessive exploitation of resources which has thrown our planet out of balance. Many different iterations of NOISE AQUARIUM (FIGURE 1) have been created over a period of three years (2017–present)—from linear videos to fully programmed interactive virtual reality (VR) and most recently, due to COVID-19 restrictions, as an online network collective meditation. In all instances, the goal remains the same: to remind everyone that the oxygen for every second breath we take is produced by plankton and that we are all part of the complex web of life. The intent is to create an ecocentric[1] point of view and immerse the audiences into active deep listening.[2]

NANO: Making the Invisible Visible, the Inaudible Audible

Although I could rewind the film back much further when considering the importance of sound, vibration and frequency as an anchor for visual experiences, the starting point of what led up to NOISE AQUARIUM was without question the NANO exhibition in Los Angeles, USA (2003–2004).[3] This ambitious collaborative project with nanoscientist James Gimzewski involved contributors from many disciplines and was the beginning of consciously imagining a series of connected art installations that involved full audience participation. The invitation to develop an exhibition in the annex of the Los Angeles County Museum of Art (LACMA) came because of institutional museum restructuring, which opened an opportunity to create a series of experimental works that allowed the audience to interact with the artworks and experience a different sense of space and movement. This exhibition took place just after completing my first fully collaborative work with Gimzewski, ZERO@WAVEFUNCTION —an interactive installation

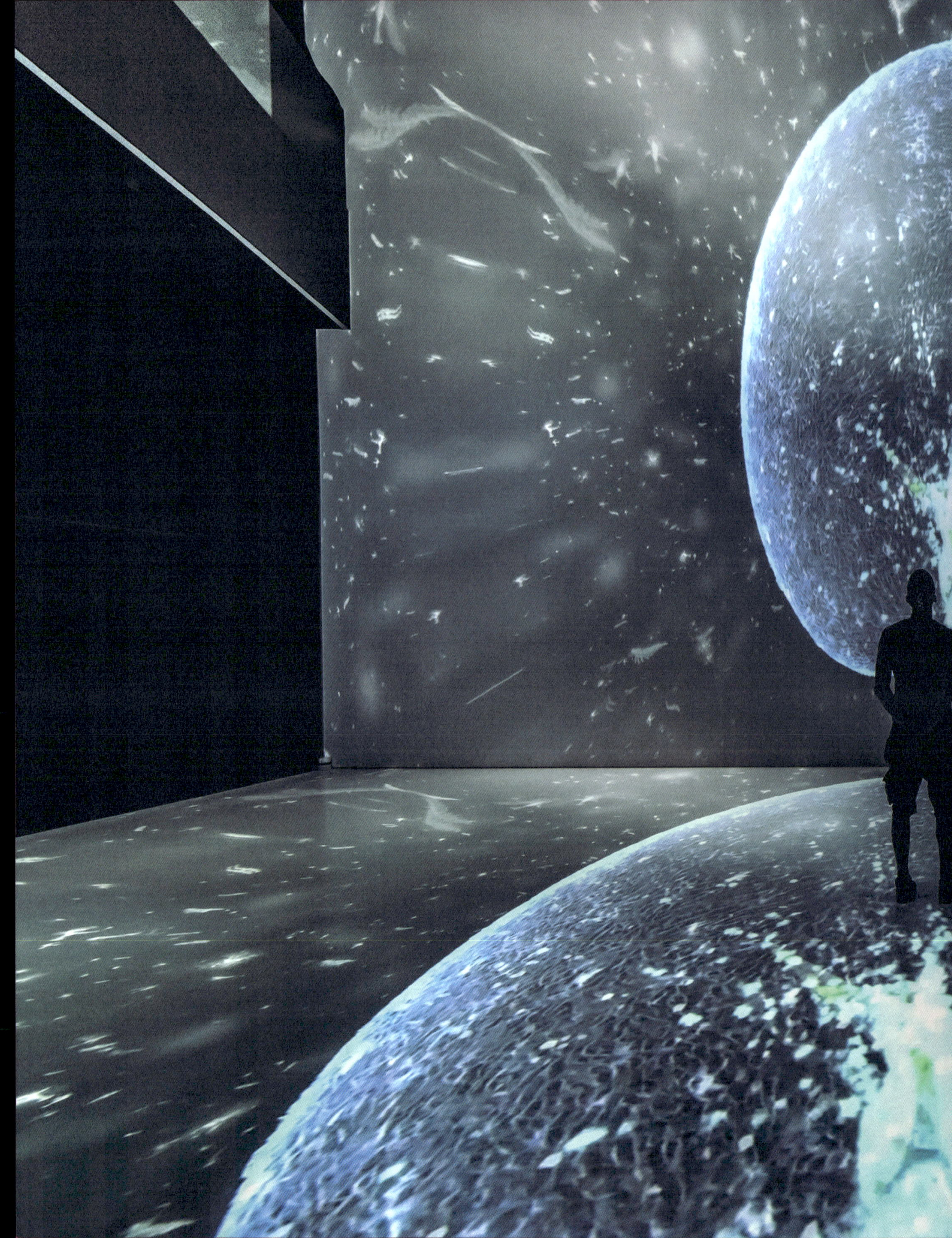

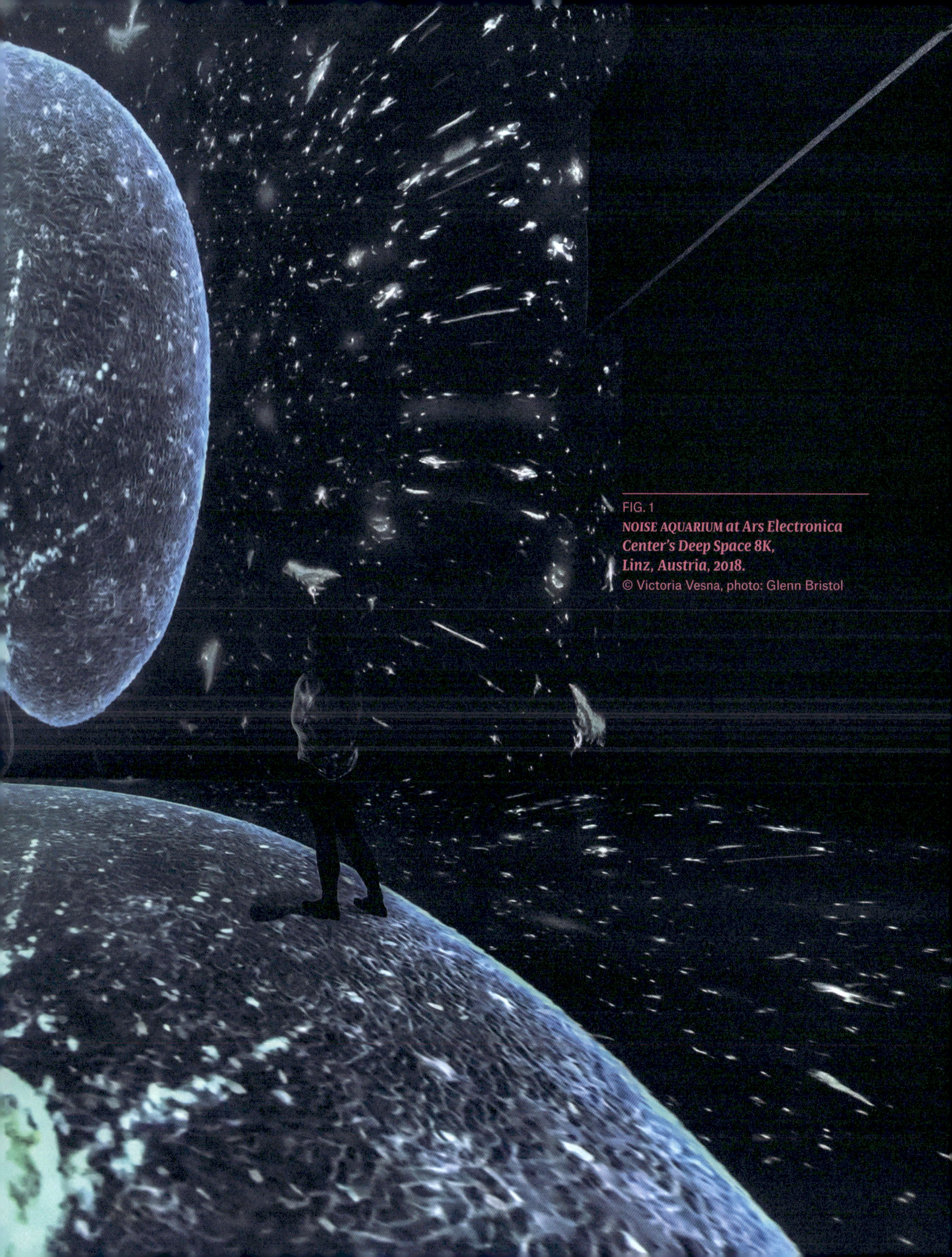

FIG. 1
NOISE AQUARIUM at Ars Electronica
Center's Deep Space 8K,
Linz, Austria, 2018.
© Victoria Vesna, photo: Glenn Bristol

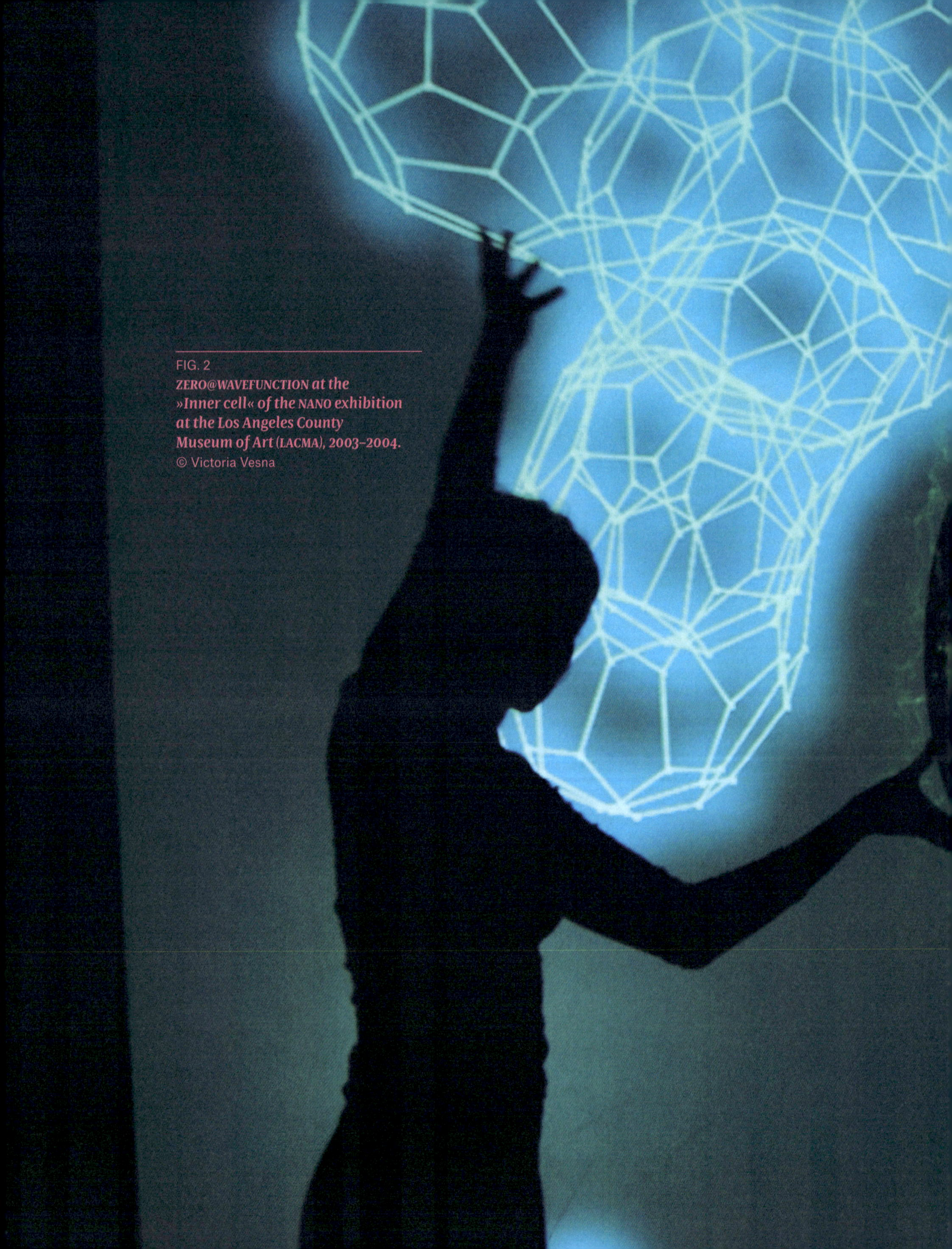

FIG. 2
ZERO@WAVEFUNCTION at the
»Inner cell« of the NANO exhibition
at the Los Angeles County
Museum of Art (LACMA), 2003–2004.
© Victoria Vesna

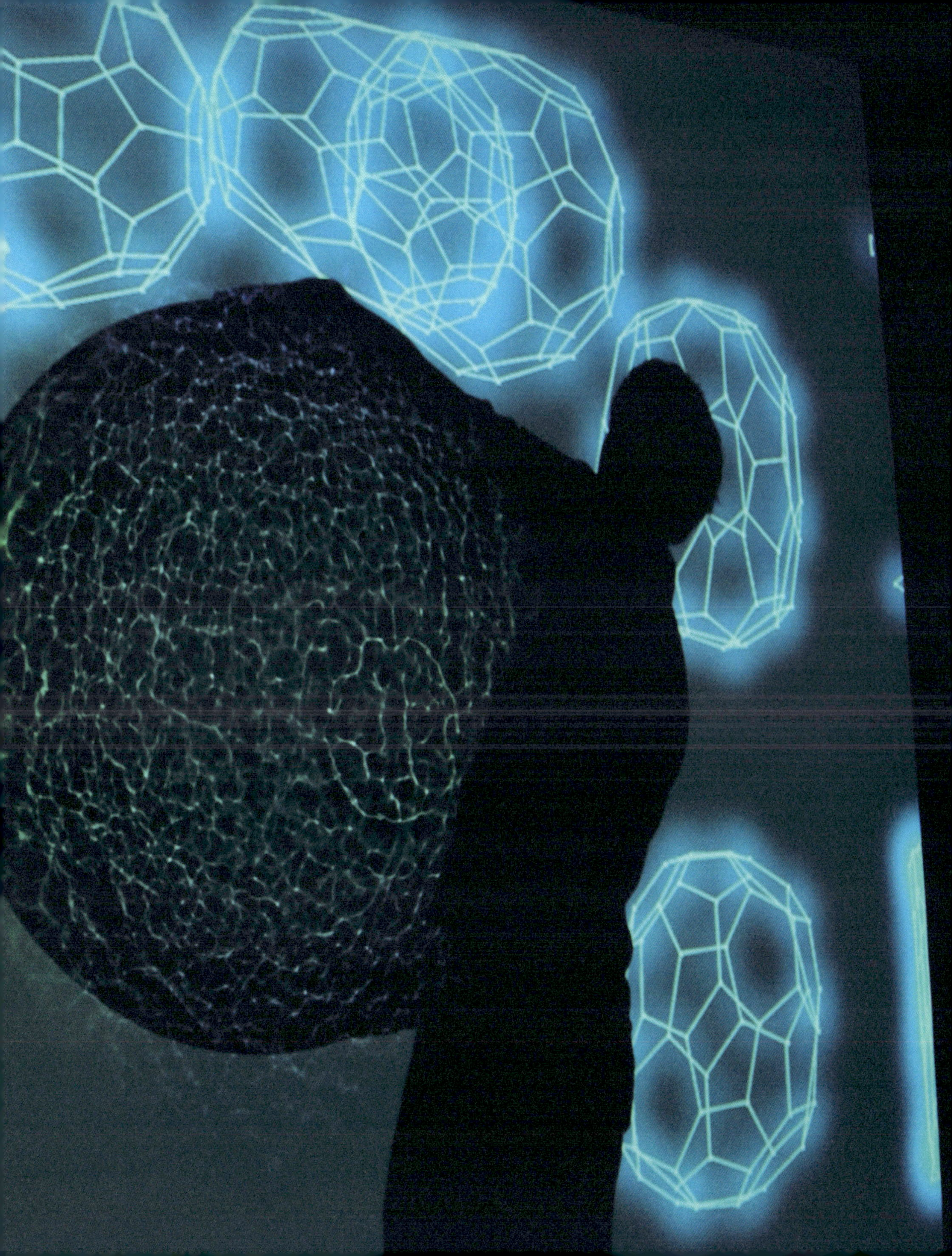

with visualizations of buckyball molecules being manipulated by the visitors' shadows (FIGURE 2).

Buckminsterfullerenes, or »buckyballs,« discovered in 1985, are named after the architect, engineer and visionary R. Buckminster Fuller, »Bucky« (1895–1983), who designed many geodesic domestructures that look similar to the molecule with the formula C_{60} — discovered two years after Fuller's death. Ideas to expand further the experience of the molecular realm were developed in discussions, amongst others with N. Katherine Hayles, a U.S. postmodern literary critic, who was Professor of English at the University of California Los Angeles (UCLA) at the time,[4] and her students who were invited to write about the process and work. On top of that, the museum connected us to architects who were also actively participating in the structural design of the exhibition — again reflecting Fuller's ideas of geometric and organic forms found in nature.[5] ZERO@WAVEFUNCTION became the central piece from which all other installations emerged, including their architecture.

As we moved further into the development of the exhibition, the museum curators became concerned that the concept was not connecting to contemporary art enough, and so I was asked to look through the collection and pick a few works that could also be included. At that point in time the highly collaborative and interdisciplinary process of developing a work of art was foreign to the museum curators and was even not perceived as art. Thus, the idea was put forward that adding some artworks from the collection would somehow help the audience to appreciate this kind of art and science work. This endeavor to put on display a selection of artworks in tune with our project was more challenging than one would imagine as it seemed forced and did not make sense in more ways than one. However, as I began to cast my net wider than the contemporary art collection, I was attracted to the East Indian collection and thought that a Buddhist work would be perfect — as James Gimzewski kept saying, we are looking into empty space when we peer through a scanning tunneling microscope (STM) since all we see there are waves and particles. When this idea was presented to the curators, they brought to our attention that a group of Tibetan monks would be at the museum to build a sand mandala and that perhaps we could work with that in one way or another. We were thrilled and wanted to talk to the monks directly as we had a shared goal — to show how everything is interconnected.

NanoMandala: CoEmergence/Transitory Nature ＿＿ Over the next three weeks, we watched the sand mandala being created. The monks were chanting throughout the process and told us that they did not know exactly what the sand mandala would look like when finished. When the curator asked when it would be completed so that they could schedule a showing, the answer — to the organizer's frustration — was that they were not sure

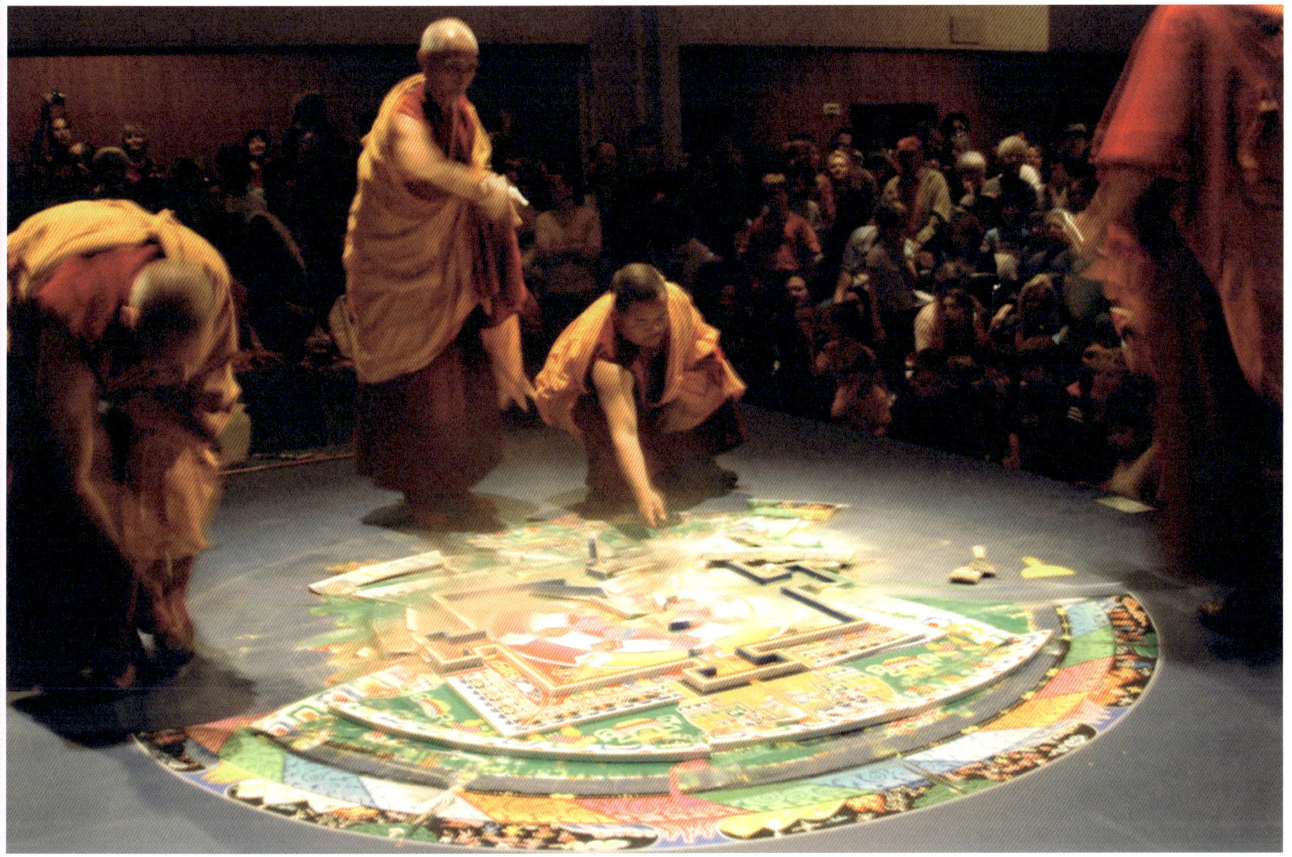

Destruction of the Chakrasamvara sand mandala by Tibetan Buddhist monks, in conjunction with the CIRCLE OF BLISS *exhibition at the Los Angeles County Museum of Art* (LACMA), *2004.*
© Victoria Vesna

and there was no way to predict or influence the creation process. It took four monks working for weeks, eight hours a day, to complete the Chakrasamvara sand mandala, and after a few weeks on display, its destruction was a major and a disturbing ritual that included a lot of movement and noise (FIGURE 3): the monks swept up the sand and, with the audience following them, they ceremoniously walked down to the beach at the Pacific Pier to throw it back into the ocean. Observing the monks' performance was a lesson for us and our own project as it demonstrated the transitory nature of life and the importance of sound in the process of creation and destruction.

Our collaboration with the Tibetan monks, which we called *NanoMandala*, immersed the audience in a meditative piece with three levels of perception: photographic, optical, and through scanning electron microscopy (SEM) (FIGURES 4—5). The *NanoMandala* is projected onto a round bed of sand that is the same size as the original sand mandala, and the images move at a scale beyond the powers of ten, beyond the visible realm. However, ultimately, what pulls one into this piece is the audio

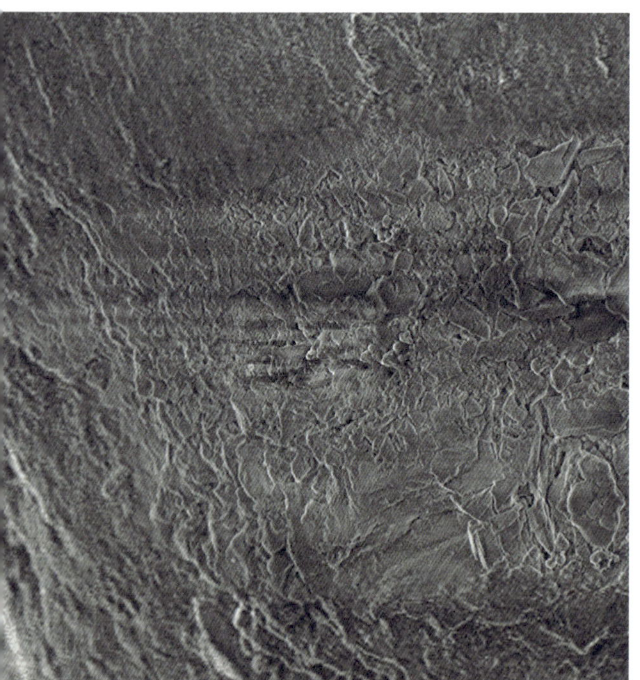

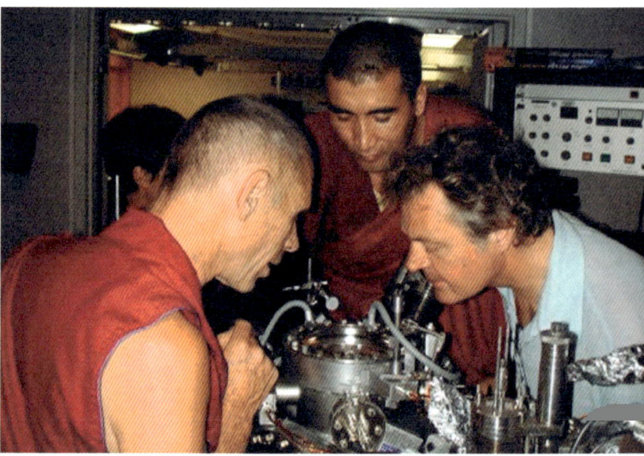

FIG. 5
*Monks being introduced to the
scanning tunneling microscope (STM)
by James Gimzewski at the UCLA
PicoLab, 2003.*
© Victoria Vesna

vibratory field: sounds of the ocean, the audio recording of the mandala being created and the human voices chanting.[6] The piece is a meditation on the importance of every particle and wave, of the interconnectivity of us all and everything surrounding us, as well as on our amazing ability to take huge data sets of information and reduce them to the essential truth in the blink of an eye. It took billions of sand grains to construct the complex sand mandala, but now we have the ability to see that each grain of sand is itself composed of many billions of molecules.

NANO: Movement through Scale and Space

In the NANO installations, one of the main rooms, the »Inner Cell« had a floor that was inspired by the surface of a graphene—an allotrope of carbon consisting of a single layer of atoms arranged in a two-dimensional honeycomb lattice. Electron waves in graphene propagate within a single-atom layer, making them sensitive to the proximity of other materials, and this laid the foundation for much of the nanotech revolution in materials science. As people moved through the space they were influencing the floor, as would happen with the needle of the STM. Here was the ZERO@WAVEFUNCTION piece, with bucky-balls projected in this central space of the installation that were influenced and transformed by the moving shadows of the audience. Very quickly, people figured out—usually by watching others—that in order to create change, they had to move slowly. This slow movement was inspired by watching the monks working patiently and slowly on the sand mandala and the scientists manipulating the molecules—in both instances, fast movement would mean destruction and thus it was important in both of these processes to move slowly and deliberately.[7]

An additional element was added to the interactivity in the form of robotic balls that were manipulated—moved—by the audience in the adjacent room from a table that had a bird's-eye view of the people in the room with the graphene floor. Here, we were mimicking the process of manipulating and moving around molecules remotely with the robotic balls moving around seemingly independently and influencing the movements and behavior of the audience in the main »Inner Cell.« All this was to show the different ways of being and how we are capable of causing change from a distance as well as how our own behavior is influenced by outside forces, whether we are aware of it or not. An important part of the series of installations was the quantum tunnel which additionally addressed the nonlocality of our thoughts, relationships and existence. The sounds utilized were those of yeast cells manipulated by Gimzewski's team, with the movement output being sound rather than a graph. This later became a larger installation, *Cell Ghosts*, and further informed and inspired the BLUE MORPH installation (see below).[8]

This virtual and metaphorical molecular space provokes the audience to discover through physical engagement, to learn by feeling. Viewers experience the act of manipulating molecules and encounter the effects of their bodily movements on the surrounding space. Instead of emphasizing how small the scale of nano is, we created an experience that reminds the audience that we live in a molecular world that is shaped by our presence: the audience becomes the performer and changes the work simply by being there.[9]

Water Bowls: Repurposing, Recycling Elements __ The robotic balls in the

ZERO@WAVEFUNCTION installation got damaged very quickly by younger audiences who were too eager to play with the objects. After the exhibition, these spheres were lying around the studio, and being aware of the fact that they were polycarbonate plastic, I was not comfortable with throwing them away. They were also quite expensive and beautiful as objects, but they no longer had a purpose and took up a lot of space. At one point, a projection light was accidentally thrown on them and the reflections were so intriguing that I was moved to start experimenting. The first thing I did was to put the balls into a bowl filled with water and project light onto them. Eventually, I placed hydrophones inside the bowl and developed a work that was in many ways a precedent to NOISE AQUARIUM. I set up four water bowls—half full or half empty—and started to extensively research water as a life-giving element, which led me to underwater noise as well

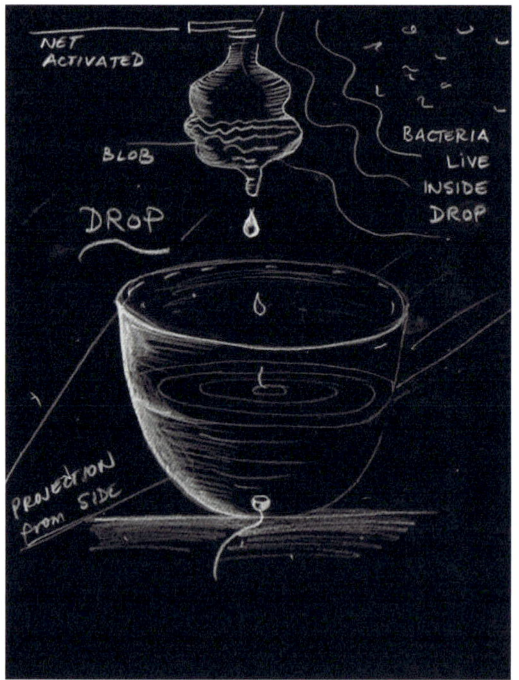

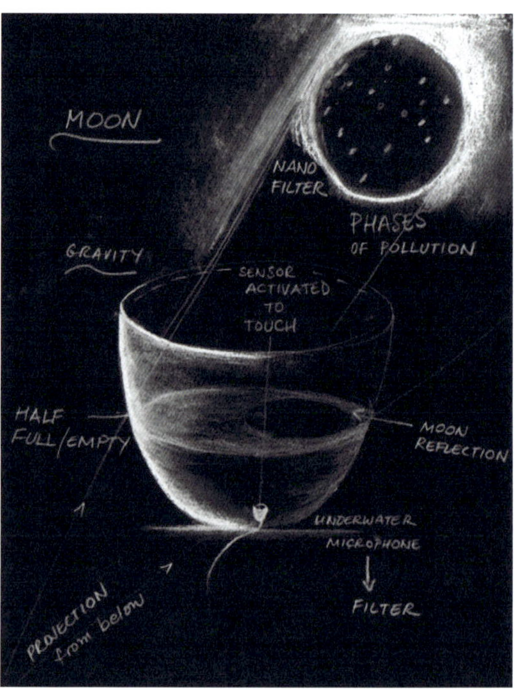

FIG. 6—9
WATER BOWLS: conceptual drawings.
© Victoria Vesna

as to oil and plastic pollution and, in general, to looking at the impending disaster of water pollution and water shortage. The installation was set up as a meditative space that, at first, seems very beautiful; but the longer you stay, the more you discover the sad state of water. The audience is invited to touch the water and make noises, create a disturbance and influence the state of the environment—we are all implicated.[10]

Water Bowls: Moon~Sound~Drop~Oil—Local/Remote Participation __

Four water bowls—half full or half empty—reflect different aspects of water related to the collective, global human condition. Some of the most common metaphorical associations of water—such as the reflection of the moon, a drop of water, the sound of water, as well as oil and water—are revisited through using some of the latest scientific observations. The bowls »Moon« and »Sound« were envisioned to be locally interactive, while »Drop« and »Oil« were interactive both locally and remotely, emphasizing the global connectivity of both water and human systems, beyond borders (FIGURES 6—10).

In »Moon,« visitors were invited to touch the water; the sounds created by their interaction were picked up by an underwater microphone and amplified. An animation of water molecules, cycling from a heavily

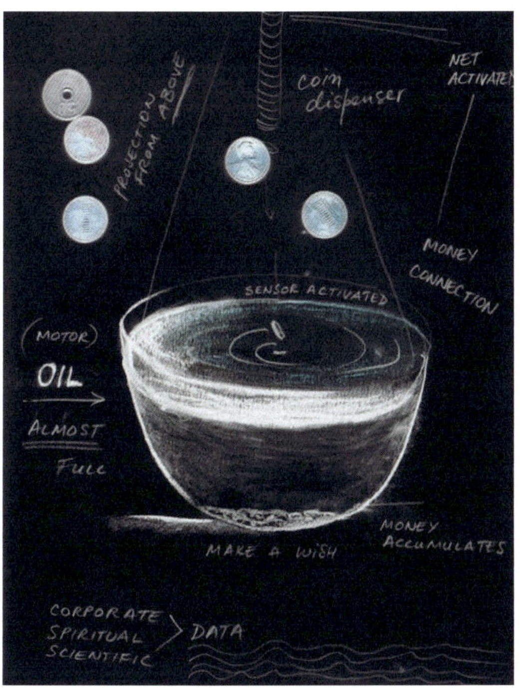

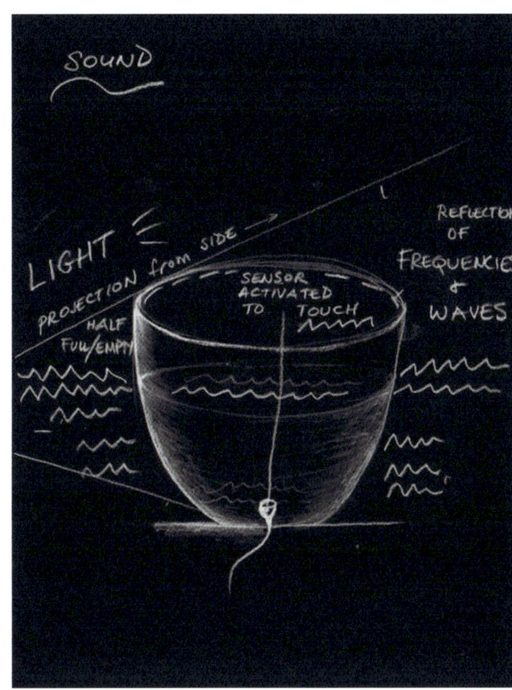

polluted state to clearing and back, was projected onto the water. »Sound« was also activated by a person's touch, which generated a disturbance of the water reflection and allowed the person to feel the vibration of sounds based on underwater pollution (such as sonar frequencies, explosions, and submarines) as well as whale sounds and cell vibrations.

In »Drop,« a drop of water released into the bowl breaks the surface and triggers images derived from maps of water bodies that ripple away. On the project website, visitors could remotely release this water drop from a dispenser suspended above the bowl in the exhibition space. Upon entering the website, visitors were asked to identify themselves with a body of water, for example, the Nile, the Ganges, the Danube, the Pacific, the Atlantic Ocean, the Mediterranean Sea, or any other water body of their choice. The online interface then pulled up a topographic map of the chosen location. The location of the participant was mapped by tracking the Internet Protocol (IP) address of their computer and pairing it with the water body they had identified with on the map. The participant could then add a drop of water to the bowl. The topographic map and the visitor's position on it were also projected onto the bowl in the exhibition space.

In »Sound,« audiences were invited to move the water with their hands, creating sounds that were underwater pollution noises – fracking, sonar, shipping, as well as naturally occurring sounds. This was really when I began to understand the dire situation that is unfolding and the importance of bringing this issue to the attention of the public. We are all participating in creating this noise and later, with *NOISE AQUARIUM I*, I was really able to bring this to the forefront. The »Oil« bowl contained both water and oil which mix over time and are not clearly separated substances—contrary to what we are often told. I would collect dirty oil from a local gas station and the coin dispenser device above was designed to throw a copper cent into the bowl that generated a disturbing explosion

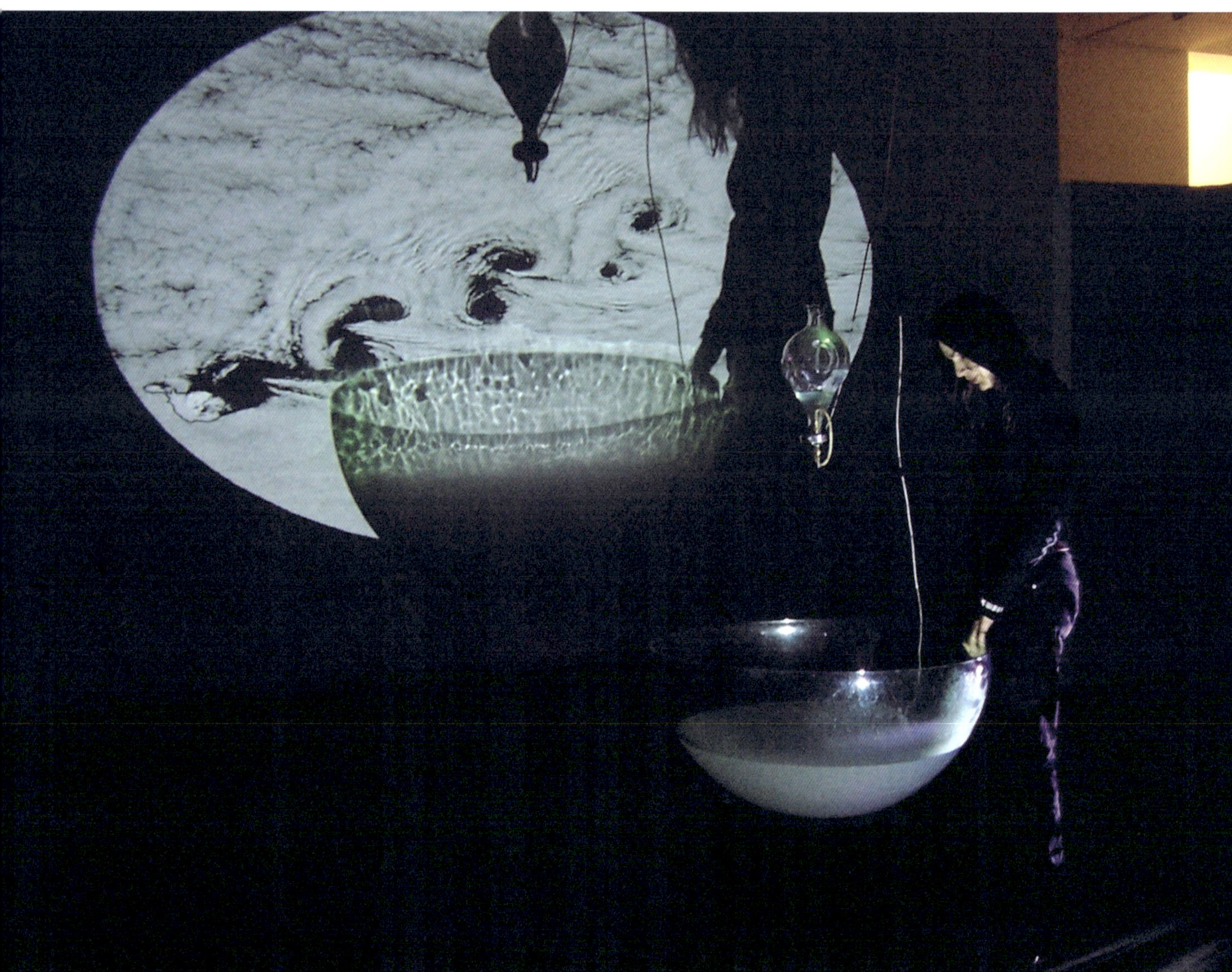

sound in the room and created a visual mix of the oil and water reflection. The idea was to additionally have people throw pennies into the oil bowl to make a wish which would generate a similar effect, and also to allow remote visitors to type in their wish and release a copper coin into the bowl from the dispenser above it. The wish would be projected onto the oil and visible on the wall behind the bowl as a visualization of dissolving particles. This bowl was almost full.

Blue Morph: Iterations, Variations/the Audience as a Performer ___ Just

like NOISE AQUARIUM, the BLUE MORPH project was initiated by an invitation from the outside and required learning and research. James Gimzewski's work on cellular vibrations attracted much attention and at one point he was asked if he could measure the vibrations of a butterfly metamorphosis. Although not of particular interest to his research or my artwork, both of us were intrigued and started looking, exploring, visualizing. What was particularly fascinating was that the beautiful iridescent wings of the Melenaus blue morpho butterfly are in fact nature's nanotechnology: it is empty space and patterns that make us see that blue shade, visible in the jungle from miles away. *Morpho melenaus* is an endangered species and has been used by many investigators of biomimicry with applications such as anti-counterfeit money and credit cards. The optics of this butterfly are amazing, but the real surprise to us was the discovery of the way cellular change takes place in the animal.[11]

Using the same technique as for the detection of nano-oscillations of yeast cells, we recorded the sounds of a metamorphosis by measuring the motion of the pupa surface as it transformed from one stage of development to another with an atomic force microscopy (AFM) detection setup. Raw data files of the chrysalis membrane »sound« vibrations were speeded up and amplified by arbitrary amounts, depending on the individual sample. Although we tend to imagine butterflies as silent, colorful creatures, they in fact generate intense noise, inaudible to humans, while in the process of changing. The microscopic encasing becomes the interface of sound intensity, while scattered light mimics the simultaneous beauty and turmoil inherent in the butterfly metamorphosis. We discovered that change does not happen gradually as we might like it to happen, but is a sudden, intense surge of energy that is destructive and creative at the same time. The visualization of the surges of metamorphosis very much resembles the ups and downs of the current financial markets in crisis. Metaphorically, we see this as a collective feeling of »butterflies in our stomachs«—a nervous anticipation of the shift in the human condition. The BLUE MORPH project emerged as an experiential installation that changed form as it moved sites and was increasingly shaped by the different audiences that interacted with it as a ritualistic experience. The installation consists of an interactive seat that one person at a time occupies, wearing a hat pulled up high

FIG. 10

into the ceiling as a space for invoking a personal metamorphosis
(FIGURE 11). In order for the microscopic butterfly wing patterns and sounds
to fully emerge, the participant has to be centered and still. The work
evolved over time in relation to site-specific installations, with the source
of the conceptual framework being the Integratron, an acoustically perfect
tabernacle and energy machine situated on a powerful geomagnetic
vortex in the Mojave Desert, California. The structure itself dictated a shift
to the audience performing their wish for change by connecting to the
frequencies of metamorphosis.[12]

BLUE MORPH came to the attention of Dr. Alfred Vendl, former head
of the Department of Technical Chemistry and Science Visualization at
the University of Applied Arts Vienna, Austria, who at that time was also
utilizing STM in his visualizations and working on visualizing plankton.
Dr. Vendl contacted James Gimzewski and visited the PicoLab and the Art
Sci Center at UCLA — the beginning of our dialogue that eventually turned
into a collaboration. This is yet another example of being open to what
is presented to us: I had little-or-no knowledge about plankton and
had it not been for this encounter, I most likely would not have gravitated
toward this subject, let alone made it the center of my work.

NOISE AQUARIUM: Immersive Sound Vibrations/Animations __ Work on

NOISE AQUARIUM started in 2016 when Dr. Vendl first introduced me to
the amazing plankton visualizations produced in close collaboration with
biologists Stephan Handschuh, University of Veterinary Medicine Vienna,
and Thomas Schwaha, based at the University of Vienna. The animations
had been commissioned for the movie *Voyage of Time* by film director
Terence Malick. In the end, however, the work was not used in the movie.
Dr. Vendl invited me to make an artwork with the beautiful animations
created by Martina R. Fröschl, his PhD student at the time, and I started
learning about plankton as fascinating micro creatures. It took me some
time to imagine how I could contribute to these amazing animations.
The images were magical. I was so intrigued by the complexity and diversity of plankton that it only took work with seven species — and very few
visualizations — to realize that there are millions of plankton species in the
ocean waters and that they produce the oxygen for every other breath
we take. »Why is this not common knowledge? How are these creatures
impacted by all the underwater noise and pollution?« I wondered.

In ZERO@WAVEFUNCTION, my wonder with regard to buckyballs
prompted me to blow up the visualizations as large as possible and give
the audience the opportunity to manipulate their shapes while hearing
the sounds of the cells. I decided to do the same with the plankton — blow
them up as huge as whales in order to bring them to the attention of
the audience and bombard visitors with the underwater noise that marine
life is subjected to (FIGURE 1). Once I had the concept of scale and noise

FIG. 11
BLUE MORPH
at St. John's Cultural Centre,
Gdańsk, Poland, 2011.
© Victoria Vesna

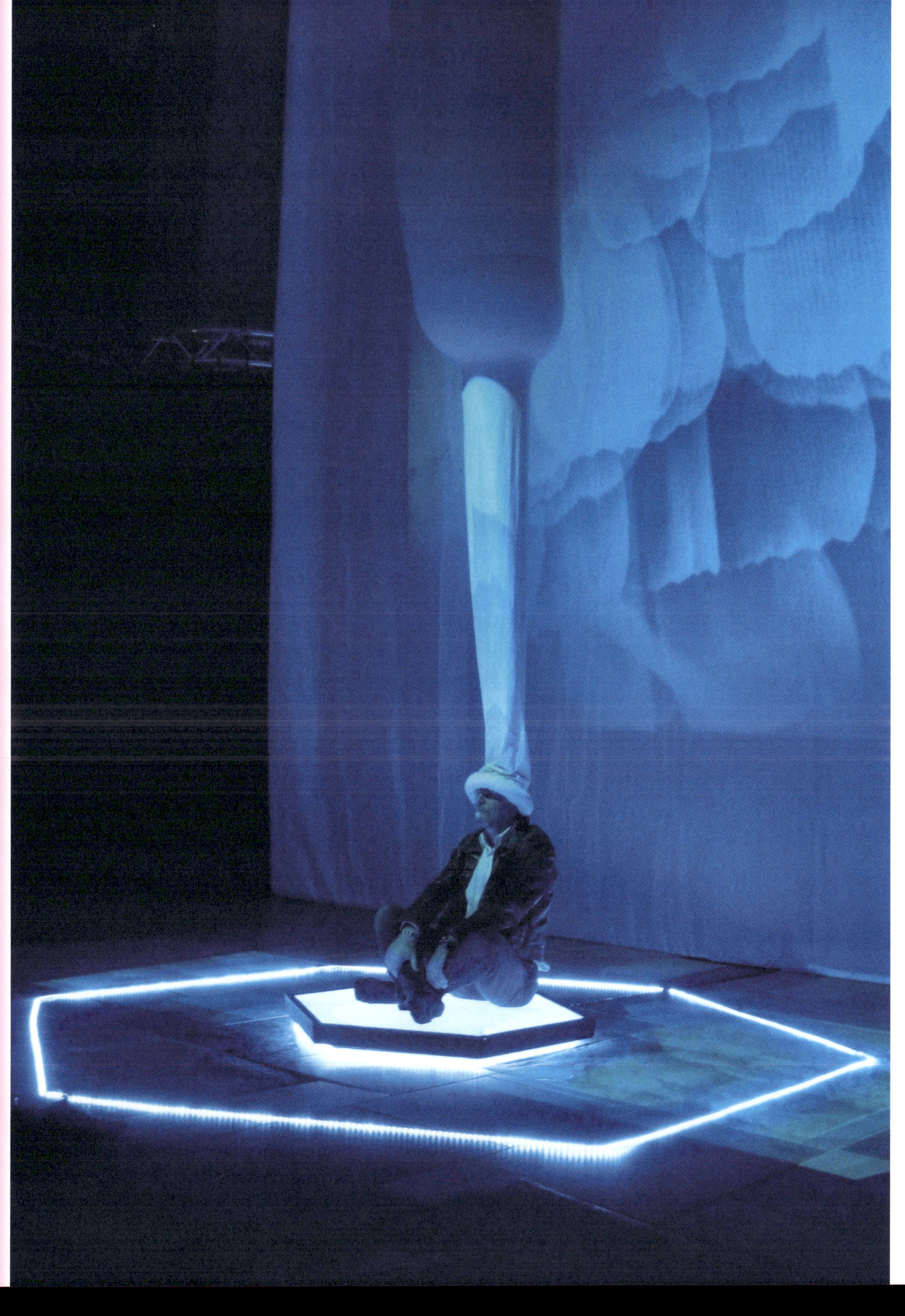

pollution in place, I reached out to Paul Geluso who specializes in immersive surround sound compositions and with whom I had already worked with on the *BLUE MORPH* project.[13]

Over the past few years, many dead whales have washed ashore and lab reports have revealed that they died due to hemorrhaging around the brain, as well as blood around the ears. Findings also revealed that the hemorrhaging was caused by many hours of intense and constant underwater sonar sounds generated by U.S. Navy warships. This was another, more urgent reason for me to blow up the plankton to the size of whales—to show the complexity of the plankton at the bottom of the food chain. And just as in *BLUE MORPH*, when the audience is quiet and centered, the noise disappears and we hear the song of the whale who depends on the plankton—just like we do.

Much of the underwater noise comes from fracking and sonar related to fossil fuel drilling—fuel that is used to produce plastic and is moved around the oceans in ships to reach consumers all over the world. As vast amounts of plastic are thrown away and end up in the ocean, the full circle is complete. When I first saw video images of microplastics being consumed by plankton, I was shocked and my sense of urgency escalated. Dr. Alfred Vendl's reaction was the same and he asked the biologists mentioned above to visualize this in their lab in order to add this important element to the project: It is the entire system of consumerism that creates all the noise we deal with underwater and in our own lives. We are all interconnected in this incredibly complex system and have lost our way in the noise that we are submerged in. Rewind back to the *WATER BOWLS* installation where the whale sounds were also heard in the far distance…

A linear video version of *NOISE AQUARIUM* was first presented in Singapore and Brisbane as part of the On|Off 1001010101 Festival. The plankton organisms were blown up to extraordinarily large dimensions, but somehow the sound was lost by the organizers and, in my opinion, this meant that the entire meaning and intent of the project was lost. The visuals alone are stunning but they do not convey the same sense of urgency of the message that sound can as the driving force of the project. So from that point onward, I insisted on having the sound as immersive and loud as possible and the visualizations as large as possible whenever *NOISE AQUARIUM* was exhibited. Without the sound, the message is lost.

Working closely with Martina R. Fröschl, we developed visualizations of various underwater noise that I mixed and recorded together with Paul Geluso, a specialist in spatialized sound composition. The same interactive computer balance platform that was used for *BLUE MORPH* was repurposed, and with the help of programmer Glenn Bristol, we created a game-like experience where one has to balance in order to get the plankton to come forward and reduce the noise. When perfectly centered, the plankton would come into full huge view, the noise recedes and

the room is filled with the song of a whale. This further inspired Bristol
to create an additional VR version of the installation where the audience
can be fully immersed while standing on the balancing board—one at
a time with others watching.

Perhaps the most impactful version of *NOISE AQUARIUM* to date
was accomplished in the Ars Electronica Center's Deep Space 8K in Linz,
Austria, in 2018, where the sound was spatially immersive, the projections
in high resolution for the audience to intensively experience visuals and
sound together. However, it was not possible to install *NOISE AQUARIUM*
in this setup at most places that were interested in showing the work,
and a more traditional approach was then taken instead, which was also
very effective. For instance, at the Historical and Maritime Museum of
Istria in Pula, Croatia, the installation consisted of Augmented Reality (AR)
prints, three aquariums with plankton and a video projection (FIGURE 12).
The sound bounced off the walls naturally, and being in a historical building made it all the more powerful.

Most recently, at the time writing this essay, the installation can
be experienced at the Łaźnia 2 Center for Contemporary Art in Gdańsk,
Poland, in a similar setup with linear video. For one day, it was also
installed as a VR experience at a sound festival in Vienna, Austria. Due
to the COVID-19 pandemic restrictions, only very local audiences, and only
a few people at a time could experience the work, which prompted me
to create a new version: an online network sound meditation where audiences from all over the world can participate and the message can spread
wider.[14] For this iteration, I asked the organizers to introduce me to a local
sound artist and they suggested Anna Nacher. We immediately connected
and started an acoustic call-and-response exchange across the ocean that
paid homage to the underwater creatures.

The COVID-19 pandemic sparked an extreme economic slowdown
in March 2020, sending cruise ships to port and oil tankers to anchor.
This is a very special moment, for the pandemic has actually rendered
the seas and oceans quiet and the underwater creatures happy; it has
also enabled scientists to make comparisons with the preindustrial age.
But on the other hand, plastic pollution is worse than ever because
single-use masks and surgical gloves are just like other plastic items,
be it plastic bags, fishing lines, drinking straws, or plastic bottles. They
take hundreds of years to break down in the environment and decompose
into tiny fragments called secondary microplastics and nanoplastics
that contain chemicals and pollutants that are released into the ocean.
These micro- and nanoplastics are mistakenly consumed by plankton
species as food.

In coastal regions, many people report that they are hearing more
birds again, which had either been drowned out by noise or their existence
forgotten. So, we do have a chance to listen, take a moment and consider

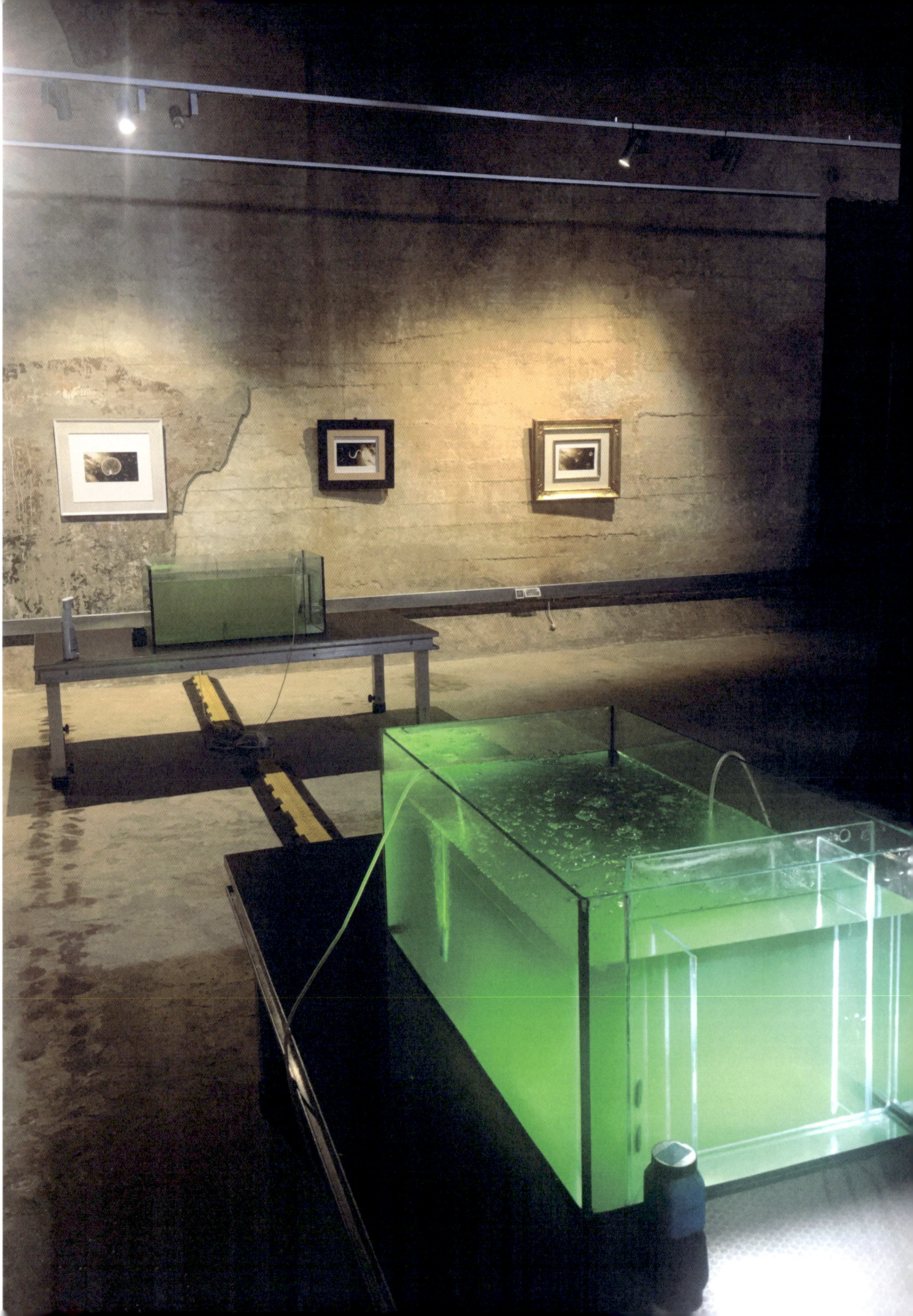

and imagine how we would like our future to look once the pandemic is over. This is the time for artists and scientists—those fortunate enough to be in safe spaces with a job—to work together, raise the frequency of our collective brain waves and move away from mass consumerism to new ways of thinking and being that respect and honor not only the scientific and technological advances but also the ancient and indigenous wisdoms.

FIG. 12
View of aquariums at the Historical and Maritime Museum of Istria in collaboration with Aquarium Pula at the Pula aMore festival, Croatia, 2019.
© Victoria Vesna

1 Joe Gray, Ian Whyte, and Patrick Curry, »Ecocentrism: What It Means and What It Implies,« *The Ecological Citizen* 1, no. 2 (2018): 130–131.

2 Pauline Oliveros, *Deep Listening: A Composer's Sound Practice*, New York: iUniverse (2005).

3 The architects of the *NANO* installations were Johnston Marklee & Associates, principals: Sharon Johnston and Mark Lee. For a full description and list of collaborators, see: http://victoriavesna.com/nano (last accessed November 3, 2020).

4 N. Katherine Hayles, ed., *NanoCulture: Implications of the New Technoscience*, Portland, OR: Intellect books (2004).

5 James Gimzewski and Victoria Vesna, »The Nanomeme Syndrome: Blurring of Fact and Fiction in the Construction of a New Science,« *Technoetic Arts* 1 (2003): 7–24.

6 The sand mandala in these images was created in 2003 by Tibetan Buddhist monks from the Gaden Lhopa Khangtsen Monastery in India and constructed on site at »The Circle of Bliss« exhibition of Nepalese and Tibetan Buddhist Art at the Los Angeles County Museum of Art. It was photographed at about ten intervals with a wide-angle lens, then switched to a macro lens at the last level. The center of the mandala was recreated in James Gimzewski's PicoLab, first imaged by an optical microscope and then with a scanning electron microscope (SEM).

7 Virtual C_{60} molecules—buckminsterfullerenes or »buckyballs«—are projected onto a large wall area. People move and distort the buckyball shapes with their shadows; slow movement rather than forceful action creates changes. This mimics the careful manipulation of molecules by scientists using the scanning tunneling microscope (STM). A camera is watching the projection, and the computer is able to isolate the participant's shadow against the gray background of the image.

8 For the complete catalog of the *NANO* exhibition, see: http://victoriavesna.com/nano/files/nano_lacma_book.pdf (last accessed December 2, 2020).

9 See the entire catalog *The Fourth State of Water: From Micro to Macro* for the description of the *Water Bowls* project and an exhibition of artists and scientists working with water: http://victoriavesna.com/ebooks/4th_State_Catalog/4th_state_catalog.html (last accessed November 3, 2020).

10 The *Water Bodies* website in collaboration with Claudia Jacques emerged as an ongoing project from the Drop Water Bowl: http://waterbodies.com (last accessed November 3, 2020).

11 *Morpho peleides* and *Danaus plexippus* wings and pupa were provided by Dr. Richard Stringer, Department of Math, Science and Allied Health, Harrisburg Area Community College. Additional technical assistance from Andrew Pelling and Paul Wilkinson in recording the heart beat of *Danaus plexippus chrysalis* is kindly acknowledged.

12 The Integratron was built in 1954 by George Van Tassel, an aircraft mechanic and flight inspector, who described it as »a machine, a high-voltage electrostatic generator that would supply a broad range of frequencies to recharge the cell structure.«

13 Using samples obtained from the U.S. National Oceanic and Atmospheric Administration, I worked with Paul Geluso to spatialize the audio at Harvestworks, New York.

14 For more information on venues, documentation and credits, visit http://noise aquarium.com (last accessed November 3, 2020).

Tomographic retrieved datasets are widely used
in the natural sciences. However, these datasets
are usually only animated without advanced
movement and physics simulations.
The artistic and cinematic visualization of such
datasets demands more complex approaches.
For an interdisciplinary art project, plankton
and underwater noise should be made visible with
attractive movements and designed in a way
that is at once factually accurate, abstracted, and
easy to understand. A combination of established
computer animation and simulation techniques
with newly researched approaches led to
authentic computer-animated scientific visualizations of noise, fluids, and floating microorganisms for the immersive interactive art
and science installation NOISE AQUARIUM.

Computer-Animated Fluidity for Stiff Datasets and the Visualization of Underwater Noise

Martina R. Fröschl and Alfred Vendl

Visualizing Microscopic Plankton ___ Water microorganisms are basically film-able with microscopic camera equipment. In practice, physical and creative limitations constrain cinematic aesthetics. The behavior of these organisms is generally unpredictable; however, artists from several genres would like them to be staged or immediately dirigible. Furthermore, there are many technical issues when it comes to filming through a microscope. Limited depth of field and lens distortions shape the look of common microscopic films, and filming through optical microscopes has constraints when it comes to camera movements. Collaborative efforts of micros-copists, biologists, computer graphics artists, and other artists, constitute an important approach and leads to data with epistemological, additional value compared to noninterdisciplinary projects. The visualization of scientific data is key to the creations of the Science Visualization Lab at the University of Applied Arts Vienna. We believe that there is a deeper knowl-edge-production process inherent in using scans of plankton organisms compared to the frequently usual and less resource-consuming efforts of model creation.

The term »model« has very different definitions in various disci-plines and contexts. In this text, »3D model« is defined as a general term in computer graphics which includes all data in current file formats for the notation of points in coordinate systems that define sculptures in digital space. The varying sources of this data may be as different as box modeling, digital sculpting or, as in the case of the 3D models, tomographic scans of real living microorganisms. These 3D models are elements which are used in digital scenes, which means software code that can be rendered into computer-generated imagery (CGI). An image has always to be seen in its context.[1] While scientific visualizations are driven by scientific data, CGI loaded with science content does not necessarily have to be. We use the term »authentic« to emphasize the scientific origins of the underlying imaging data. Recently, NASA's Scientific Visualization Studio (SVS) has coined the term »cinematic scientific visualizations«; in a similar context, but with less emphasis on cinematic productions, we use the term »computer-animated scientific visualizations.«[2]

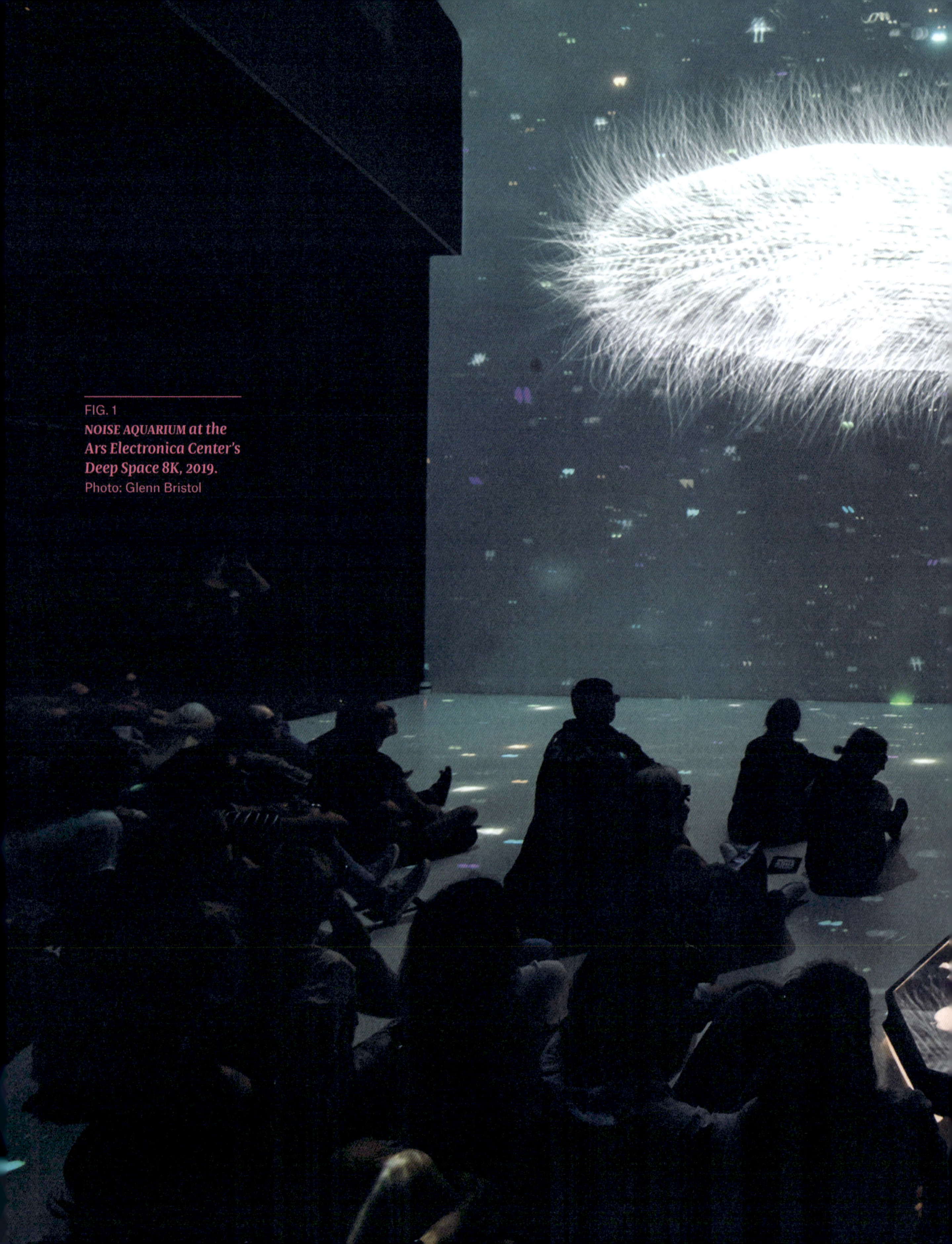

NOISE AQUARIUM at the Ars Electronica Center's Deep Space 8K, 2019.
Photo: Glenn Bristol

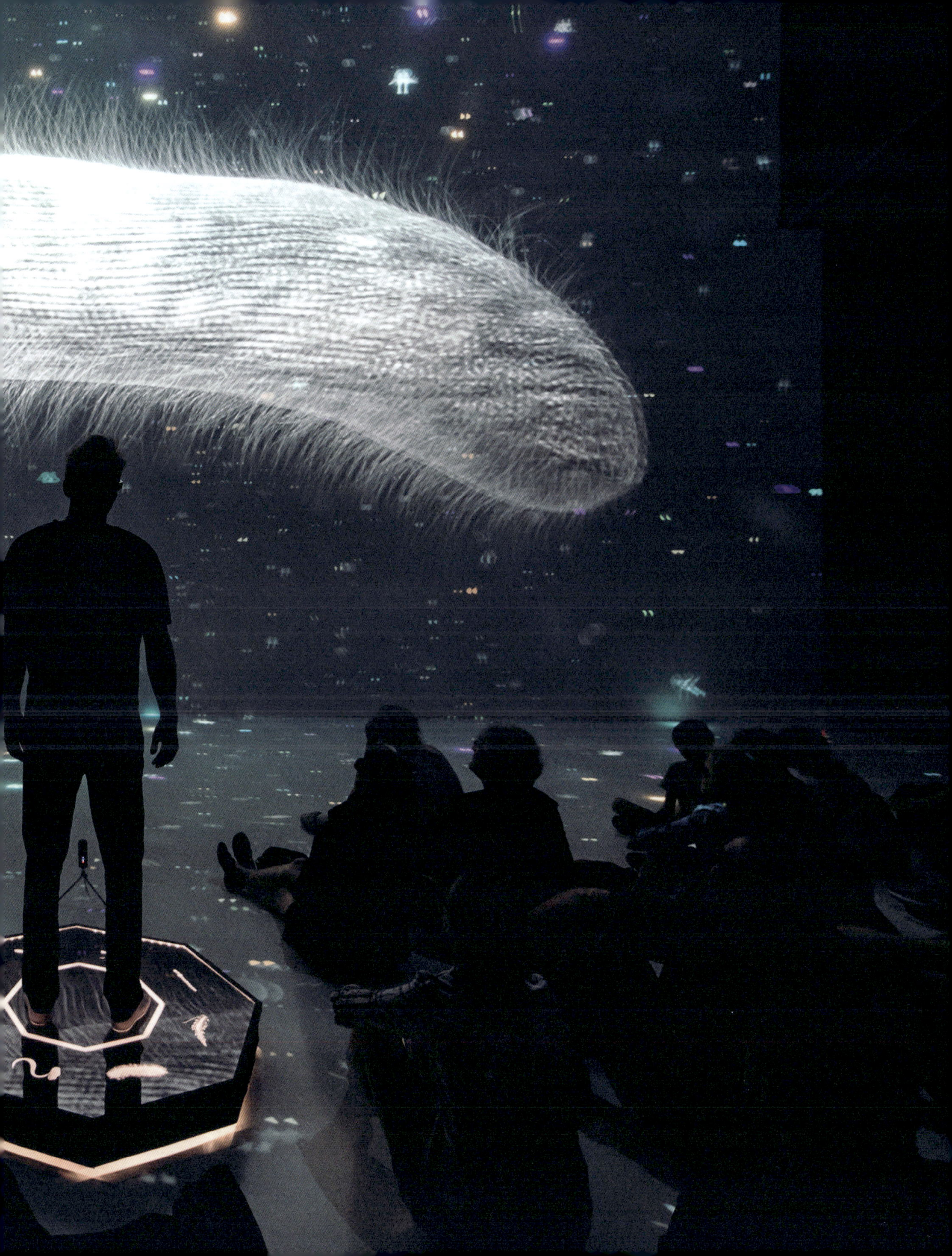

Preparation, scanning and, last but not least, the generation of 3D model animations of the plankton are elaborate, yet result in unique fluid particle simulations and animations. In contrast to most workflows in computer animation and scientific visualization, organisms are not only a reference for digital sculptures, but are also the artistic base material and structure for isosurface geometry. The samples are taken out of their natural environments and transferred to the digital realm. The process is not an imitation but the actual transfer of three-dimensional information. The basis of all 3D models are scanned real organisms that were once alive. Most of the datasets are tomographically retrieved. This approach is widely used for research in the natural sciences; however, hardly any projects use the stiff 3D models obtained by scientific imaging techniques to create computer-animated scientific visualizations as for the immersive art installation *NOISE AQUARIUM* (FIGURE 1).[3]

Fluids in computer animations are interpretations or approximations of more or less realistic behaviors in physics described virtually. The basis of life on Earth is fluids; in other words, fluid behavior surrounds us humans, from the air we breathe to the oceans that cover the planet. Fluids enable captivating effects in the media arts. Both digitally simulated and in nonvirtual realms, it is possible to make powerful statements. The expressive aesthetics of the underwater world and of fluids attracts numerous contemporary artists. The field of art and science is often the chosen field for these creatives. By providing cinematic scientific visualizations of microscopic plankton, the Science Visualization Lab at the University of Applied Arts Vienna created the inspiring basis for an interdisciplinary project on the complex relationship of biodiversity, noise and plastic pollution in the water bodies of our world.

At the Science Visualization Lab, we are keen to create visualizations of scientific nexuses which cannot be experienced without technical augmentation, for instance, microscopes or special cameras. In the processes and the outcomes, we focus on key advantages of scientific visualization, as stated in our white paper.[4] Particularly when it comes to promoting environmental protection and issues related to habitats of animals and plants, impressive pictures are useful. Especially underwater sceneries can have an intriguing, mesmerizing effect.

Compared to our size as humans, for species orders of a magnitude of less than one millimeter, water is a substance with stronger surface tension; therefore, its behavior is comparable to large pieces of jelly. At the same time, it is possible to see the gelatinous interior of many organisms in the microworld, since the organisms are often partially transparent. The reflections of the water surface and phenomena like caustics and floating and drifting add to these alien and interesting sceneries. With digital tracking shots, it is possible to dive through these semitransparent creatures. Diving through tissues is possible because of visual effects shots and

computer animation. This way, it becomes possible to dive up front to
a microscopic creature and see it glimmering in the refracted sun of
the water waves. Moreover, there are water organisms which drift through
water with tremulous, beautiful movements, or swim and still get drawn
away by the currents. Plankton organisms are drifters in the water without
any distinct direction.

NOISE AQUARIUM assumes that the increasing sound pollution of the
water bodies on our planet has a negative impact on microscopic plankton
life. There is sound scientific evidence that this applies to at least a few
researched species. Furthermore, the drifting microorganisms in the project could be seen as a metaphor for general problems of noise pollution
in the world's water bodies. Therefore, it is very important indeed to urge
action for the protection of these ecologically important living animals.
Additionally, the art project's recipients should be confronted with the
problem of the tremendous amount of plastic particles polluting the
oceans. The topic of plastic pollution spread all over the world was only
just beginning to gain attention when the project *NOISE AQUARIUM* started
in 2016. Microplastics and the toxins they contain have been detected
everywhere.[5] Plastic pollutants can be found in the largest and smallest
creatures on the planet, and they are found in the remotest areas of Earth.

Our contribution deals with an art and science project during which
scientific datasets were transformed into CGI and advanced animations.
Seven microscopic plankton organisms based on scans (for further details,
see the chapter by Stephan Handschuh and Thomas Schwaha, this volume)
were transferred to the digital 3D space. All creatures in *NOISE AQUARIUM* are
soft-tissued organisms. Therefore, threshold segmentation can only be
a starting point for an approach to isolate specific animal parts; this means
that experts should manually or semiautomatically segment the features
of interest from those that should not be in the resulting 3D model.

The incredible extent of marine biodiversity in microscopic plankton is hinted at using seven examples which are very different in shape and
behavior. The seven organisms featured are *AMOEBA*, *PARAMECIUM*, *CYLIN-
DROSPERMUM*, *TOMOPTERIS*, *ACTINOTROCHA*, *OIKOPLEURA*, and *NOCTILUCA*. Due to
preparation and scanning by the biologists, tomographic 3D data was available which serves as starting material for a large number of animations and
simulations of microscopic fluid systems and plankton. The implementation of tomographic scanned microscopic plankton, however, does pose
a variety of technical and aesthetic challenges.[6]

Up to now, visualizations in biology are usually only animated without advanced computer animation techniques. The artistic and cinematic
visualization of biological datasets demands more complex approaches.
For *NOISE AQUARIUM*, the goal was to make plankton and underwater noise
visible with intriguing movements and designed in a way that is both
abstracted, physically correct and easy to understand.

The triangulated 3D models are imported into the 3D software Blender (Blender Foundation, Amsterdam, The Netherlands). The organisms are either rigged with digital bones or animated with other deformers. The digital geometry complexity is reduced and edge loops retopologized to a processable state as a prerequisite for various advanced animation and simulation processes.

The combination of established computer animation and simulation techniques and newly researched approaches resulted in internationally acclaimed computer-animated scientific visualizations of noise, fluids and floating microorganisms for NOISE AQUARIUM. After brief portraits of the different plankton genera, we will discuss the visualization of noise and plastic pollution and provide a general discussion of the project.

Amoeba

An important inspiration to start the endeavor of depicting microscopic plankton organisms in computer animation was the shape-shifting and fluid appearance of AMOEBA. The fluid aesthetics and the locomotion of this unicellular organism and how it feeds on other life to gain energy is a great inspiration and prerequisite for future projects. The fluidity of the organelles and the refraction of light in the details play a significant role. AMOEBA can be viewed from the outside and the inside, and the cell nucleus and other organelles drift past the viewer like precious shiny gemstones in a tour of a liquid treasure trove (FIGURE 2).

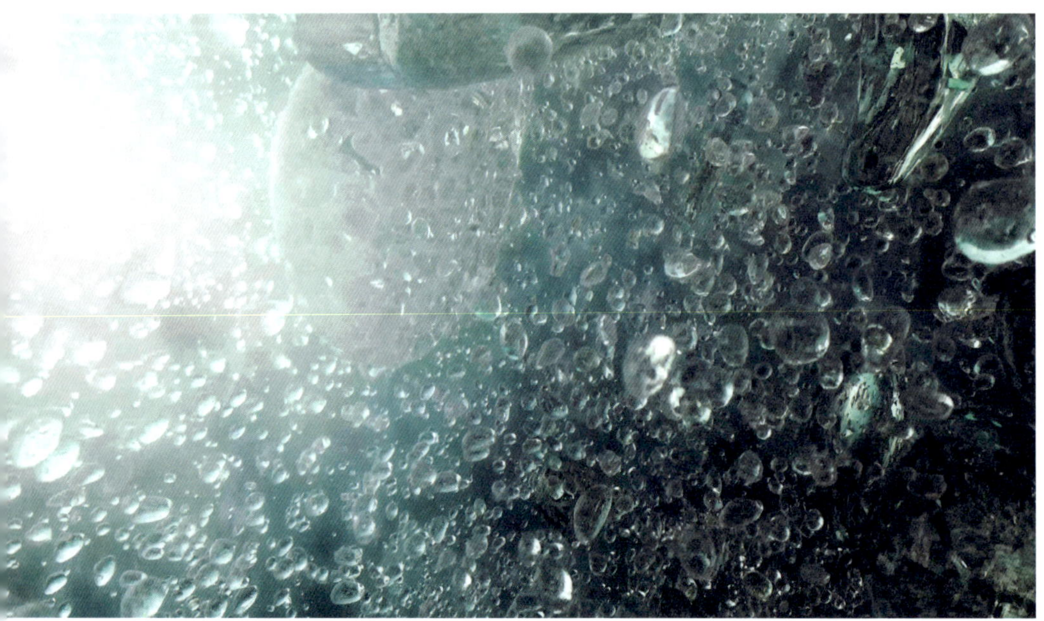

FIG. 2
*CGI dive through
an AMOEBA.*
© Science Visualization
Lab, University of
Applied Arts Vienna

Paramecium

This organism is covered in many tiny hair-like structures called cilia.
Because of this amount of cilia, which make the creature look like a plush
slipper, it has been given the nickname »slippershoe animal.« The model
used in NOISE AQUARIUM contains a displacement map from scientific
imaging, particle and hair simulations, and animated interior organelles.
The 3D model of PARAMECIUM includes a shape animation that shows
the fluid processes inside the organism (FIGURE 3).

185

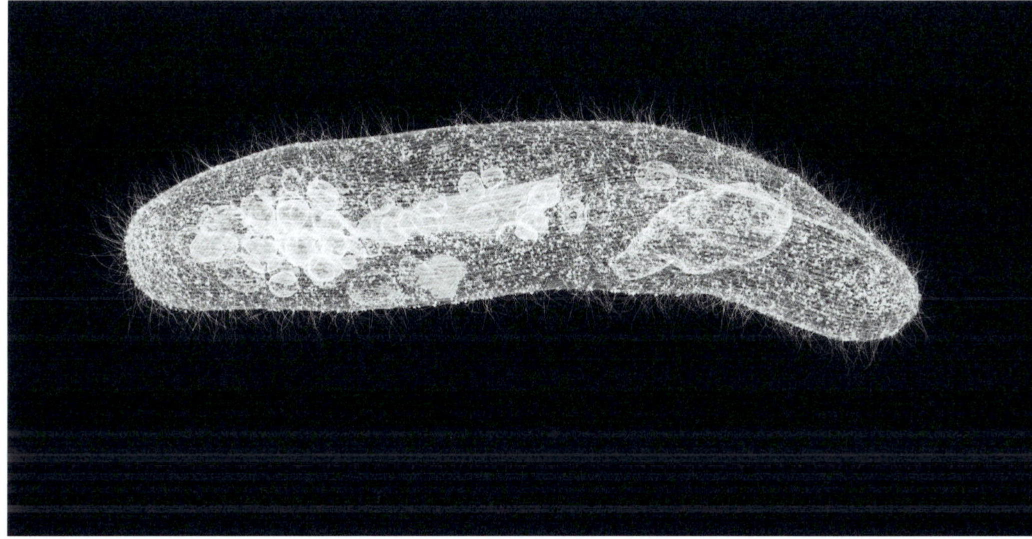

FIG. 3
*CGI of PARAMECIUM and
inner simulations.*
© Science Visualization
Lab, University of
Applied Arts Vienna

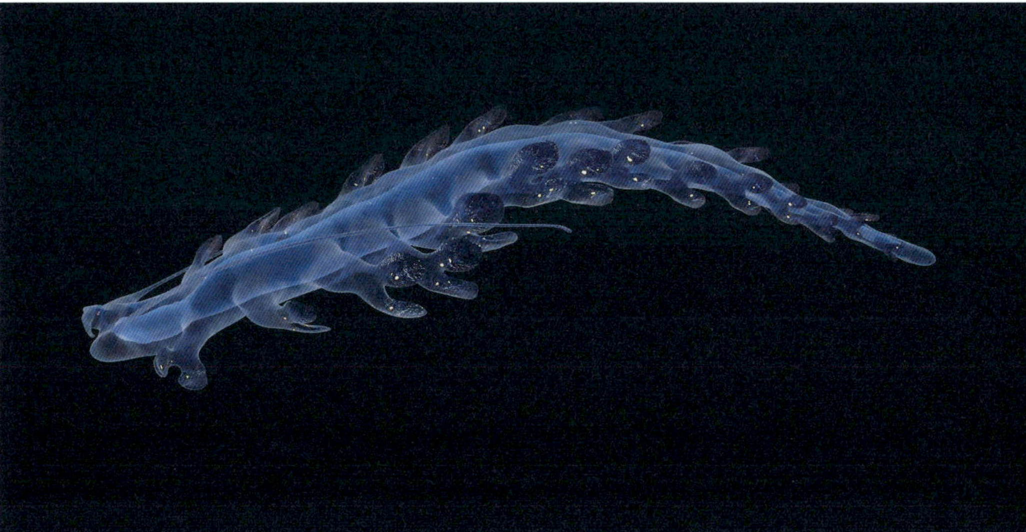

FIG. 4
CGI of TOMOPTERIS.
© Science Visualization
Lab, University of
Applied Arts Vienna

Tomopteris

TOMOPTERIS is a marine worm with deeply divided, forked, fin-like pods,
a pair of lensed eyes and long front tentacular cirri. These animals are
among the few marine animals with yellow bioluminescence. This
plankton genus has a predatory lifestyle and locomotes intriguingly
and elegantly using wave-like movements (FIGURE 4).

Cylindrospermum

This microbe usually consists of chains of single cells. The CYLINDROSPER-
MUM 3D model incorporates a combination of transmission electron
microscopy, scanning electron microscopy and light microscopic image
stacks, and shows both the photosynthetic interior of the organism
and the outer membrane (FIGURE 5).

FIG. 5
CGI of CYLINDROSPERMUM.
© Science Visualization Lab,
University of Applied Arts Vienna

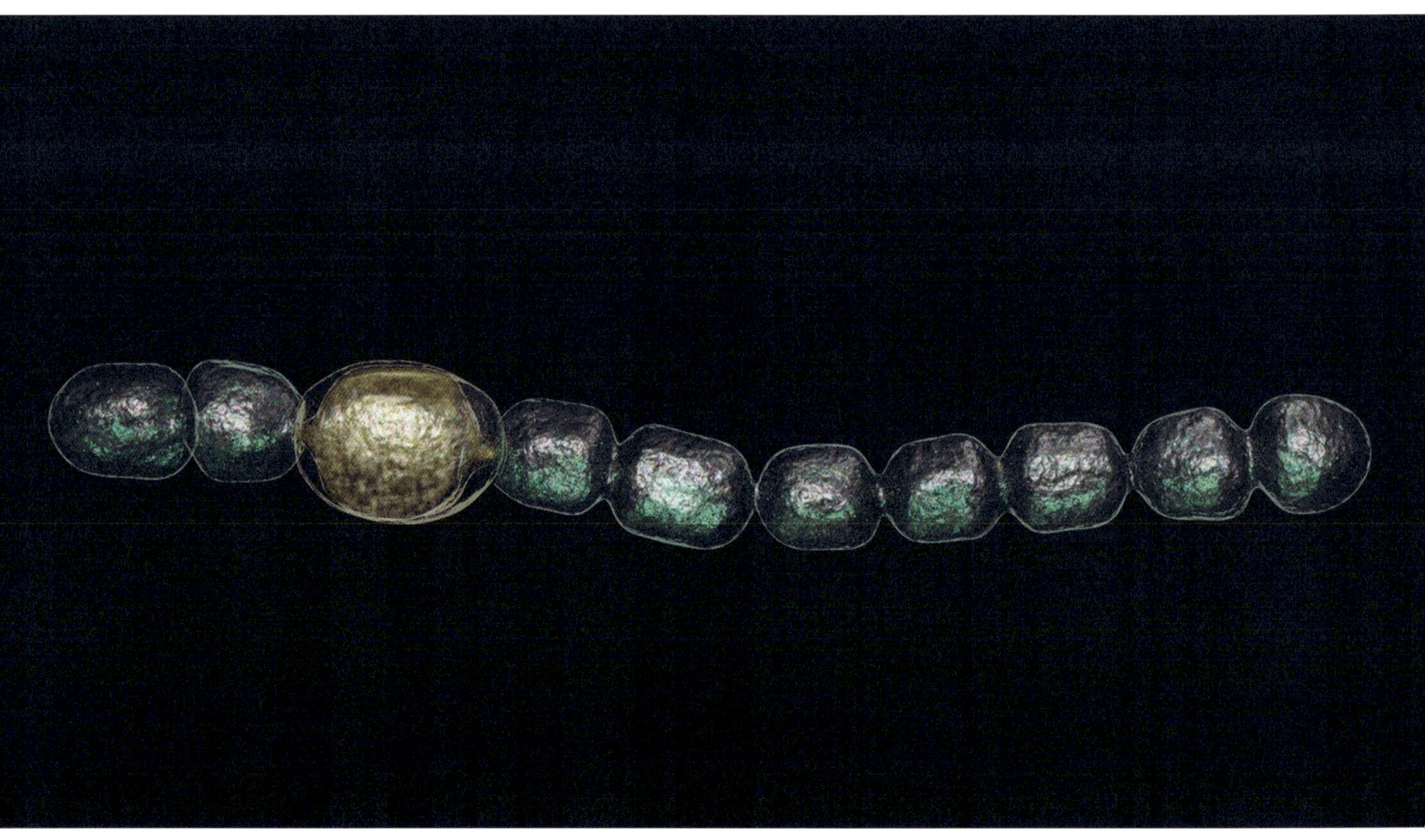

Noctiluca

NOCTILUCA is responsible for the spectacular phenomenon of shining
blue coastlines at night in several areas of the world. The cytoplasm-filled
balloon-shaped cell causes this brilliant bioluminescence. *NOCTILUCA*
species live mostly in coastal waters and digest organic matter, but they
can also synthesize their own nutrients (FIGURE 6).

187

Oikopleura

OIKOPLEURA species form a complex, gelatinous cave called a »house, which
they use to filter food particle streams to their ›mouth.‹« The definition of
the light reflection properties for the rendering of the *OIKOPLEURA* 3D model
was particularly challenging because of its many transparent intestine
layers and an iridescent tail that shimmers when light shines onto it at a
certain angle (FIGURE 7).

FIG. 6
CGI rendering of NOCTILUCA.
© Science Visualization Lab,
University of Applied Arts Vienna

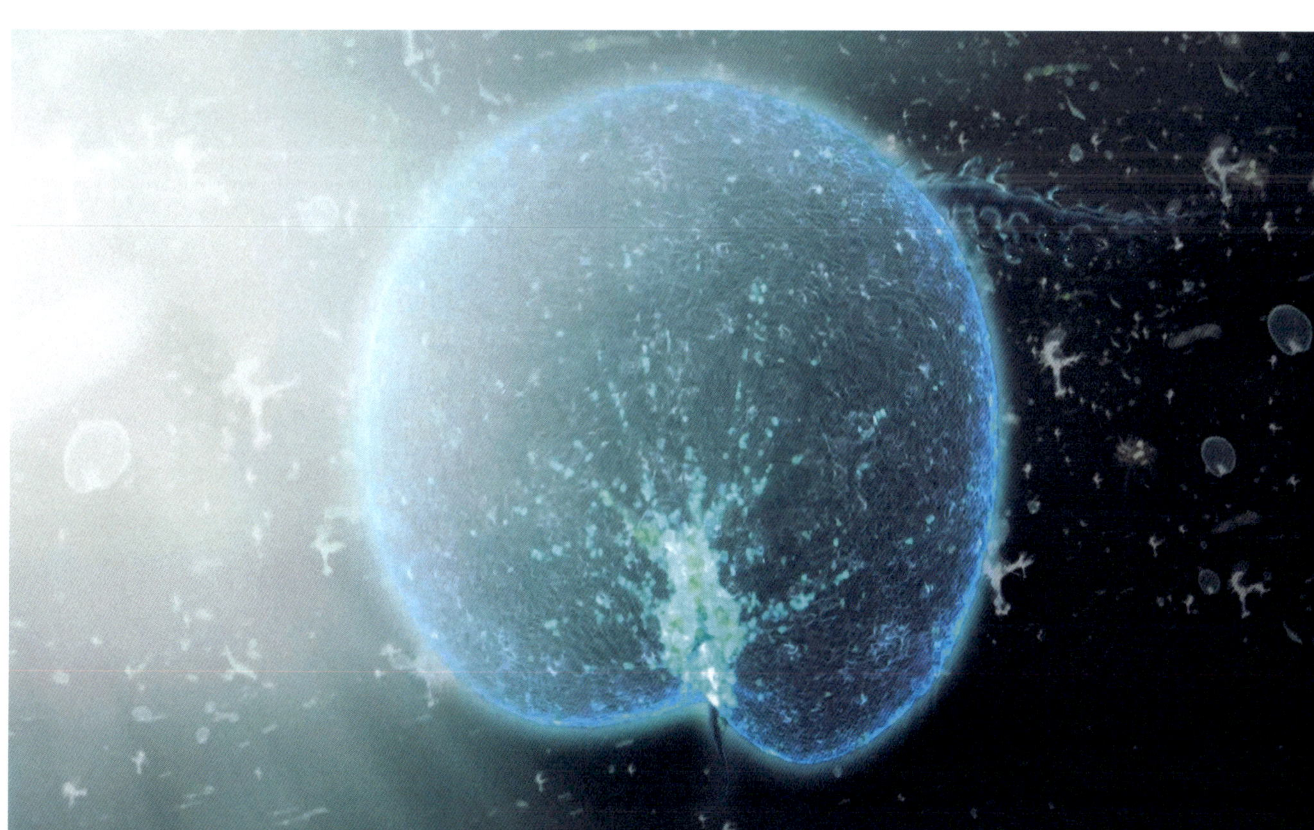

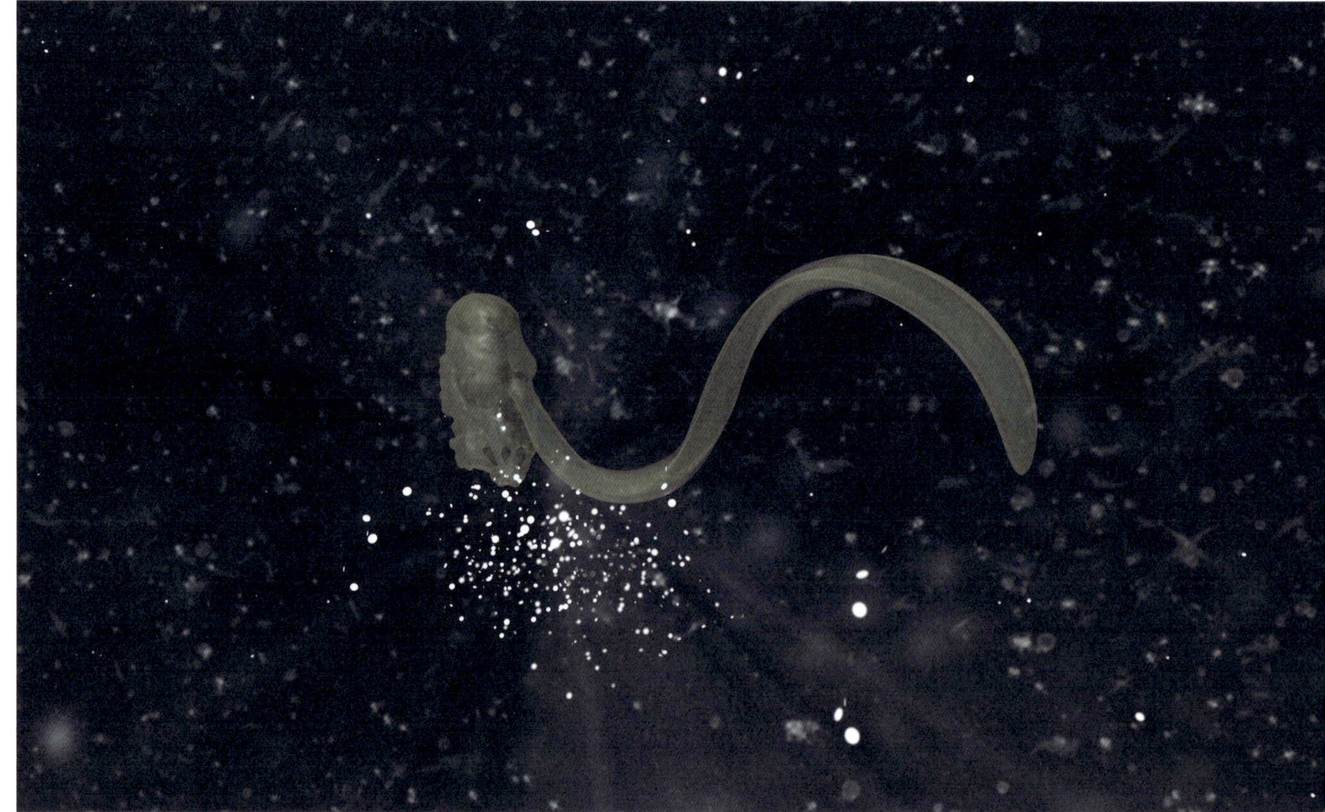

FIG. 7
CGI rendering of OIKOPLEURA.
© Science Visualization Lab,
University of Applied Arts Vienna

Actinotroch

ACTINOTROCH organisms are larvae of one type of *Phoronids* and they are
sometimes called horseshoe worms. Their look reminds some people
of squids. The depicted ACTINOTROCHS have typical convulsions and the
cilia move in patterns which are clearly visible in various video references
(FIGURE 8).

Art and Science Collaboration, Underwater Noise, and Plastic Pollution __ __

The visuals of the pressure waves for NOISE AQUARIUM were developed gradually and through various experiments. Much emphasis is on the depiction
of sound waves. One challenge was to visualize the densification of the
water authentically and therefore visualize the sound waves in a compre-
hensible manner. The first publicly presented video version utilized the
Blender Wave Deformer animated by sound samples in 3D space to visualize the noise pollution. The chosen solution shows planes which are
deformed by waves. The sound waves have a common origin which adds
a kind of fake central perspective to the underwater view. This has the

qualities of a »sonic rosette,« which pulls the viewer into the visual.[7] For later versions of the linear NOISE AQUARIUM videos, the sound waves were implemented with the CC Ripple Pulse effect[8] in the compositing software After Effects (Adobe, Inc., San Jose, CA, USA) (FIGURE 9).

Additional virtual and augmented reality versions of NOISE AQUARIUM, were often combined in large installations with videos and extra information on the subject to motivate visitors to think about the ocean, microscopic plankton life as well as noise and plastic pollution. The perception of the whole project strongly depends on the context of the presentation. The project was mainly exhibited at art venues, but in most cases information was available on the scientific data origin of the artistic CGI. According to statements and a short survey among visitors, the underlying scientific data add additional information and emotional value to the art installation.[9] Especially virtual and augmented reality versions of NOISE AQUARIUM, often combined in large setups with videos and additional information on the subject matter, prompted visitors to perceive the ocean, microscopic plankton life and noise pollution in alternative ways. They promote thinking about acute environmental issues.

The plastic pollution in the virtual reality version of the art installation is visible through modeled plastic particles. They are sometimes perceived by visitors as a sky of colorful stars underwater, consequently people start thinking about the conflict of plastic being useful and often aesthetically pleasing, but at the same time, poisonous trash. In the tomographic plankton scans used as well as the additional video references, no identifiable plastic particles were found. Nevertheless, we decided to experiment with *Copepod* plankton organisms and standardized fluorescent nano plastic beads. The experiments showed how shockingly vulnerable plankton is to intake of this form of anthropogenic waste.

NOISE AQUARIUM is a mind-opening experience for many people. After the CGI of the first three plankton species were finished, Alfred Vendl asked U.S. media artist Victoria Vesna, Director of the Art|Sci Center at the University of California Los Angeles (UCLA) to collaborate on an art project featuring these cinematic scientific visualizations, which were originally created but not used for Terrence Malick's *Voyage of Time: Life's Journey* (2016).[10] The collaboration of the Science Visualization Lab in Vienna, the UCLA Art|Sci Center, the Veterinary Medicine University Vienna, and the University of Vienna is leading to significantly more than artistically distributed scientific visualizations. The members of the NOISE AQUARIUM collective experience the benefits of a so-called »Renaissance Team,«[11] which means experts from different fields contribute to a greater good, inspire each other and try to overcome the historical split between the arts and the sciences. The project began with large immersive projections of a video with the plankton CGI, and this caught the attention of Ars Electronica 2019 (Linz, Austria) for which a first interactive version was

189

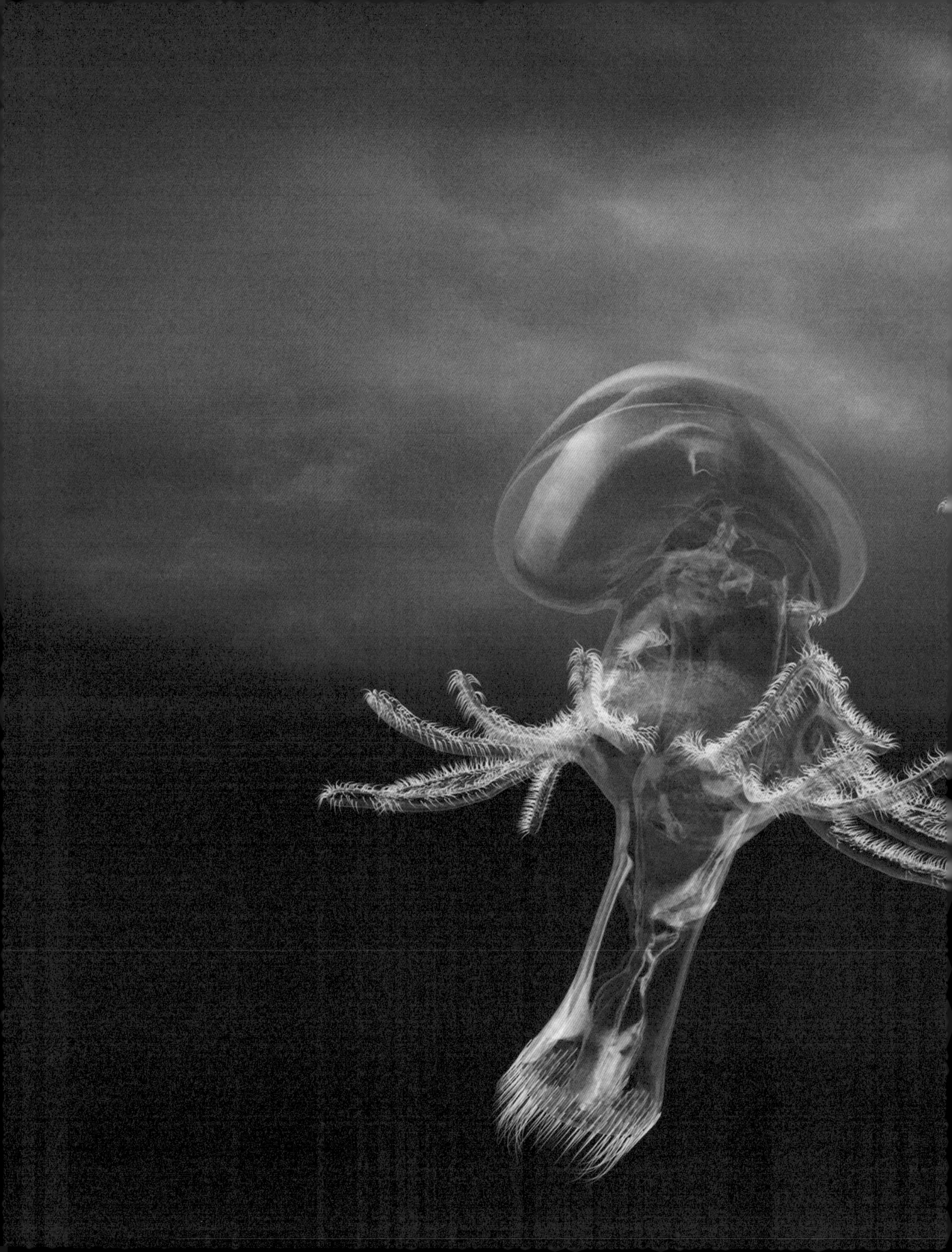

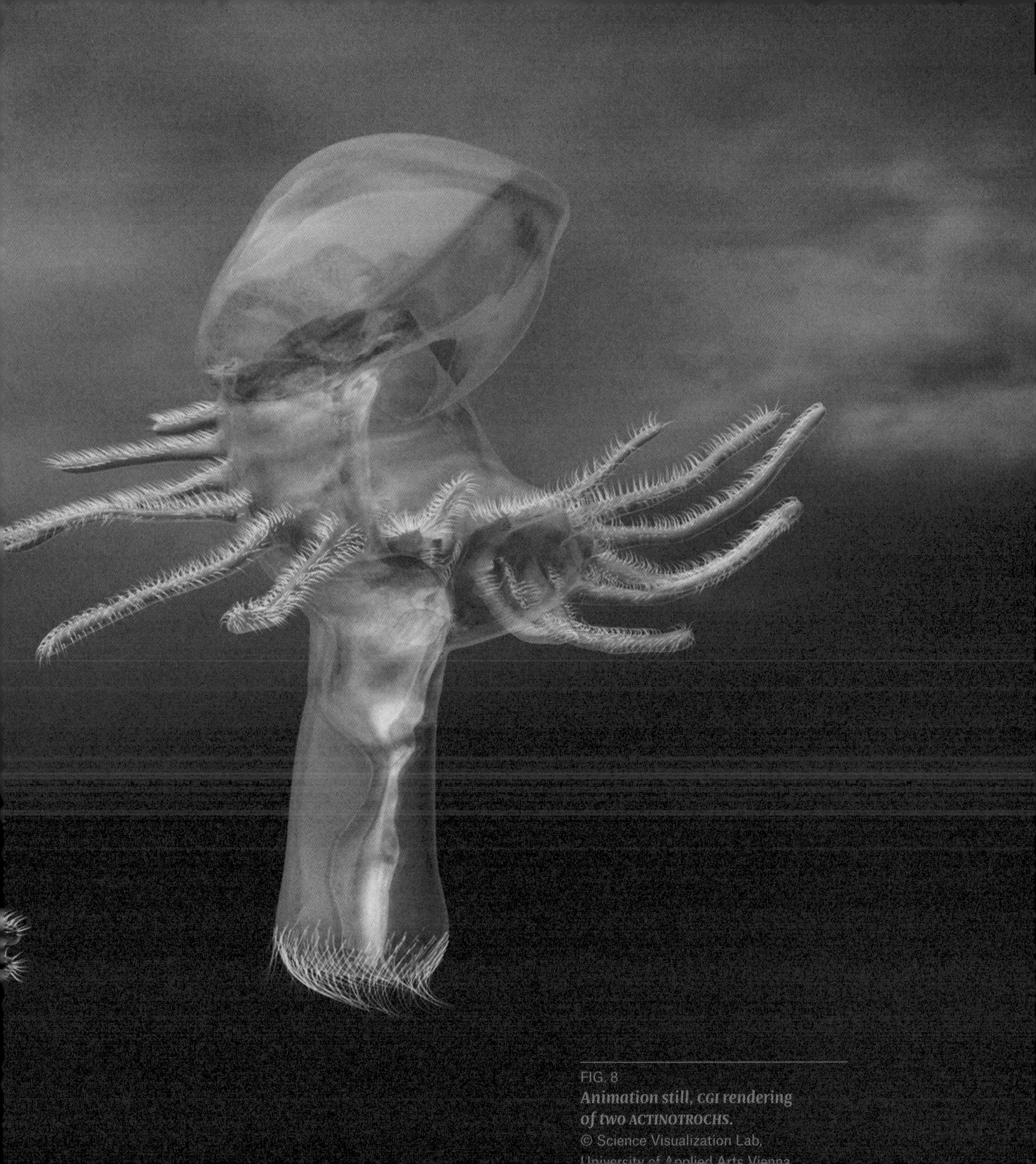

Animation still, CGI rendering
of two ACTINOTROCHS.
© Science Visualization Lab,
University of Applied Arts Vienna

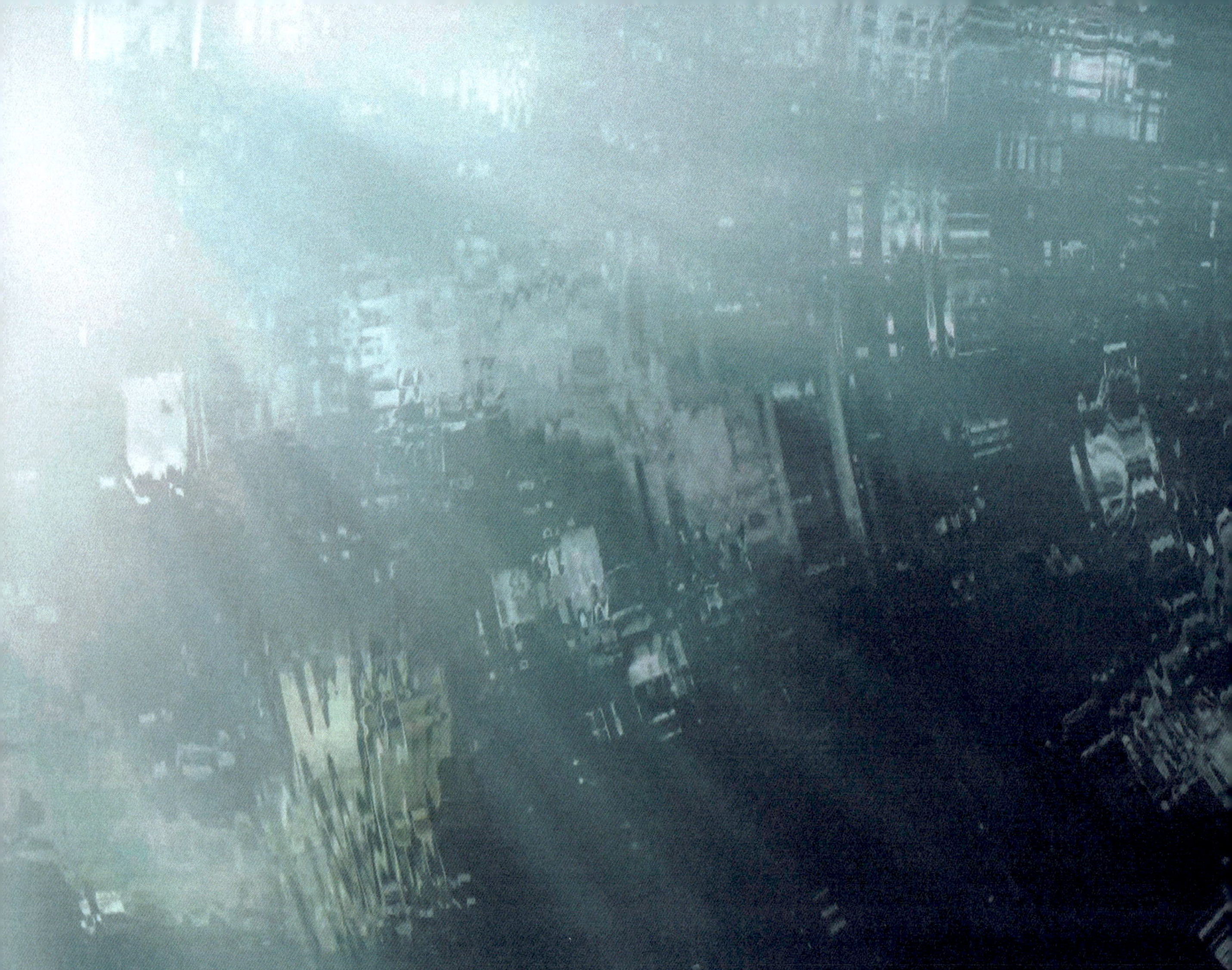

developed. Victoria Vesna had the idea of an octagonal interface and created a framework of artistic questions about noise pollution and microscopic plankton life in the ocean.

The collaborative work on the project enhanced creativity, expanded the international target audience and made computer-animated fluidity for stiff datasets and the visualization of underwater noise possible. Innovative attempts to visualize biological tomographic datasets of plankton organisms as in the case example of NOISE AQUARIUM may very well be interesting for future collaborations in science as well as in art. Creating computer-animated scientific visualizations involving scientific imaging of the natural sciences is a promising future direction for a multitude of different presentation formats, both in the arts and the sciences, to encourage thoughts about microscopic organisms and their importance for our ecosystems.

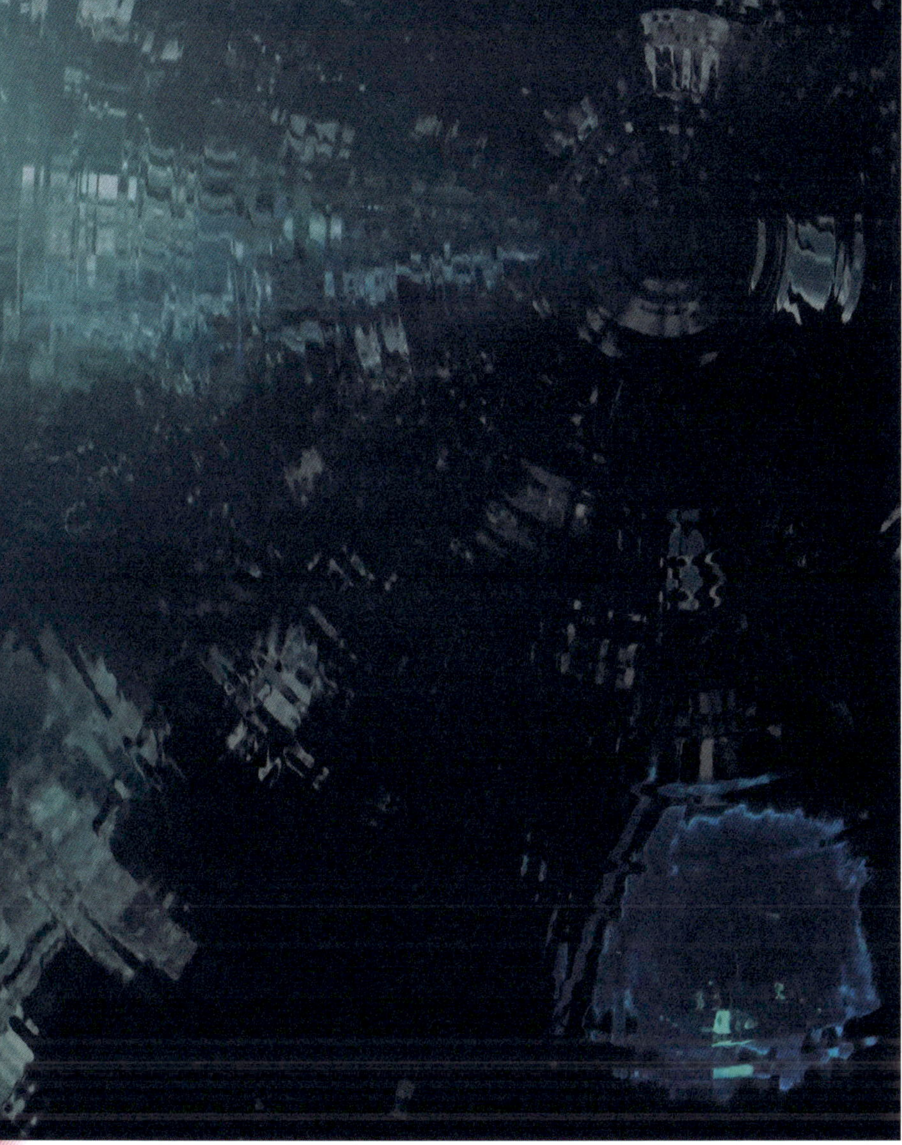

1 Ingeborg Reichle et al., »Die Familien-ähnlichkeit der Bilder,« in *Verwandte Bilder: Die Fragen der Bildwissenschaft*, eds. Ingeborg Reichle, Steffen Siegel, and Achim Spelten, Berlin: Kulturverlag Kadmos, 2nd ed., (2008): 7–11.

2 Kalina Borkiewicz et al., »Introduction to Cinematic Scientific Visualization,« *ACM SIGGRAPH 2020 Courses* (2020), 267.

3 http://noiseaquarium.com (last accessed May 15, 2020).

4 www.researchgate.net/publication/334291503_Science_Visualization_Lab_Angewandte_Whitepaper (last accessed April 3, 2020).

5 Melanie Bergmann et al., »White and Wonderful? Microplastics Prevail in Snow from the Alps to the Arctic,« *Science Advances* 5, no. 8 (2019). Online: https://advances.sciencemag.org/content/advances/5/8/eaax1157.full.pdf (last accessed April 23, 2020).

6 Martina R. Froeschl, *Computer-Animated Scientific Visualizations of Tomographic Scanned Microscopic Organic Entities*, PhD dissertation, University of Applied Arts Vienna (2019); Martina R. Froeschl, »Computer-Animated Scientific Visualizations in the Immersive Art Installation *NOISE AQUARIUM*,« *2020 IEEE Conference on Virtual Reality and 3D User Interfaces Abstracts and Workshops (VRW)*, (2020): 183–187.

7 Martina R. Froeschl, *Computer-Animated Scientific Visualizations of Tomographic Scanned Microscopic Organic Entities*, PhD dissertation, University of Applied Arts Vienna (2019).

8 www.cycorefx.com/downloads/cfx_hd_std/CycoreFX%20HD%201.7.1%20Manual.pdf (last accessed June 5, 2020).

9 Martina R. Froeschl, »Computer-Animated Scientific Visualizations in the Immersive Art Installation *NOISE AQUARIUM*,« *2020 IEEE Conference on Virtual Reality and 3D User Interfaces Abstracts and Workshops (VRW)*, (2020): 183–187.

10 *Voyage of Time: Life's Journey* is a 2016 American documentary film written and directed by Terrence Malick. It was released in two versions: a forty-minute IMAX version narrated by Brad Pitt and a 35-millimeter feature-length edition narrated by Cate Blanchett.

11 Donna J. Cox, »The Art and Science of Visualization: Metaphorical Maps and Cultural Models,« *Technoetic Arts* 2, no. 2 (2004): 71–80.

This chapter reviews the workflow for creating accurate digital 3D models from planktonic organisms. We describe how living planktonic organisms are collected, chemically fixed, and prepared for imaging. Different imaging modalities such as light microscopy and microscopic X-ray computed tomography are capable of acquiring 3D image volumes depicting the anatomy of organisms. Finally, we demonstrate how image segmentation tools are used to transform these image volumes into 3D geometry models of specimens. Digital artists can use such models for animating specimens in projects such as NOISE AQUARIUM, which is treated in another chapter of this volume.

From Live Imaging to 3D Modeling: A Guide to Documentation and Processing of Planktonic Organisms

Thomas Schwaha and Stephan Handschuh

The Fascination of Plankton ___ The biologic diversity of living organisms is immense and ranges from simple viruses to multicellular plants, fungi and animals. The term »plankton« summarizes aquatic organisms that predominantly move passively in a water body – as opposed to the term »nekton« that is used for active swimmers including larger squids, fish, and marine mammals.[1] Plankton includes a vast array of different organisms that occur in the free water column of water bodies. Planktonic organisms are fascinating, mesmerizing and puzzling at the same time, and the microscopic study of plankton has intrigued researchers and naturalists for more than a century owing to the challenges when unknown bi- or multi-phasic life cycles are involved. Many marine larvae were initially described as new species, but much later they turned out to be juvenile stages of already known adult organisms.

Plankton comes in all ranges of sizes, from a few micrometers as in unicellular green algae, a few hundred microns to a millimeter as in many invertebrate larvae (e.g. larvae of starfish or phoronid worms), to several millimeters or even a few centimeters as in larger polychaetes, crustaceans (e.g. krill), or fish larvae. Classifications such as nano-, meso-, or megaplankton have been used for categorizing different planktonic organisms according to their size. It should be noted that organisms grow in their ontogeny, and thus different developmental stages of the same species may be assigned to different size classes of plankton. Another classification can be made according to the systematic affinity of the organism, for instance phytoplankton for planktonic plants, or zooplankton for planktonic animals.[2]

Plankton have even inspired naturalists to write numerous poems, most notably the famous *Larval Forms and Other Zoological Verses*, especially »The Ballad of the Veliger,« by British zoologist Walter Garstang (1868–1949), first published in 1951, two years after his death. Similarly, the dissertation *De Salpa* by the German poet and botanist of French origin Adelbert von Chamisso (1781–1838) of 1819 revealed and analyzed the life cycle of thaliacean urochordates (a group closely related to vertebrates).[3]

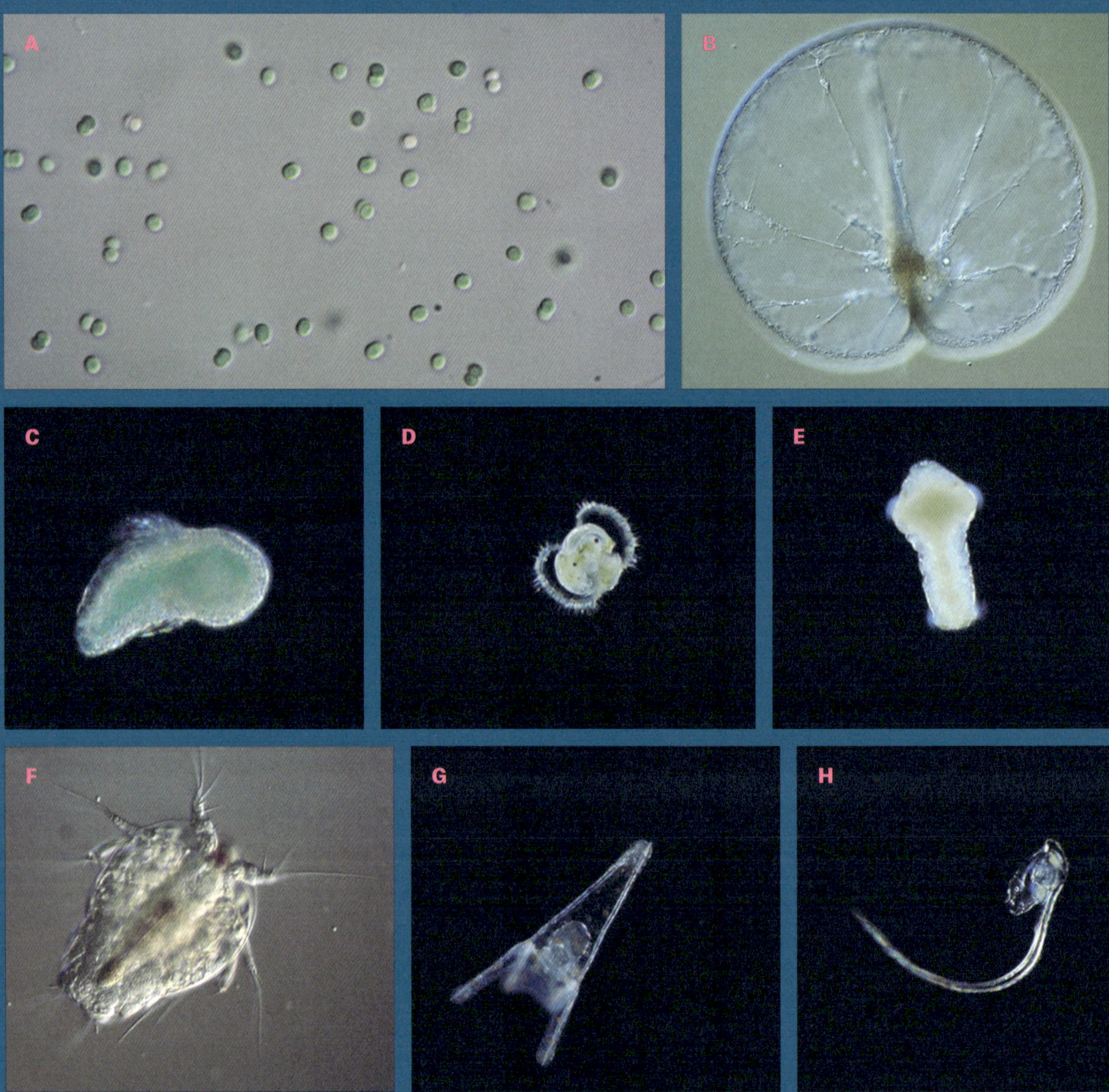

FIG. 1

Diversity of selected plankton organisms

A *Synechococcus* sp., a live cyanobacterium of the size of a few microns
B *Noctiluca* sp., a dinoflagellate
C A molluscan trochophore of about 300 µm
D A typical gastropod veliger of 300 µm
E A polychaete larva of 400 µm
F Crustacean larva of *Tisbe* sp. specimen, approximately 300 µm in size
G Pluteus larva of an echinoderm
H Larvacean *Oikopleura dioica*

© Thomas Schwaha and Stephan Handschuh

Biphasic Metazoan Life Cycle: Diversity and Ecology of Marine Larvae ___

Most marine invertebrate phyla (a taxonomic rank below kingdom level such as Animalia or Plantae) have a biphasic life cycle that involves a planktonic larva living in the free water column and often a benthic (sea-floor associated) or even sessile adult stage. Most marine larval forms carry cilia (»hairs«), including larvae of sponges, cnidarians (corals, hydroids, jellyfish, etc.), spiralians (e.g., molluscs, annelids) and even echinoderms (e.g., starfish, sea urchins, sea cucumbers) or ascidians (sea squirts).[4]

Larval planktonic forms are incredibly diverse and include simple ovoid larvae as in sponges or cnidarians, and more complex trochophore or trochophore-like larvae of lophotrochozoans (FIGURE 1). Such larvae are characterized by a ciliated ring called a prototroch that uses coordinated ciliary beating for propulsion in the water column. Some special forms, for instance the veliger larvae of gastropods, have enlarged this area to a multilobular velum (FIGURE 1D). Other larvae such as ascidian larvae rely on muscular activity for locomotion. Crustacean larvae generally have a chitinous cuticle, lack external ciliation and also rely on muscular movements (FIGURE 1F). Many planktonic organisms have enlarged appendages such as fins or paddle-shaped legs for increased buoyancy and increased propulsion during swimming.

In terms of feeding, there are two main strategies of planktonic larvae: those that are self-nourishing with a functional gut (termed »planktotrophic«), and those supplied with yolk and lacking a gut (termed »lecithotrophic«). In general, planktotrophic larvae last longer as they are not dependent on given reserve substances, which also enables a higher dispersal rate, whereas lecithotrophic larvae rely on their provided nutrients. This implies that their lifetime in the free water column is limited and they have to settle in a shorter time. Most planktotrophic larvae also utilize ciliary bands for creating feeding currents.[5]

Biomechanical Constraints ___
All organisms are subject to different mechanical forces they need to overcome for changing their position in space. The Reynolds number is a dimensionless quantity implying the ratio of inertial versus viscous forces and is mainly defined by the viscosity of the medium, velocity, and specific size (length) of the organisms. Smaller larvae are mainly subject to viscous forces while larger larvae are mainly subject to inertial forces. This is usually reflected in the shape and mode of propulsion: small larvae tend to be more roundish, reducing the surface area, and use ciliary locomotion, whereas larger ones are more streamlined, avoiding pressure drag, and frequently use muscular activity for locomotion.[6]

Workflow for Generating Accurate 3D Models of Planktonic Organisms

___ The wide range of sizes of planktonic organisms is very challenging for imaging and for creating 3D models, as each of the applied morphological techniques has advantages, restrictions and limitations. In many cases, several different techniques have to be used on different levels in order to create realistic and detailed impressions of the organisms modeled (FIGURE 2).

Sampling/Collection

Plankton is usually collected with the help of plankton nets with various mesh sizes (down to a few microns) that allow the concentration of aquatic organisms in small volumes, which enables easier analysis (FIGURE 3). Plankton samples can be taken manually with a handheld plankton net trawled in more superficial layers, or attached to a boat. Depending on the boat, plankton nets can be released to a certain depth and trawled horizontally or vertically. After retrieval of the net, it should be rinsed with seawater (or freshwater, respectively) to ensure that organisms close to the net

200

FIG. 2
Workflow of 3D model generation from sampling until surface generation

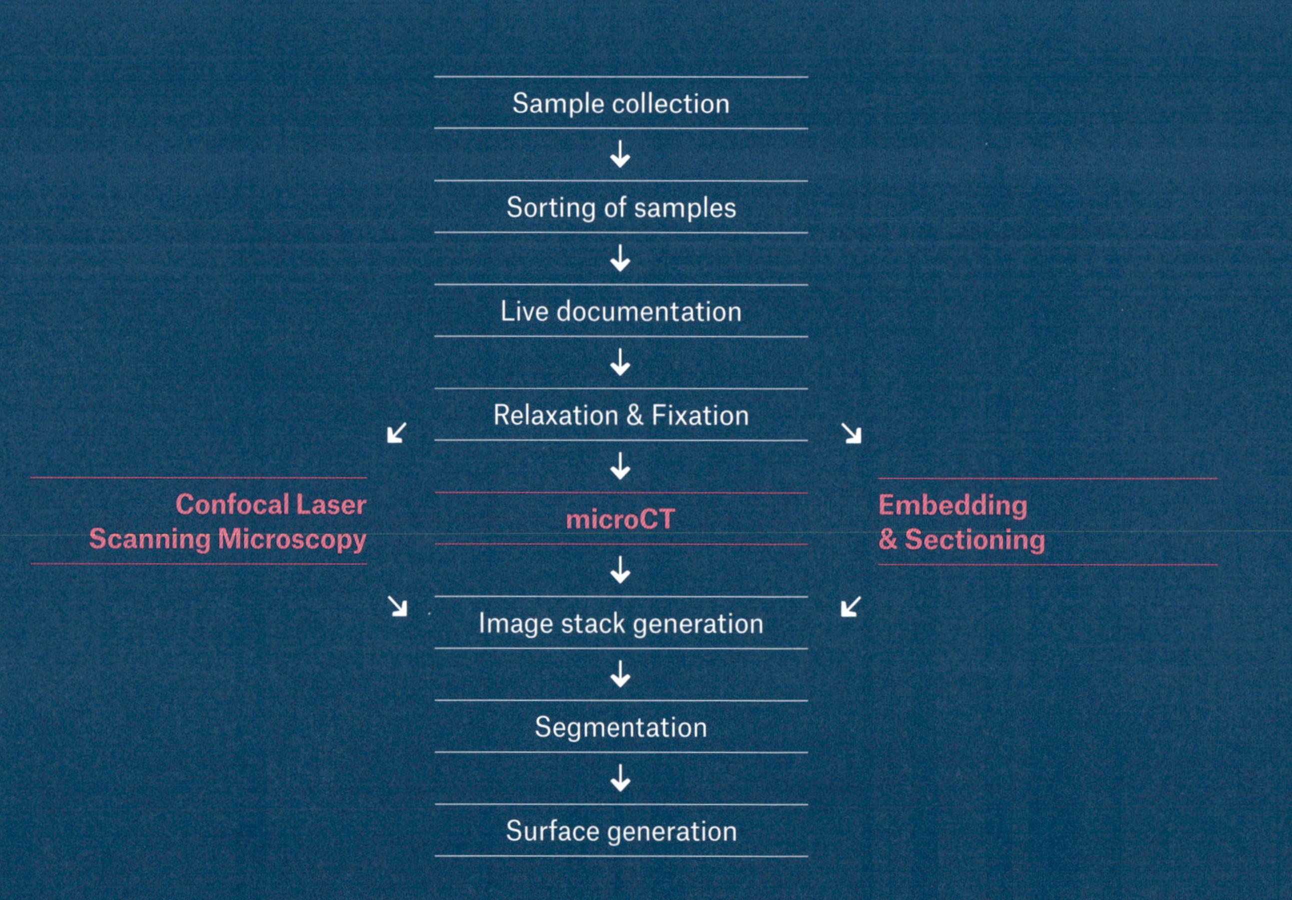

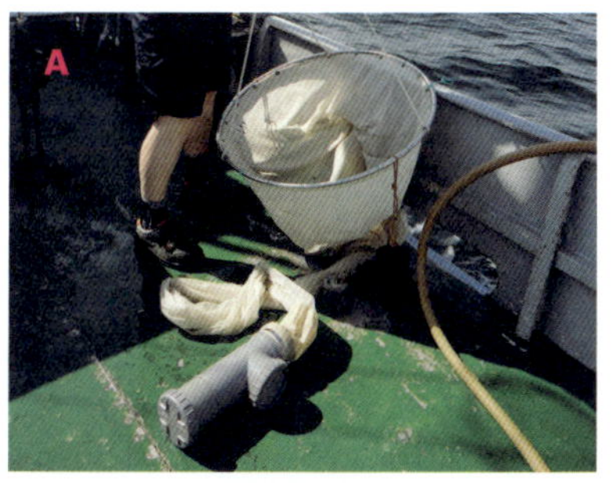

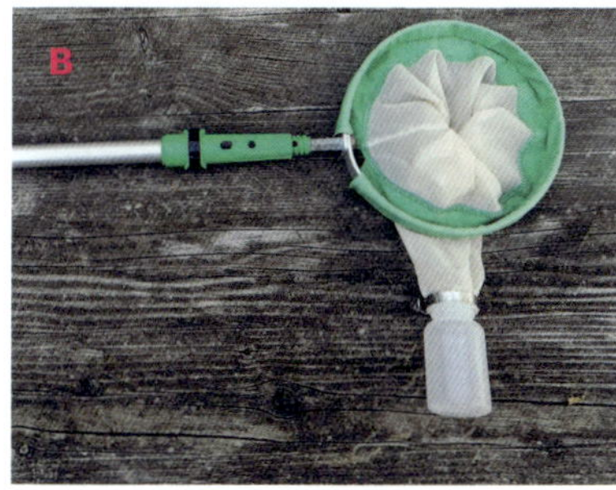

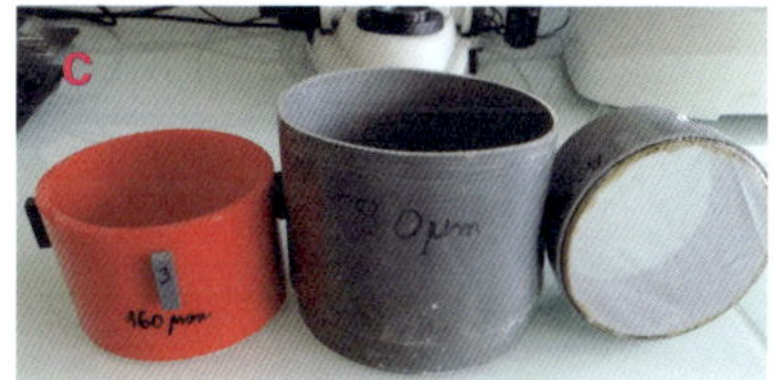

FIG. 3
Sampling of plankton
A Large plankton net on a boat
B Handheld plankton net
C Sieves of different mesh sizes for
concentrating planktonic samples
© Thomas Schwaha and
Stephan Handschuh

are flushed into the collecting bin. Collected plankton can also be further concentrated by additional filtering with various mesh sizes in the lab (FIGURE 3C).

After sample collection, plankton — especially zooplankton — is examined through a conventional stereomicroscope equipped with transmitted or oblique incidental lighting. Sorting of plankton is essential in order to separate different organisms or species from the bulk of concentrated plankton. Ranging down the mesh size of the plankton net can facilitate sampling of specific organisms, for example, using a 250 µm mesh size for 300–400 µm sized organisms. Overly concentrated plankton will not last long in the lab unless sorted quickly. It is also highly recommended to keep samples for sorting cool (16–18°C) to ensure longer vitality.

Live Observation/Video Microscopy

Concentrated plankton can be quite overwhelming in dense samples. Macroscopic samples such as krill or fish larvae can be separated quite easily, but the separation of smaller specimens requires thorough sorting in order to identify for instance specific larvae for subsequent analysis. Ideally, the specimen to be live-imaged is transferred to a small dish or microscope slide with filtered (sea)water to ensure a clean periphery. Documentation with the stereomicroscope is best conducted with lateral lighting, oblique transmitted light, or transmitted darkfield illumination. In stereomicroscopic video documentation, it is best to have samples in small dishes with a small amount of water, as the depth of focus is limited with any optical device (higher magnifications yield a smaller depth of focus). Especially with small larvae that locomote in all three dimensions, this is particularly challenging as the fast movements require continuous adjustment of stage position and focus during video documentation.

Samples with specimens mounted on microscope slides for detailed microscopic observations pose the same problems. Larger magnifications yield more limited depth of field, causing restrictions and difficulties in documentation. Limitation of space in all axes eases observations. For example, slight squeezing of specimens with a coverslip restricts specimen movement and thus allows steady motion analysis of ciliary beating or other muscular or intracellular activity. Particularly phase or differential interference contrast microscopy, but also darkfield illumination may reveal fine anatomical details during live observations.

Relaxation and Fixation of Specimens

Relaxation is necessary prior to fixation in most samples in order to prevent morphological changes of body shape, which may occur owing to fixation-induced muscle contractions that obscure any natural and realistic shape of an organism. The most common relaxation agent used in marine organisms is magnesium chloride.[7]

201

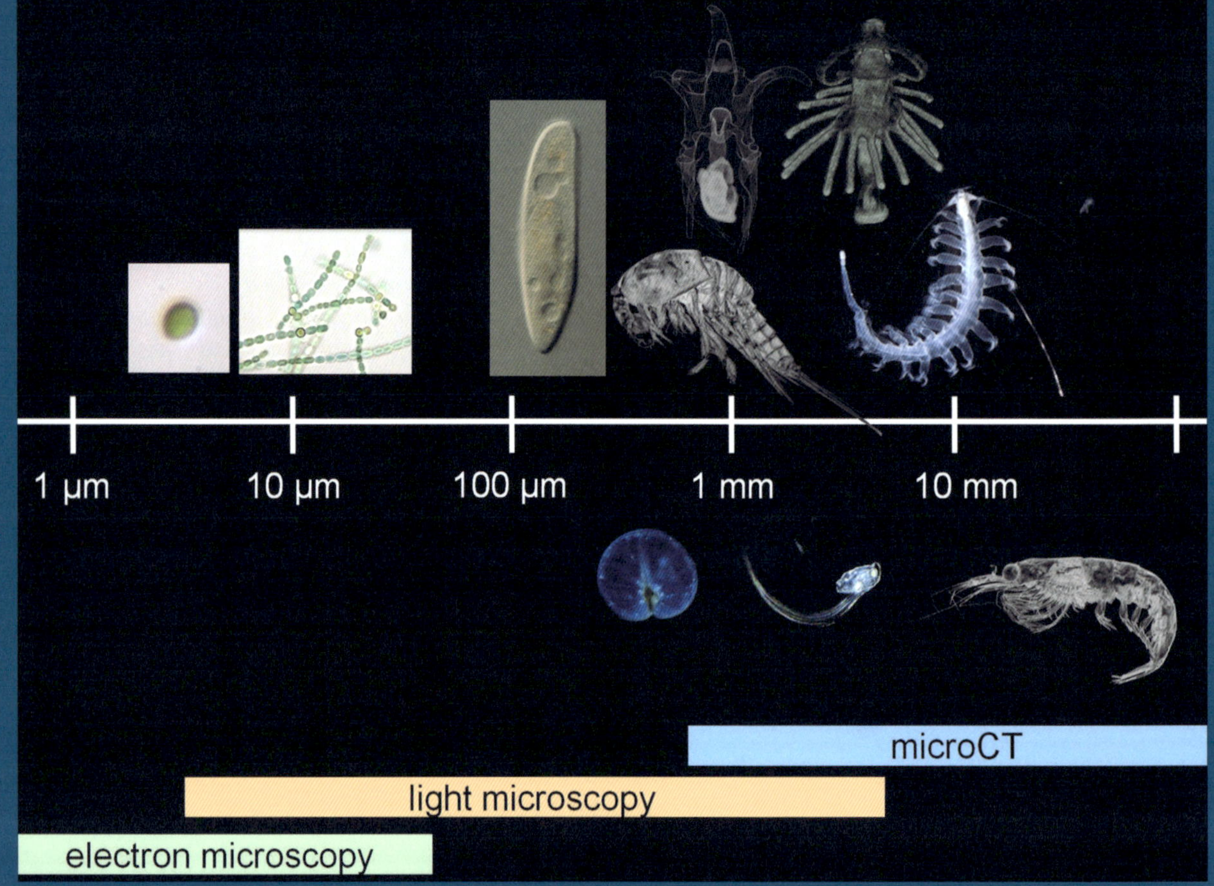

FIG. 4
*Body size variability of planktonic organisms
and size range of common imaging techniques*
Total body size of planktonic organisms spans
several length scales, ranging from a few microns
(e.g. unicellular green algae) to roughly a millimeter
(e.g. phoronid or starfish larvae, larvaceans, or
small crustaceans such as copepods) to several centimeters (e.g. larger crustaeans such as krill). Depending
on specimen size, different modalities are used for
imaging. While microCT is most efficient for imaging
of specimens of 1 mm or more, light and electron
microscopy provide finer anatomical detail for smaller
specimens.
© Thomas Schwaha and Stephan Handschuh

Fixation is the process of stabilizing and preserving biological samples for subsequent analysis. Common fixation strategies use either protein crosslinking by aldehyde solutions, or dehydration of tissues in alcoholic solutions. High-quality preservation of fine morphological structures can be achieved by aldehyde fixation using either paraformaldehyde or glutaraldehyde in various buffered solutions. Paraformaldehyde is used for immunocytochemical analyses such as antibody stainings. Glutaraldehyde is used for finer morphological preservation which is required especially for electron microscopic investigations.

Various other fixation protocols are available.[8] Fixation is a crucial step that also has severe effects on the morphology and can cause shrinkage, particularly when using dehydrating fluids like alcoholic solutions. In soft specimens such as marine larvae this might even result in drastic shrinkage, rendering the specimens useless for morphological analyses. Arthropod specimens with a hard and sometimes mineralized cuticle are less susceptible to such massive changes.

Postfixation Sample Processing and Imaging

This section briefly introduces the most commonly used modalities for generating 3D image volumes from planktonic organisms. Depending on specimen size, different modalities may be favorable for imaging (FIGURE 4). Tomographic techniques such as microCT are most efficient for imaging medium-sized and larger species (FIGURE 5), while light microscopy offers a lot of fine anatomical details for small planktonic organisms below 1 mm body size (FIGURE 6).

Microscopic X-Ray Computed Tomography (Micro CT)

MicroCT is a nondestructive tomographic imaging technique that can be used for imaging dense and optically nontransparent samples ranging in size from around 500 µm to around 30 cm or even more. During a microCT scan, about 1,000 to 3,000 X-ray projections are recorded over a 360° specimen rotation. From these projections, virtual tomographic slices are reconstructed via a filtered backprojection algorithm. Image contrast in the reconstructed slices is based on X-ray attenuation in samples. Nonmineralized biological tissues show very low inherent X-ray attenuation and, as a consequence, low image contrast. Thus, X-ray-dense contrast agents such as iodine, phosphotungstic acid or osmium tetroxide are routinely used to enhance sample contrast.[9] After microCT imaging, the same samples can be further investigated using light microscopy and/or electron microscopy in order to add finer anatomical details.[10]

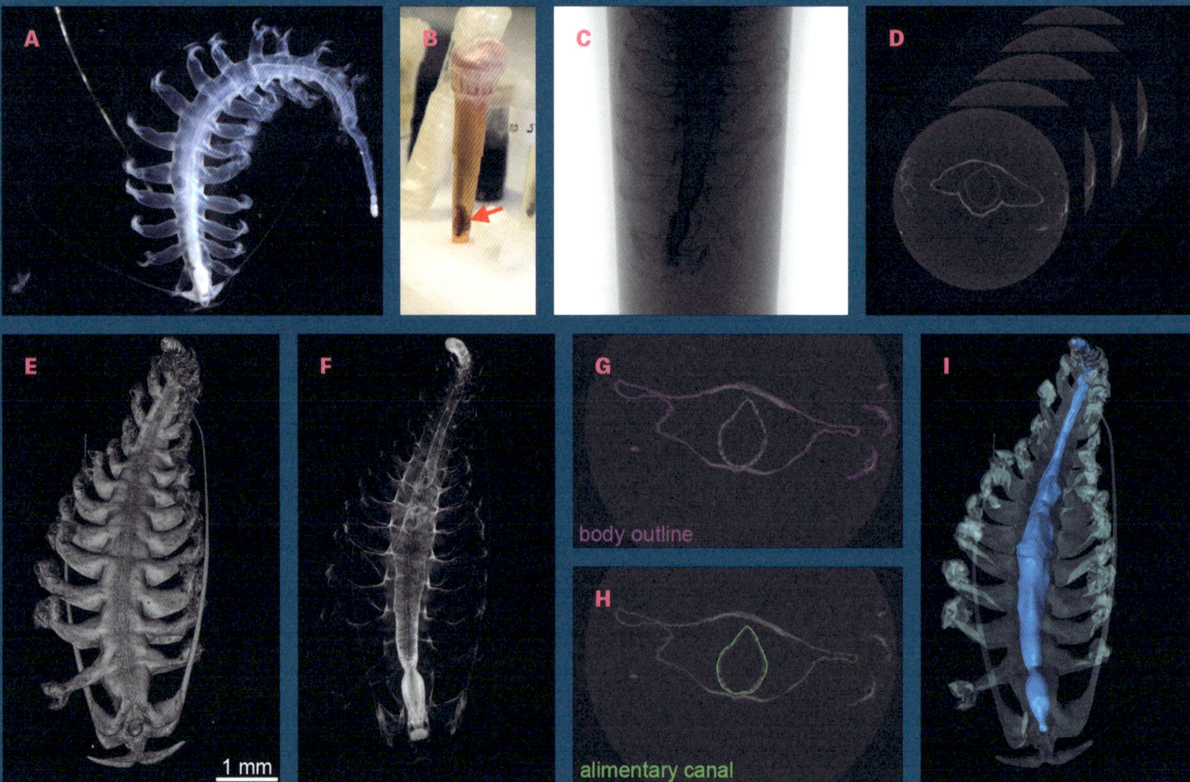

FIG. 5

Workflow for the generation of 3D models for medium-sized and larger planktonic organisms, demonstrated on the polychaete TOMOPTERIS *sp.*

Tomopteris species are holoplanktonic marine worms exhibiting bioluminescence.

A Fixed specimen imaged by stereomicroscopy

B Iodine-stained specimen mounted in a polypropylene pipette tip for microCT scanning; © image courtesy of M. Glösmann

C X-ray projection image of the anterior body part; roughly 1,500 such X-ray images are recorded over a 360° sample rotation

D Virtual slices reconstructed from X-ray projections (**C**)

E Volume rendering of microCT data set, solid view depicting external body features

F Volume rendering of microCT data set, transparent view depicting internal body features

G and **H** Based on the microCT data set, image segmentation tools are used to contour selected body regions

I Polygon mesh surface models triangulated from image segmentation masks (shown in (**G**) and (**H**))

© Thomas Schwaha and Stephan Handschuh

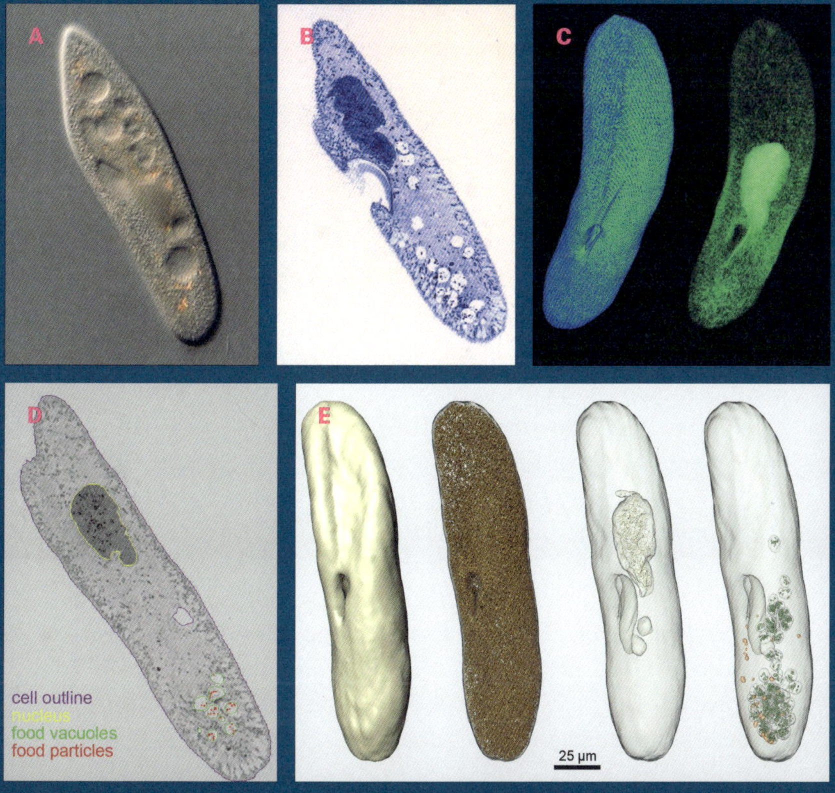

FIG. 6

Workflow for the generation of 3D models for small planktonic organisms, demonstrated on the unicellular ciliate PARAMECIUM *sp.*

Paramecia occur in freshwater, brackish and marine environments. They are fascinating because—despite being unicellular organisms—they possess a functional mouth and their cell organelles fulfill multiple different functions similar to organs in multicellular species.

A Light microscopic DIC image of a living paramecium; after live imaging, small specimens are fixed, embedded in epoxy resin and physically sectioned on a microtome

B Micrograph of a toluidine blue-stained 0.5 µm thick section of a paramecium; routinely, several hundred of such serial sections are cut, stained and imaged

C Volume renderings of aligned serial section micrographs

D Like for larger specimens, image segmentation tools are used to segment relevant anatomical structures

E Polygon mesh surface models triangulated from image segmentation masks (shown in (D))

© Thomas Schwaha and Stephan Handschuh

Histology/Physical Sectioning

Physical serial sectioning yields thin slices of three-dimensional objects that can be imaged with high optical resolution. However, the process is time consuming and prone to sectioning artifacts (i.e. geometric distortion, inhomogeneous staining of sections, etc.). In addition, processing for embedding prior to sectioning often involves complete dehydration, which in many soft-bodied organisms can lead to severe shrinkage.

For histology, small fixed samples are best postfixed with osmium tetroxide. Especially minute samples intended for sectioning require some contrast for sample preparation; otherwise, they become hard to detect once infiltrated with the embedding medium. Osmium postfixation will turn samples gray to black, depending on the density of their tissues. Alternatively, other dyes can be applied to stain small larvae prior to embedding. Prior to embedding into media not miscible with aqueous solutions, samples are gradually dehydrated by various agents such as ethanol or acetone and subsequently infiltrated and embedded in epoxy resin.[11] Resin blocks are afterwards sectioned with a microtome that is capable of producing serial sections of 0.5–1 µm thickness. For sectioning of resin blocks, diamond knives are best suited as they remain sharp for a long time and yield the best serial sections. Serial sections are mounted on glass slides, stained and sealed, and further analyzed by microscopic photography.

Confocal Laser Scanning Microscopy (CLSM)

CLSM is a technique suitable for creating optical section series of small objects. The upper size limit for proper stack generation is usually around 200 µm. CLSM depends on fluorescence signals using lasers of different wavelengths to excite native or induced fluorescence within tissues. Hence, the excitation laser has to penetrate the tissue to get a signal back, which is easily done for small soft-bodied and transparent organisms. For harder and especially pigmented structures such as some cuticles penetration will only be superficial. CLSM is an elegant and fast way of obtaining 3D data sets.

Most useful for imaging planktonic organisms is autofluorescence; that is, the general fluorescence of tissues, which can be native or induced by aldehyde fixations.[12] Native autofluorescence is, for instance, given by cuticular structures and is thus very helpful in analyzing crustacean cuticles. Often, particular details like cuticular bristles or setae of individual legs where the micro CT might lack the specific resolution, can be imaged via CLSM.

Image Stack Generation, Alignment, and Processing

Volumes generated from optical sections (CLSM) or virtual sections (microCT) do not require any alignment procedures. However, these stacks usually require some processing steps, including, for instance, the correction of refractive index mismatch in CLSM volumes or image filtering for noise reduction in microCT volumes. MicroCT volumes are inherently isotropic, which means that image resolution is the same for all three image axes (X–Y–Z). This becomes obvious when visualizing microCT volumes by direct volume rendering (FIGURE 5E, F). In comparison, owing to physical limitations, CLSM images show higher lateral (X–Y) resolution compared to a lower axial (Z) resolution.

Serial physical (histological) sections are imaged with a conventional transmitted light microscope using either a color or a grayscale camera. After image capture, all micrographs of an image stack must be aligned with reference to the neighboring sections because the spatial context of consecutive sections is lost during sectioning. Alignment is mostly performed using automatic algorithms such as the least squares algorithm. Aligned image volumes from serial physical sections are usually post-processed by image filters[13] and can be readily visualized by direct volume rendering (FIGURE 6C). Image stacks from physical sections are usually anisotropic, showing distinctly higher lateral image resolution (X–Y) compared to the section thickness (Z).

Image Segmentation

Image stacks, also often called »image volumes,« are three-dimensional data sets consisting of voxels (volume pixels). Each voxel contains a certain intensity or color value. Image segmentation is the process of partitioning a single image stack into multiple segments. Each of these segments can then be used as a mask for image analysis or for the generation of 3D models. Many different image segmentation tools are available for segmentation of biological data sets, including manual segmentation tools, intensity-based segmentation tools as well as machine learning and deep learning approaches. One commercial program that is frequently used in biomedical research is the software Amira from ThermoFisher Scientific. This software includes a highly versatile segmentation editor, which combines manual and intensity-based segmentation tools, machine learning, and morphological image operations such as grow/shrink, fill holes, remove islands, or smooth contours. These tools allow interactive and efficient labeling of multiple materials of interest in a data set (FIGURES 5G, 5H, AND 6D). Image segmentation needs to be completed for the whole image volume and, depending on the complexity of the data set and the number of segmented materials/tissues, this step may take up 10–40 hours per data set or even more, thus usually being the most time-consuming step in the whole workflow.[14]

208

Surface Generation

After image segmentation is completed, segmentation masks can be used for the generation of polygon mesh models by using the marching cubes algorithm.[15] In this step, voxel data sets (segmentation masks) are converted to triangle surfaces. This process can be thought of as digitally wrapping the segmented objects in some virtual polygon mesh foil. Each surface can be processed and edited, including triangle reduction, smoothing and remeshing.[16] During these editing steps, polygon meshes are visualized by simple surface rendering, depicting single materials with uniform colors and variable transparency level rendering (FIGURES 5I AND 6E). Edited surfaces are then exported to some standard file format (such as *.stl) and delivered to digital artists.

Conclusions ___ The presented workflows are well established and widely utilized in biomedical research. Polygon surface models may be used for the visualization or quantitative analyses of biological structures (including measurement of volume, surface area, and tissue density, among many other parameters) as well as for complex analyses such as geometric morphometrics, finite element analysis (FEA), multibody dynamic analysis (MDA), or flow simulations. In the context of arts and science, biological models can further be edited by artists, for instance through rigging and/or animation.[17] Eventually, informative and valuable biological 3D models can be used in art projects such as Victoria Vesna's NOISE AQUARIUM,[18] which is also addressed in a chapter of the present volume. Such projects also aid in conveying knowledge about little-known organisms to a wider nonscientific audience and can spark interest in the fascinating diversity of life on our planet.

1 Ernst Haeckel, »Plankton-Studien,« *Jenaische Zeitschrift für Naturwissenschaft* 25 (Neue Folge) 18 (1891): 232–336.

2 Claudia Castellani and Martin Edwards, *Marine Plankton: A Practical Guide to Ecology, Methodology, and Taxonomy*, Oxford: Oxford University Press (2017).

3 Matthias Glaubrecht and Wolfgang Dohle, »Discovering the Alternation of Generations in Salps (Tunicata, Thaliacea): Adelbert von Chamisso's Dissertation *De Salpa* 1819, Its Material, Origin and Reception in the Early Nineteenth Century,« *Zoosystematics and Evolution* 88, no. 2 (2012): 317–363.

4 Reinhard M. Rieger, »The Biphasic Life-Cycle: A Central Theme of Metazoan Evolution,« *American Zoologist* 34, no. 4 (1994): 484–491; Richard C. Brusca and Gary J. Brusca, *Invertebrates*, 2nd edition, Sunderland, MA: Sinauer Associates Inc. (2003).

5 Richard R. Strathmann, »Feeding and Nonfeeding Larval Development and Life-History Evolution in Marine Invertebrates,« *Annual Review of Ecology and Systematics* 16 (1985): 339–361.

6 Steven Vogel, *Comparative Biomechanics: Life's Physical World*, 2nd edition, Princeton, NJ: Princeton University Press (2013).

7 Bernhard Ruthensteiner, »Soft Part 3D Visualization by Serial Sectioning and Computer Reconstruction,« *Zoosymposia* 1 (2008): 63–100.

8 Peter Böck, ed., *Romeis, Mikroskopische Technik*, 17th edited and extended edition, Munich: Urban & Schwarzenberg (1989).

9 Brian D. Metscher, »MicroCT for Comparative Morphology: Simple Staining Methods Allow High-contrast 3D Imaging of Diverse Non-mineralized Animal Tissues,« *BMC Physiology* 9 (2009), 11; Brian D. Metscher, »MicroCT for Developmental Biology: A Versatile Tool for High-Contrast 3D Imaging at Histological Resolutions,« *Developmental Dynamics* 238, no. 3 (2009): 632–640.

10 Stephan Handschuh, Natalie Baeumler, Thomas Schwaha, and Bernhard Ruthensteiner, »A Correlative Approach for Combining MicroCT, Light and Transmission Electron Microscopy in a Single 3D Scenario,« *Frontiers in Zoology* 10 (2013), 44.

11 Bernhard Ruthensteiner, »Soft Part 3D Visualization by Serial Sectioning and Computer Reconstruction,« *Zoosymposia* 1 (2008): 63–100.

12 Frederick W.D. Rost, *Fluorescence Microscopy*, Cambridge: Cambridge University Press (1992).

13 Stephan Handschuh, Thomas Schwaha, and Brian D. Metscher, »Showing Their True Colors: A Practical Approach to Volume Rendering from Serial Sections,« *BMC Developmental Biology* 10 (2010), 41.

14 Bernhard Ruthensteiner, »Soft Part 3D Visualization by Serial Sectioning and Computer Reconstruction,« *Zoosymposia* 1 (2008): 63–100.

15 William E. Lorensen and Harvey E. Cline, »Marching Cubes: A High Resolution 3D Surface Construction Algorithm,« *ACM SIGGRAPH Computer Graphics* 21, no. 4 (1987): 163–169.

16 Bernhard Ruthensteiner, »Soft Part 3D Visualization by Serial Sectioning and Computer Reconstruction,« *Zoosymposia* 1 (2008): 63–100.

17 Martina R. Froeschl, Stephan Handschuh, Rudolf Erlach, Thomas Schwaha, Helmuth Goldammer, Reinhold Fragner, and Manfred G. Walzl, »Computer-Generated Images of Microscopic Soil Organisms for Documentary Films,« 9th International Seminar on Apterygota, Görlitz, Germany, Sept. 7–9, 2014, *Soil Organisms* 86, no. 2 (2014): 95–102.

18 Victoria Vesna, Alfred Vendl, Martina R. Froeschl, Glenn Bristol, Paul Geluso, Stephan Handschuh, and Thomas Schwaha, *Noise Aquarium*, *Leonardo* 52, no. 4 (2019): 408–409.

The objects shown in this chapter were made during
a course on basic principles of three-dimensional design
that I offered to art students at the Burg Giebichenstein
University of Art and Design Halle during the summer
term of 2019. To foster the students' ability to design
complex three-dimensional forms while simultaneously
gaining insights into the characteristics of form
solutions that natural systems present, I provided them
with iconic images of microscopic organisms like
radiolaria, which are found as zooplankton throughout
the world's oceans. Besides images of radiolaria,
I also showed images of microalgae, especially unicellular diatoms, which live in the oceans, waterways,
and soils of the world. The images were provided by
Dr. Christian Hamm, head of the Bioinspired Lightweight
Design group of the Helmholtz Center for Polar and
Marine Research at the Alfred Wegener Institute in
Bremerhaven, Germany, who investigates diatom micro-
and nanostructures in the field of functional morphology and their systematic application for sustainable
lightweight products in automotive, aerospace,
mechanical engineering, and other industries.

I AM A RADIOLARIAN

Reiner Maria Matysik

embarrassing objects – north sea – fresh breeze – stiff breeze –
try about aquatic sculpture – afflicted by… – violence of the deep –
focus on the sea as a habitat – the mighty power of water –
the artificial separation of water and land – microscopically small
radiolarians – fragile objects – glass – water sculptures – the tidal
rangeforce of suffering – vicious – why are you blaming yourself? –
community – related role models – nonparticipation – affection –
ill will – unwillingness – unplastic – huh – »away from you all.
just where?« – damnation – stillness – sigh of pleasure – irrevocably
separated – to be able to speak – gender – inveterate – exhaustion –
involuntarily – open your eyes – conditionality – divisive – sense
of wholeness – embodied – uselessness – draft – heartlessness –
uniform – light up – follow up – transfiguration – working alone –
most generously unscathed – longed for – unalterable – irreversible –
animated – presence of mind – sweaty – that which is going into
the uncertain – alive dead – tired of creating – amusement – becoming
accessible – refuge plastic – that can be touched – becoming one –
get into each other – »suspension bridge of the me« – distress –
eyelid flip – idiot – the sculptures form themselves without being
mediated by the hands. things formed without action, completely
in the absence of activity, out of themselves. for me, like you, all
»gebilde« were already formed at the same time. – best possible –
educational detour – tension – central sculpture – sculpture according
to darwin – »every flower is half a brothel.« – active idleness –
adorable sacral secular – art is something that finds its origin in
language, still in the mind or intellect. art arises outside of language.
we can try to circumnavigate you with language. not more. –
from hand to ear: from start to finish – it can no longer be about
the presentation of objects. we have to activate it. better still,
it needs active structures that touch us.

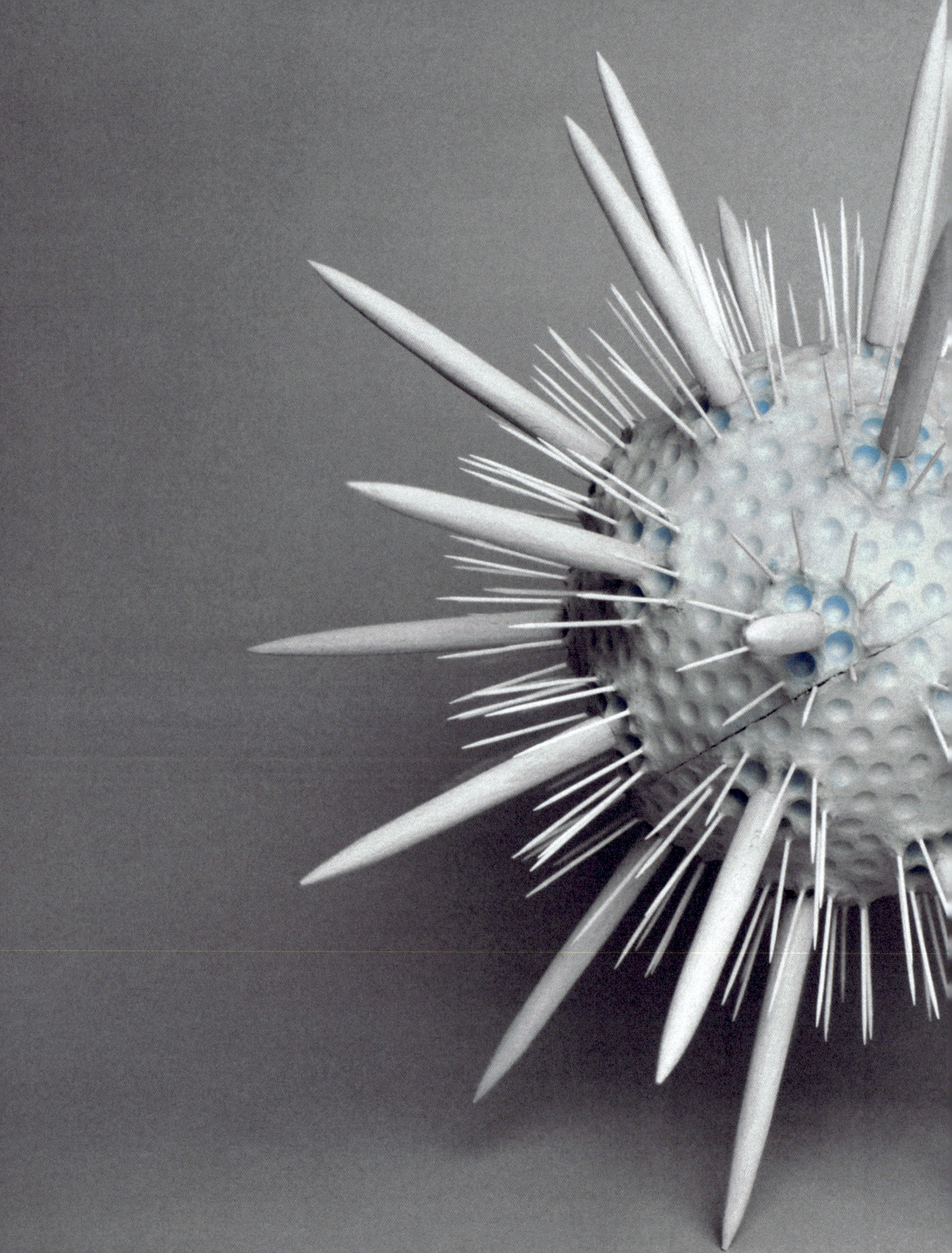

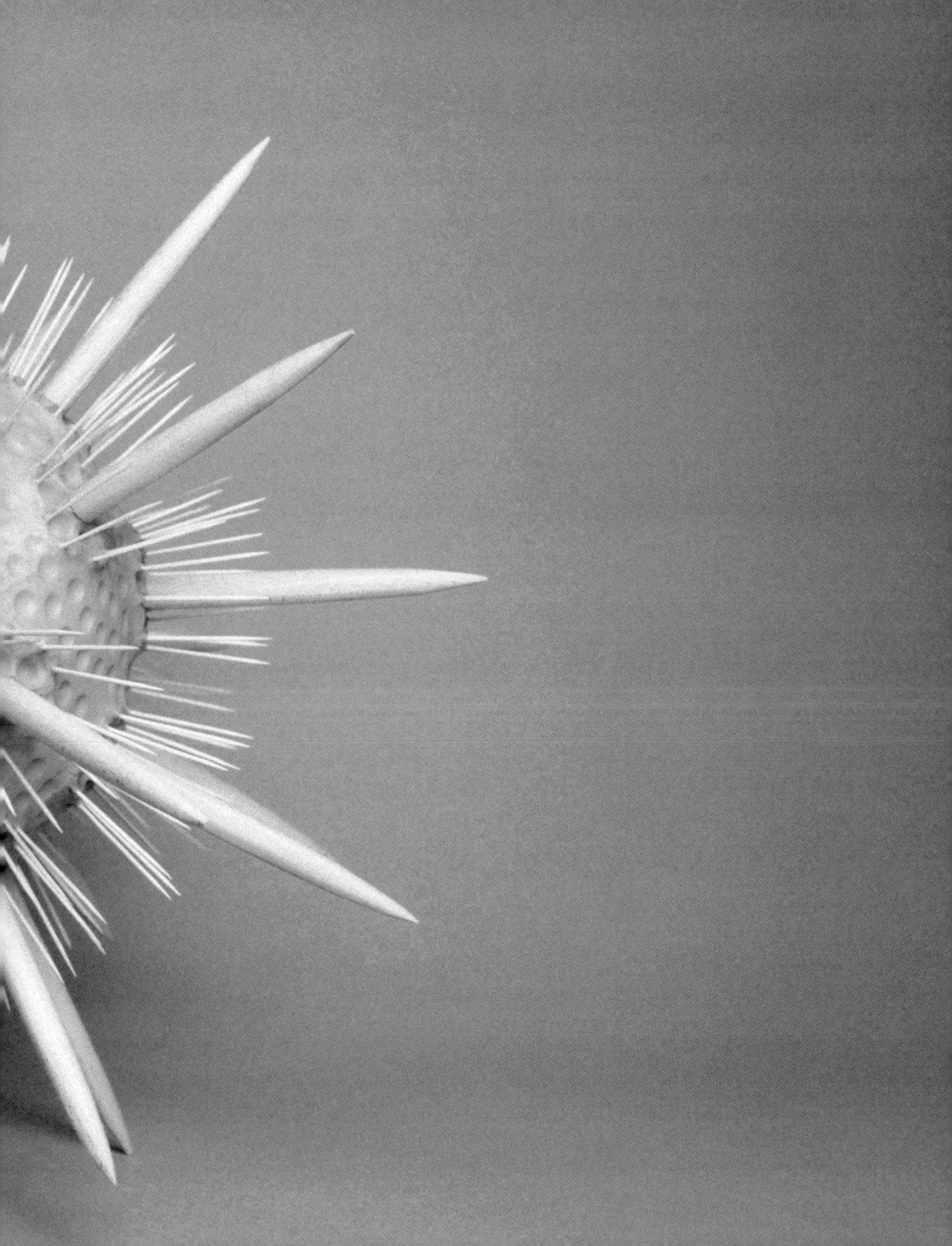

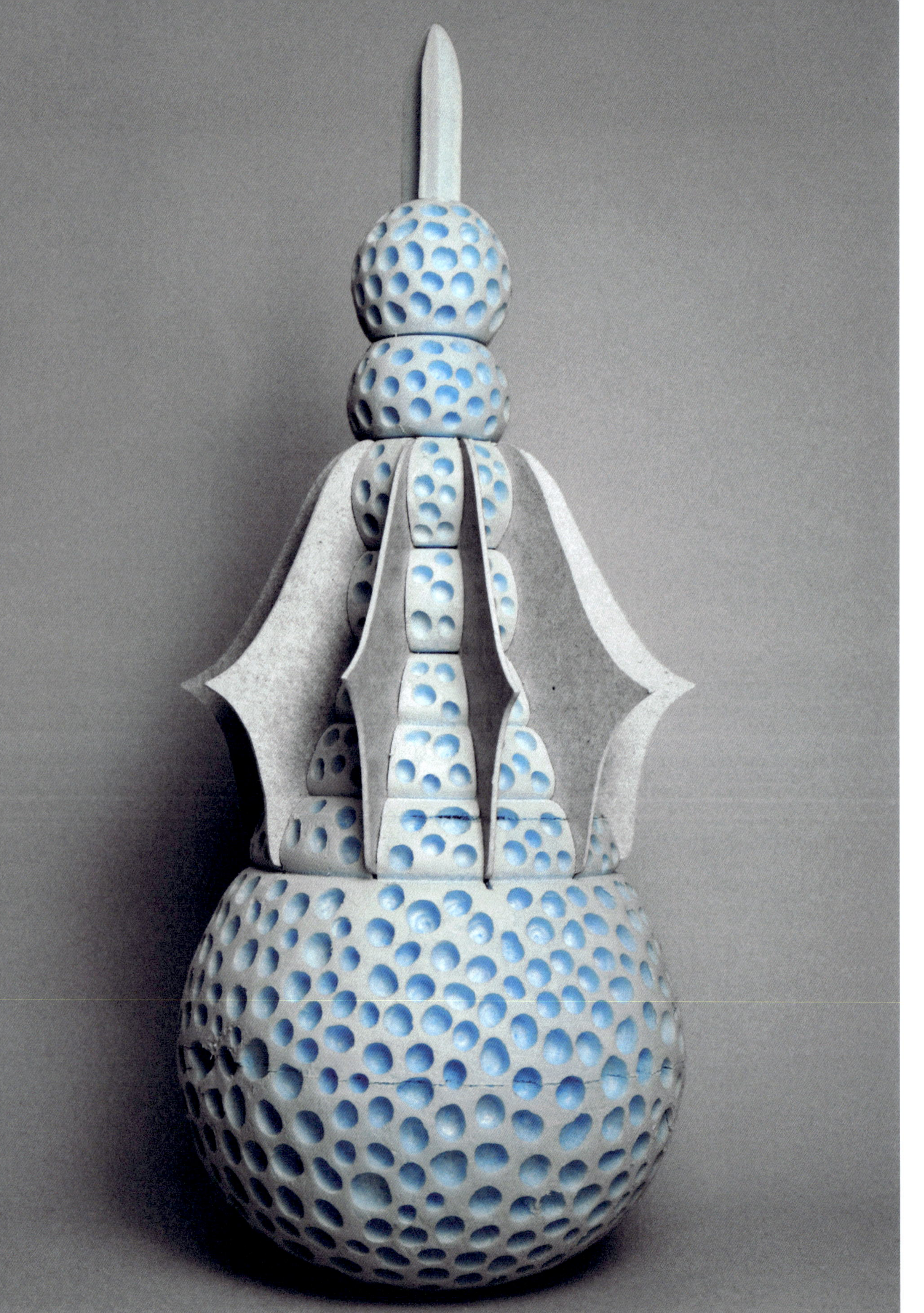

Radiolarians are microscopic creatures that were first introduced in 1834 by German physician and botanist Franz Julius Ferdinand Meyen (1804–1840) and extensively researched by 19th-century marine biologists such as German marine biologist and physician Ernst Haeckel (1834–1919). The models of radiolarians represent morphological structures of these creatures, which are usually no more than 0.5 millimeters in size. The three-dimensional shape is carved out of a cubic block. The deepening in the aesthetic design of radiolarians coupled with questions about the necessity of their shaping enables an intensive haptic–plastic experience. In order to understand the forms formed, questions such as how skeletons function must be clarified. The observations and findings are used to create the physical manifestation of a model based on the depiction of the living being from a geometric block. Sculptural design is an activity that explores what already exists and translates it into spatial appearances and three-dimensional results. Sculpture, therefore, means analyzing the origin, the model, the original, and using it to form objects. In basic plastic processes, experience with the entire body is placed on an equal footing with seeing.

FIG. 1

The scanning electron microscope image shows the radiolarian HEXACONTIUM SPEC.
Photo: Nils Niebuhr,
Alfred Wegener Institute, 2010

FIG. 2

The scanning electron microscope image shows the diatom ASTEROMPHALUS HOOKERI.
Photo: Friedel Hinz,
Alfred Wegener Institute, 2004

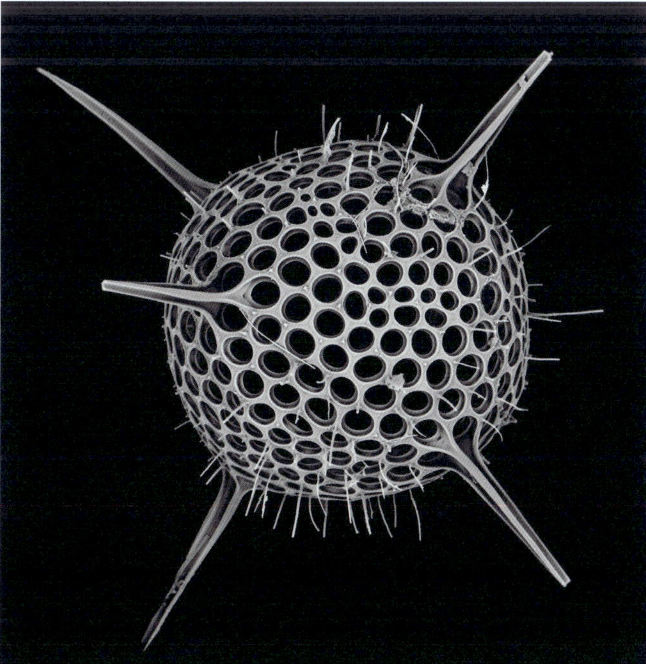

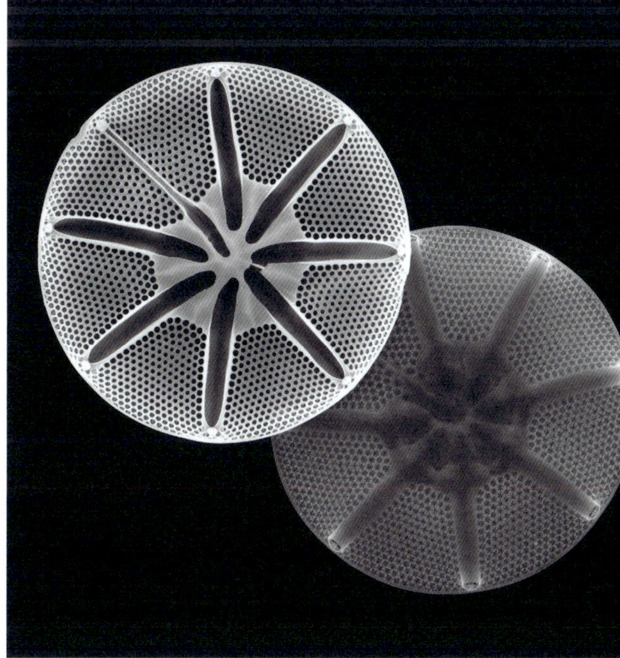

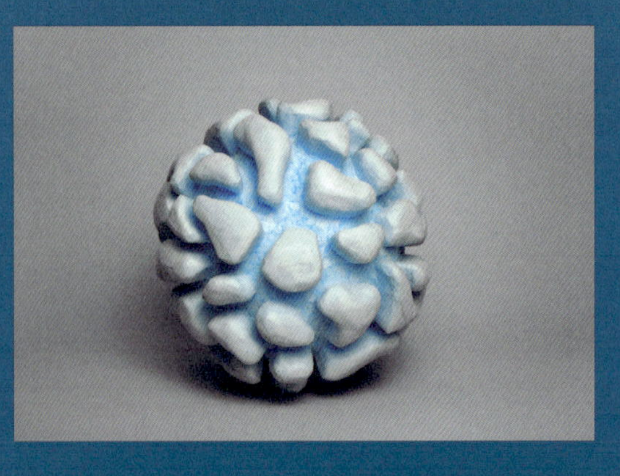
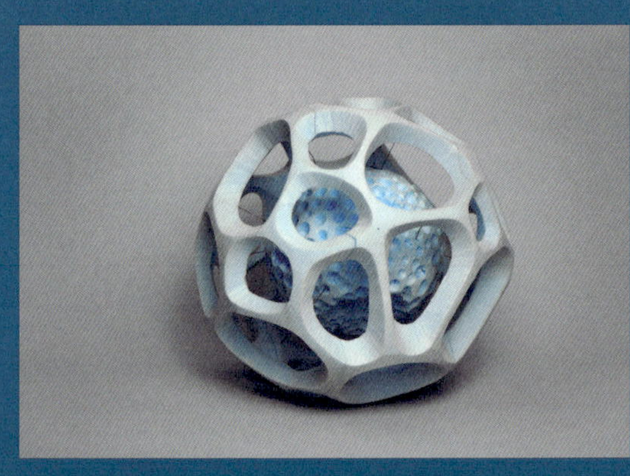
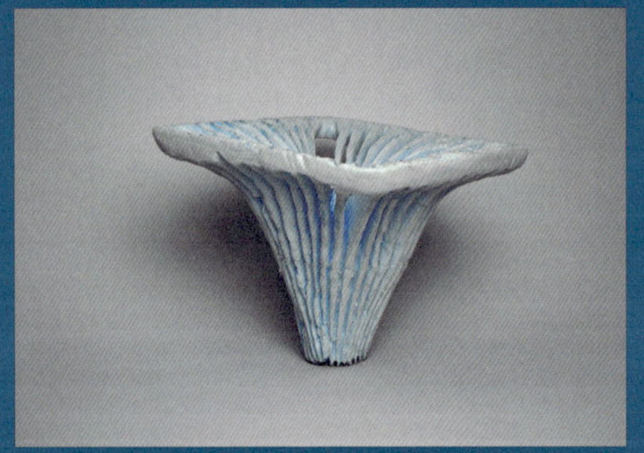
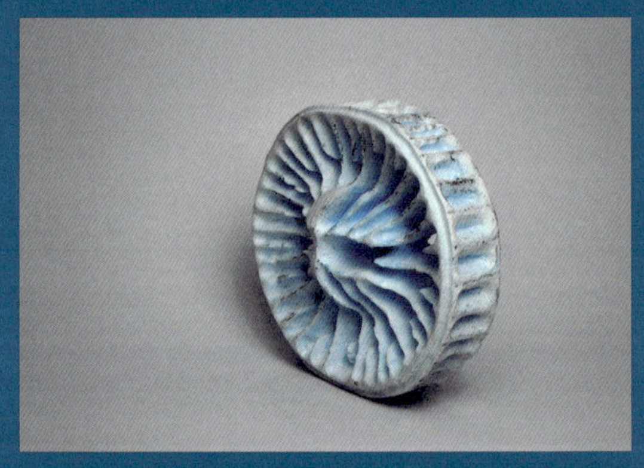
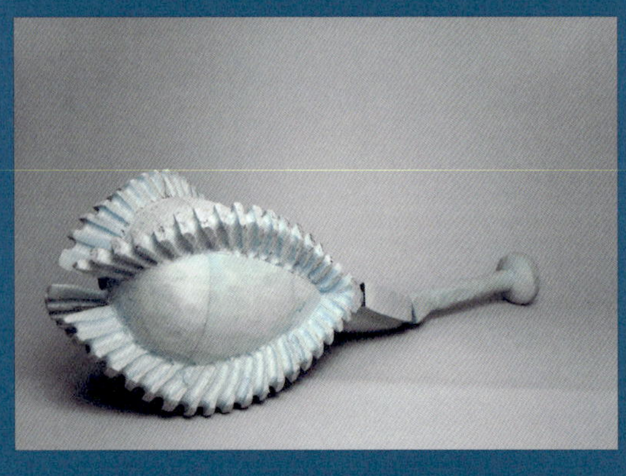
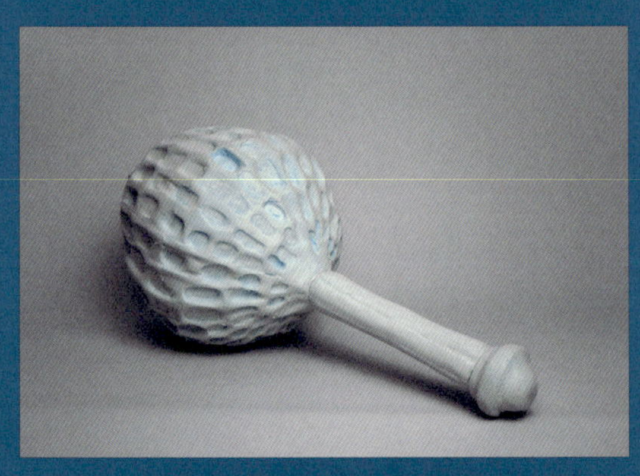

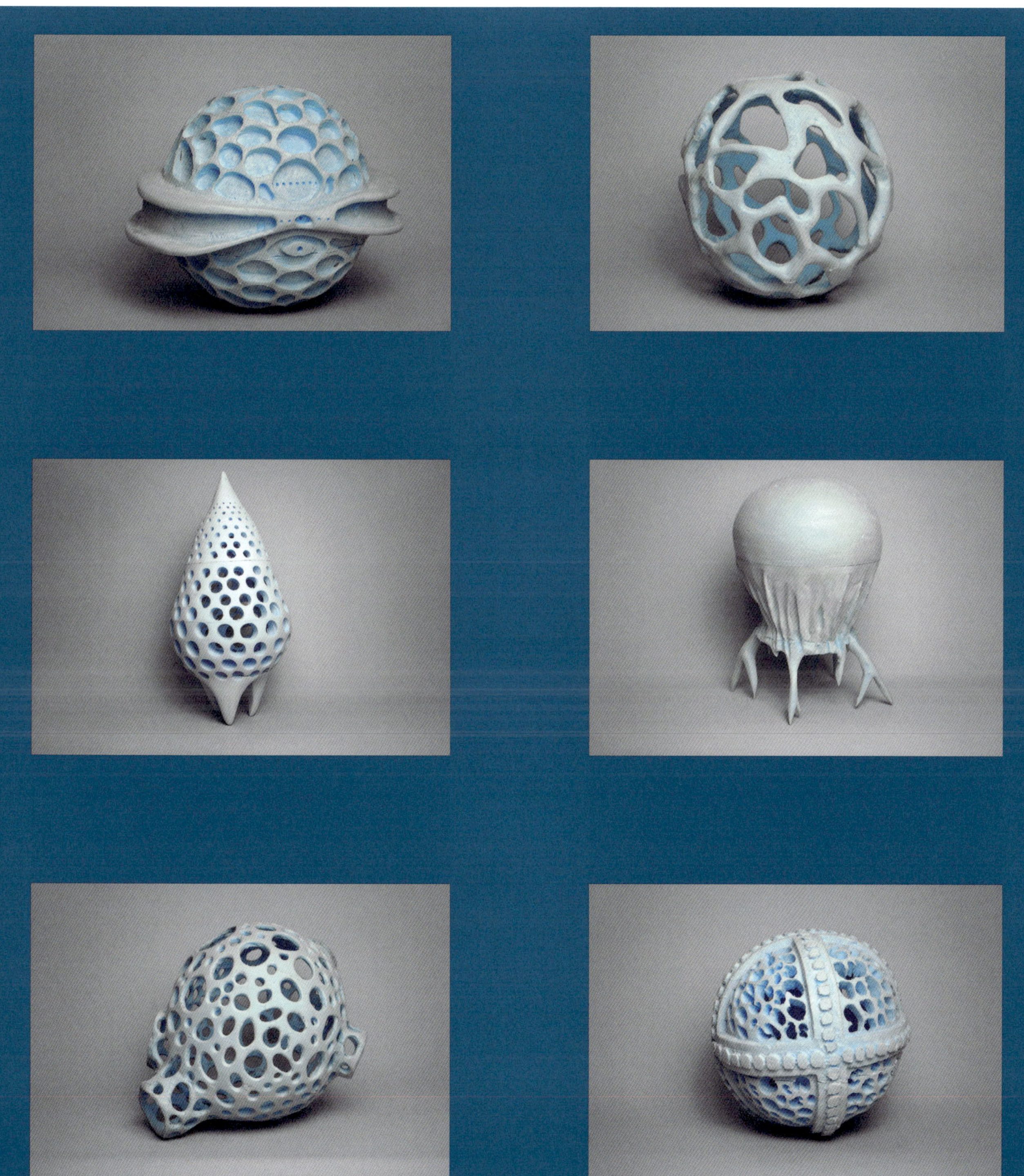

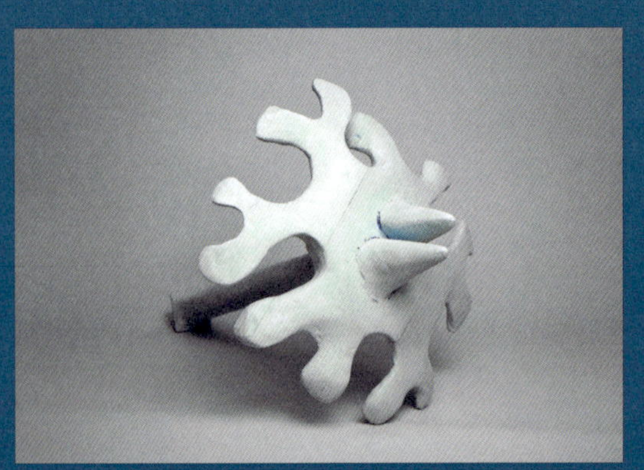
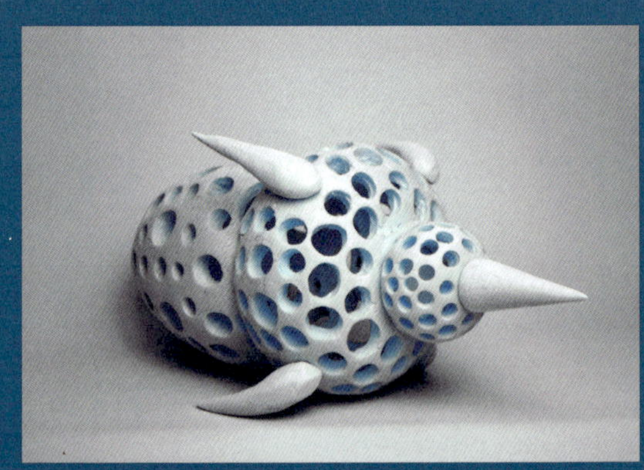
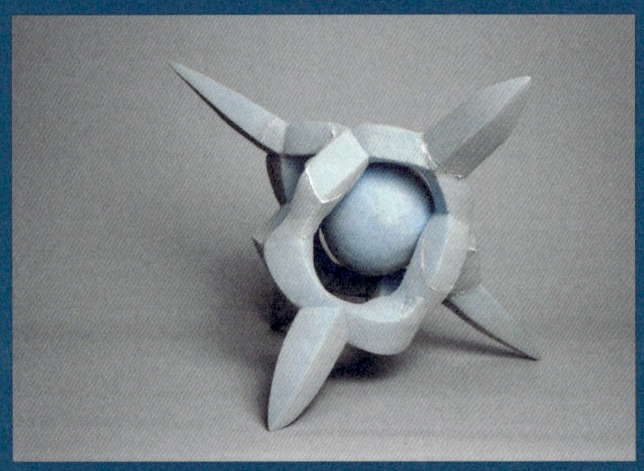
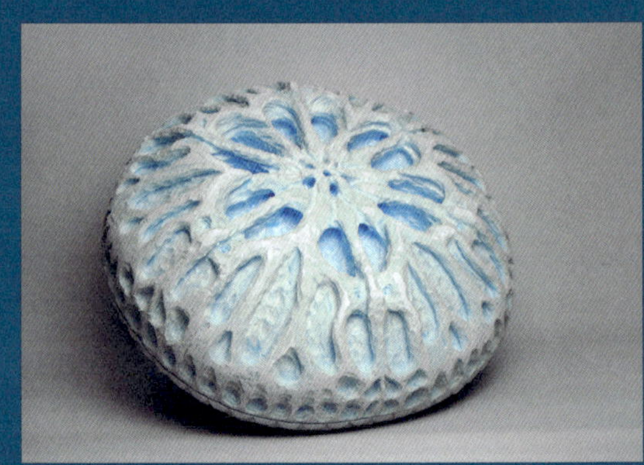
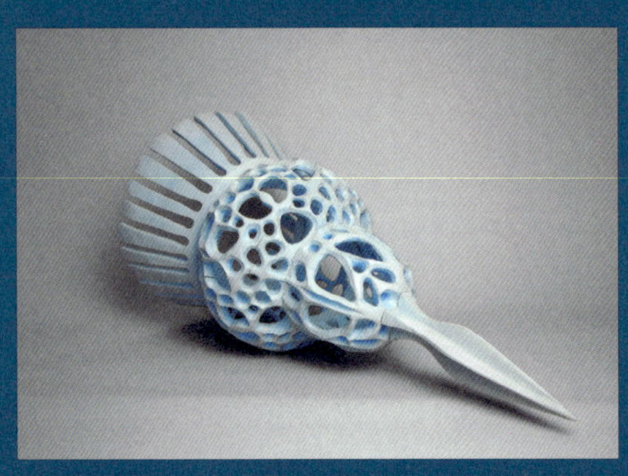
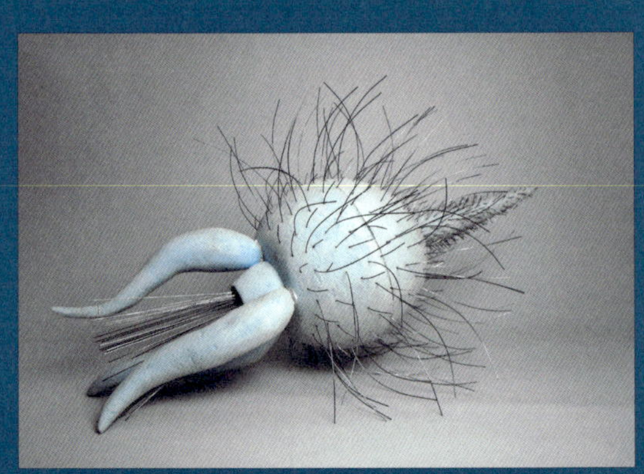

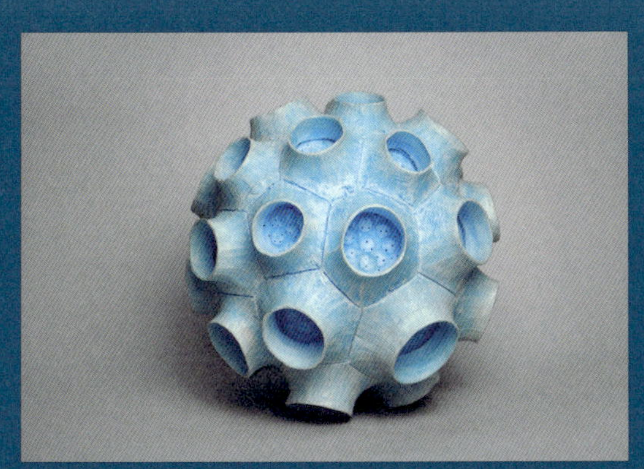
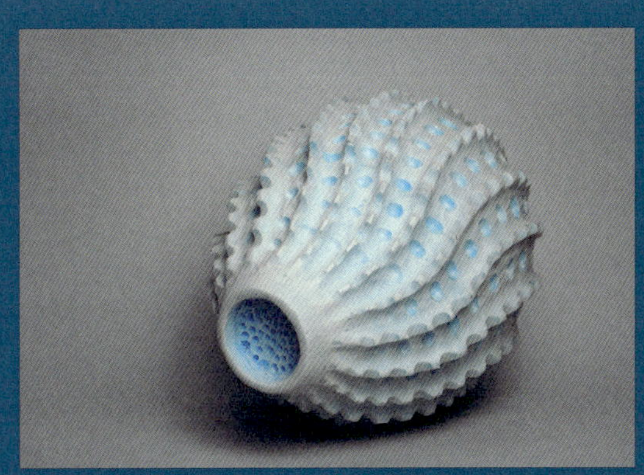
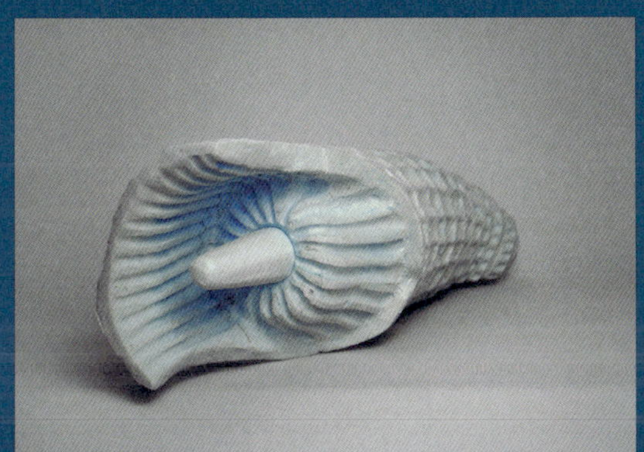
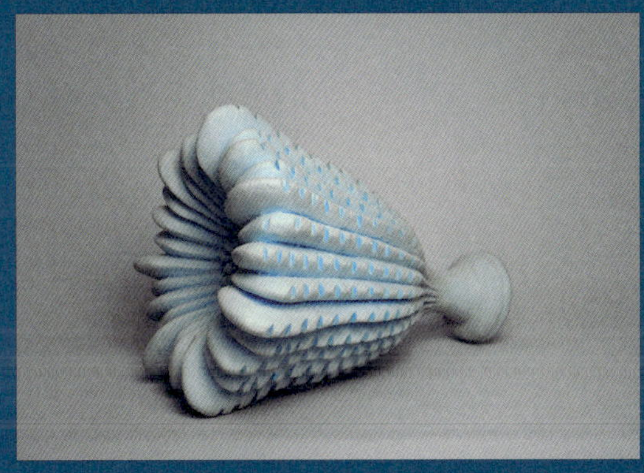
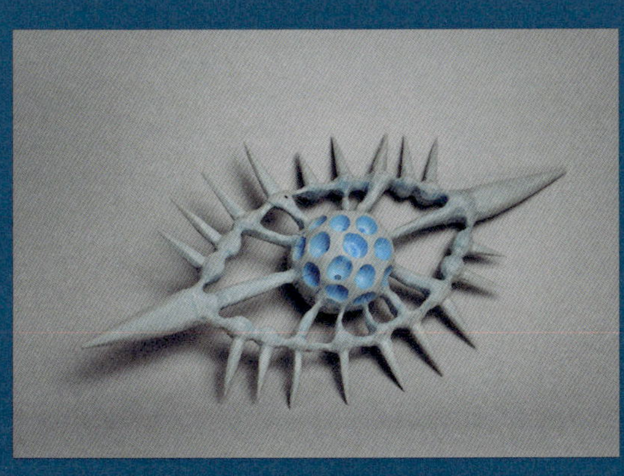
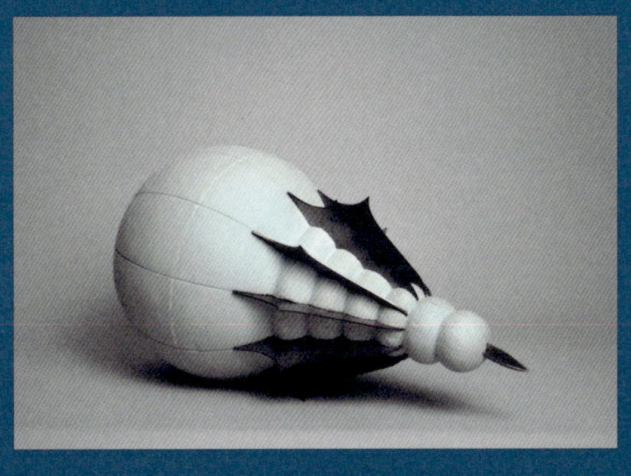

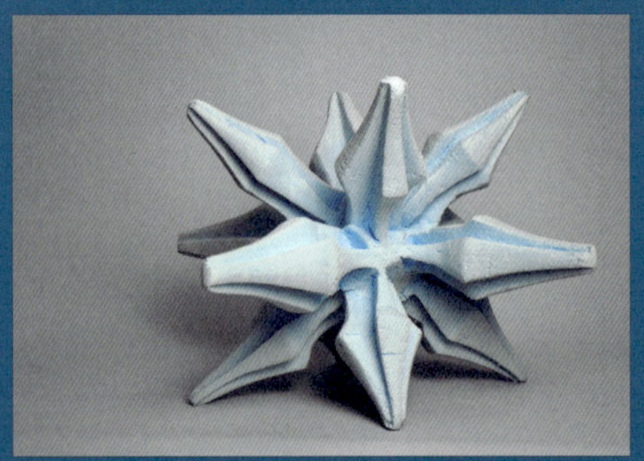
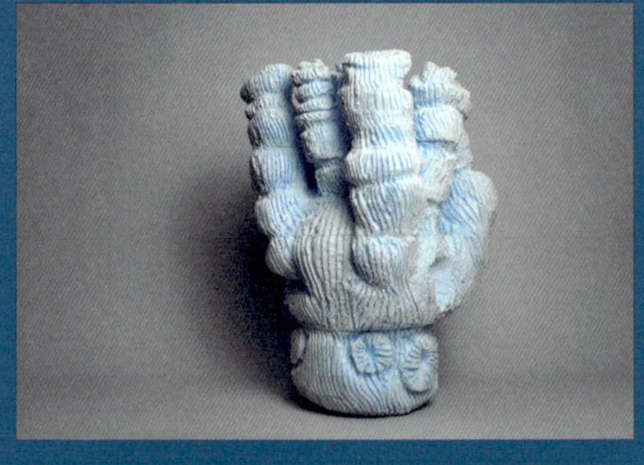
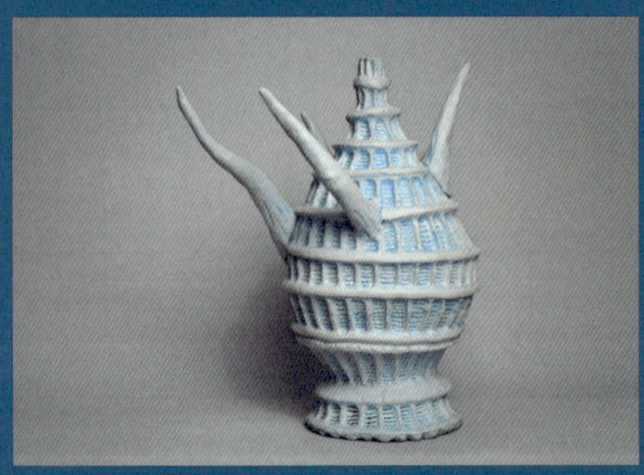
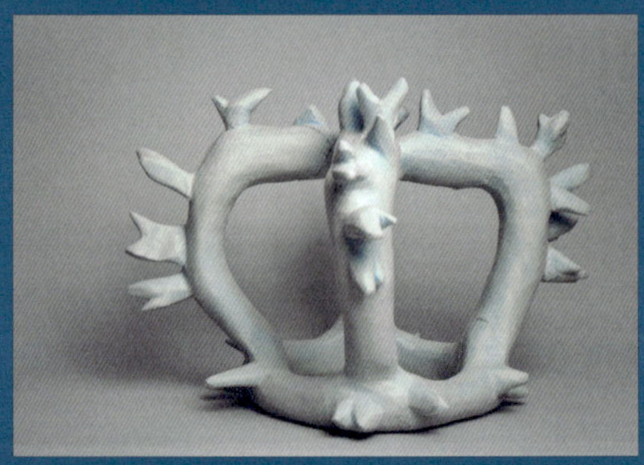
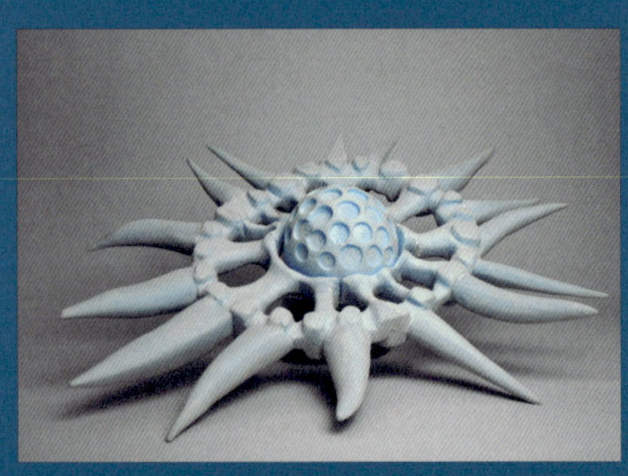
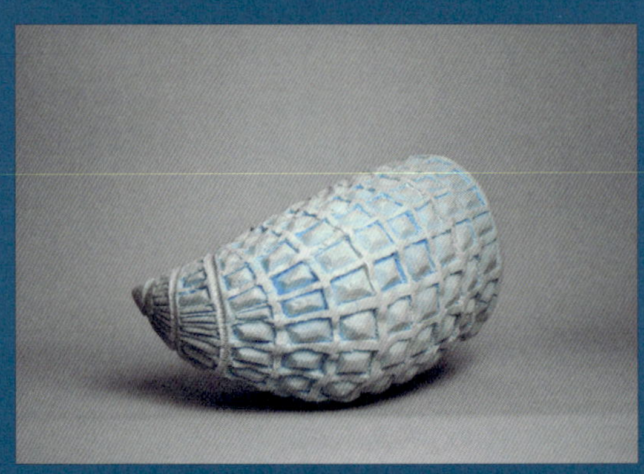

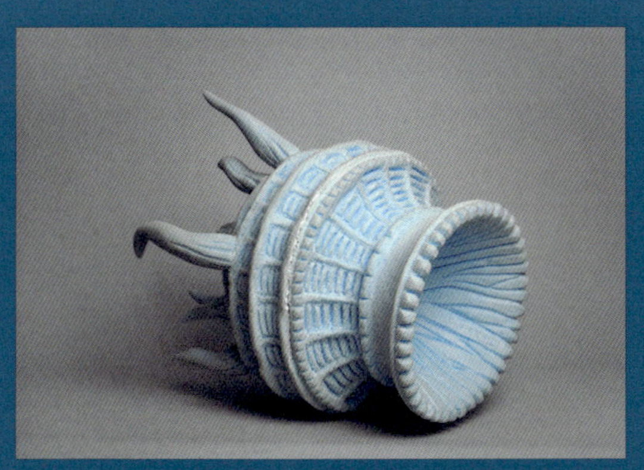
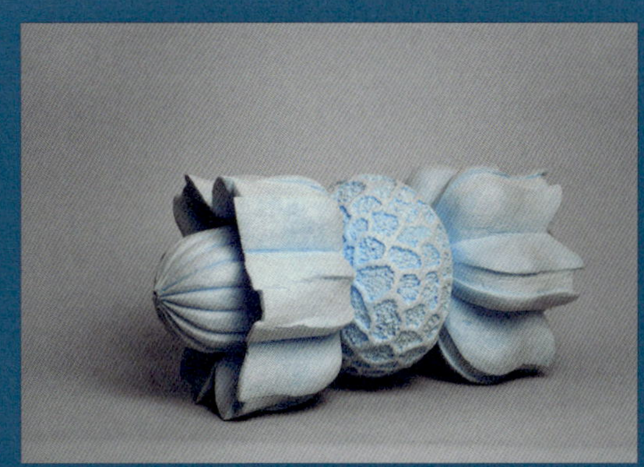
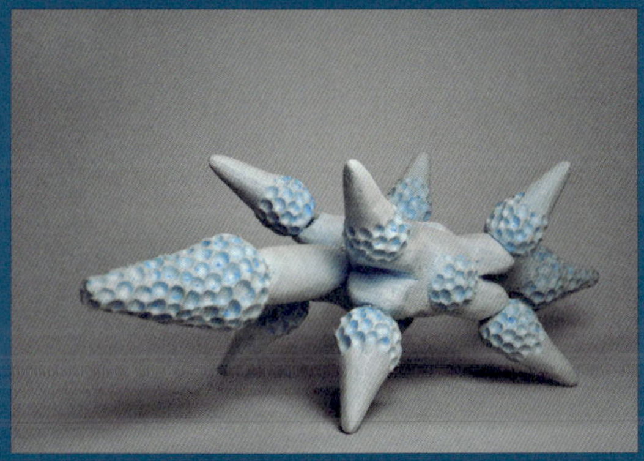
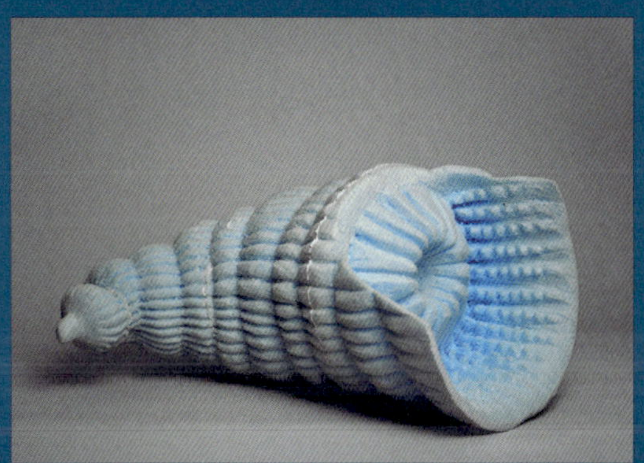
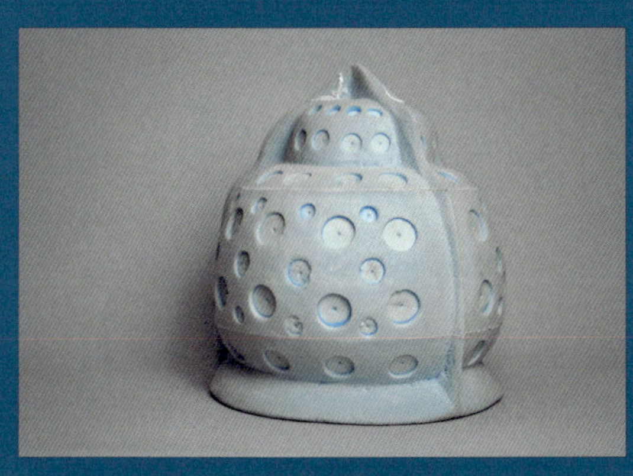
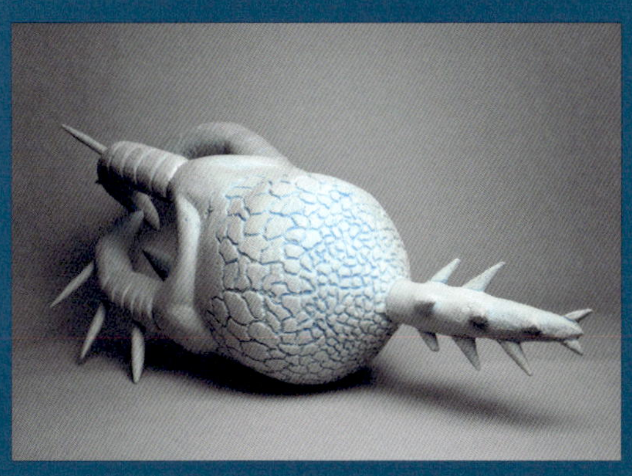

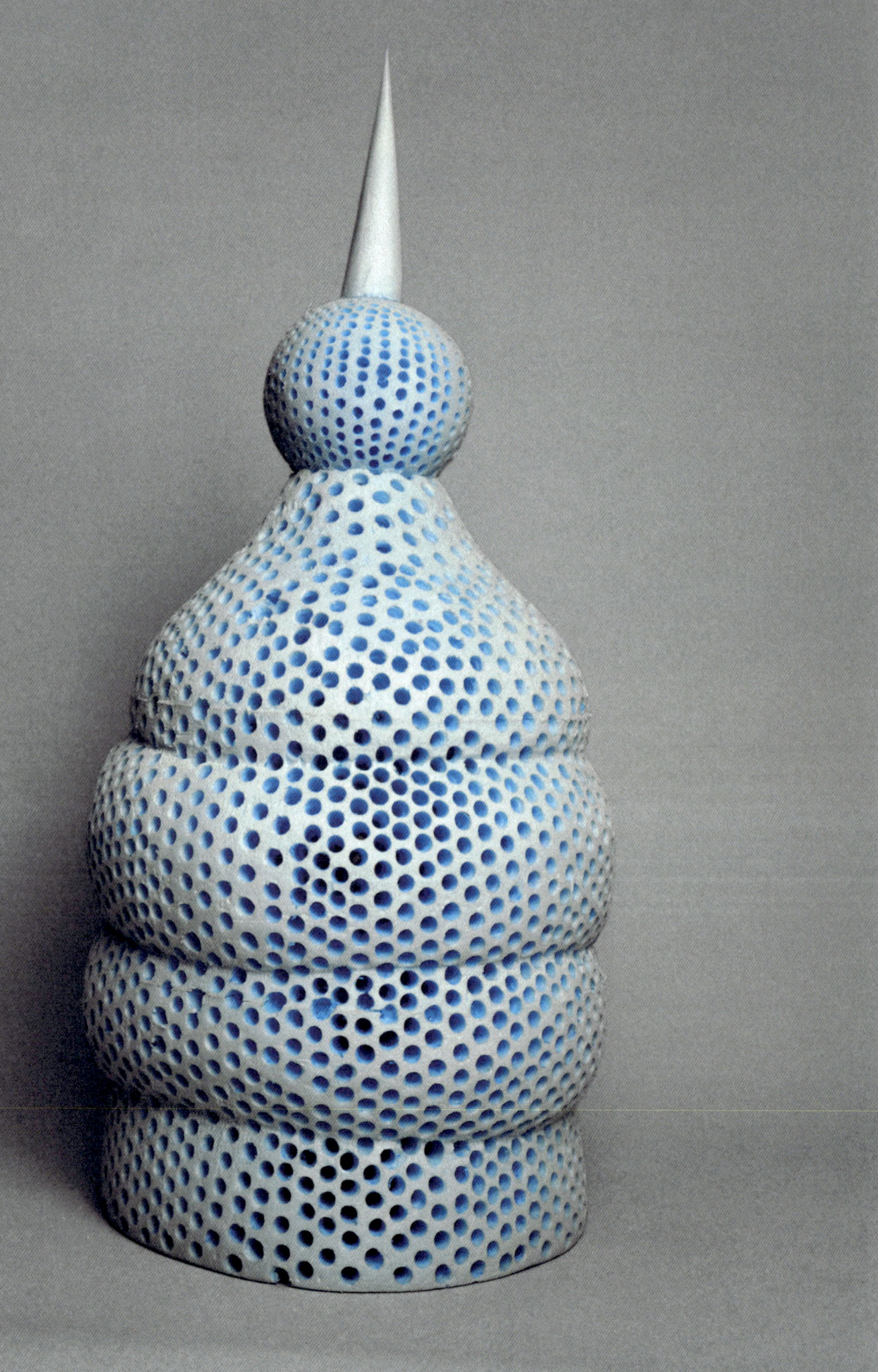

The »Radiolarienstudien« by Reiner Maria Matysik were developed in 2019 by the following students from Burg Giebichenstein University of Art and Design Halle: »Styrodur sculptures« by Robert Walther, Sang Hee Kang, Nanette Henschke, Johann Kogge, Bianca Poppe, Theresa Güldenberg, Zoe Haufler, Vincent Leicht, Anna Skuratovski, Luis Braun, Benjamin Neubauer, Jakob Trepel, Greta Ruppert, Georg Stahlbock, Isabell Bilfinger Unzueta, Franziska Hofmann, Vincent Aubin, Hannes Koch, Daniel Grahl, Paula Rieß, Jule Eretier, Katharina Michnik, Nadja Schulze, Felix Cordes, Louis Rohmer, Andreas Yakub Poda, Marie-Sophie Runge, Andreas Schwab, Lore Elstermann, Fanny Liebhardt, Paula Ködding, Lena Eichhorn, Leah Frey, Justus Büchner, Nadine Begenat, Lotte Büdel, Johanna Schmidtke, Miriam Hantzko, Salima Jalala, Josephine Kraus, Katja Undeutsch, Friedrich Wördehoff, Frieda-Marie Knödler, Isabelle Selwat, Gege Ji, Jisu Park, Lara Herrmann, Magdalena Meißner, Maximilian Pflugk, Leah Messerschmidt, Florentine Bahls, Isabell Bilfinger Unzueta, Chaewon Song, Alessa Scivoli, Sophie-Charlotte Bolinski, Sae Kaneko, Max Greiner, Luc Sohrmann, Lena Endner, Anton Grabolle, Victoria Woywodt, Lena Schliemann, Sarah Mende, Christiane Wöhrmann, Pia Eggers, Dayeon Auh, Michelle Cerny, Paula Holzhauser, Nadine Härting, Julia Müll, Ole Peters, Hyelim Kim, Marc Goldbach, Yakub Poda, Karl Schinkel, Amelie von Hausen, Sophia Reichmann, Hannah Köchy, Ina Gercke, Loris Stephan, Lena Hasler, Janine Harnisch, Jannis Scheerbarth, Hanna Mühlbach, Sculpture made of plaster, Kim Cordes, 2020.

225

FIG. 3
The scanning electron microscope image shows the diatom
SURIRELLA SPEC.
Photo: Friedel Hinz,
Alfred Wegener Institute, 2004

FIG. 4
The scanning electron microscope image shows the diatom
ARACHNOIDISCUS SPEC.
Photo: Lars Friedrichs,
Alfred Wegener Institute, 2008

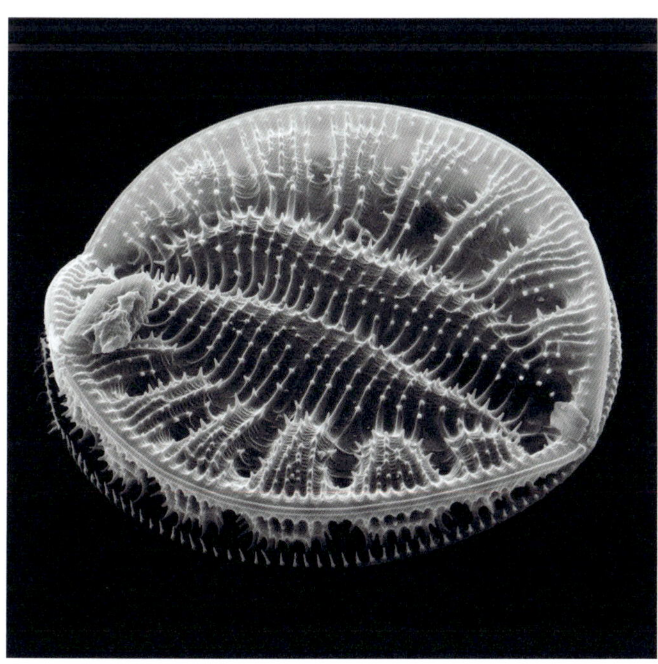

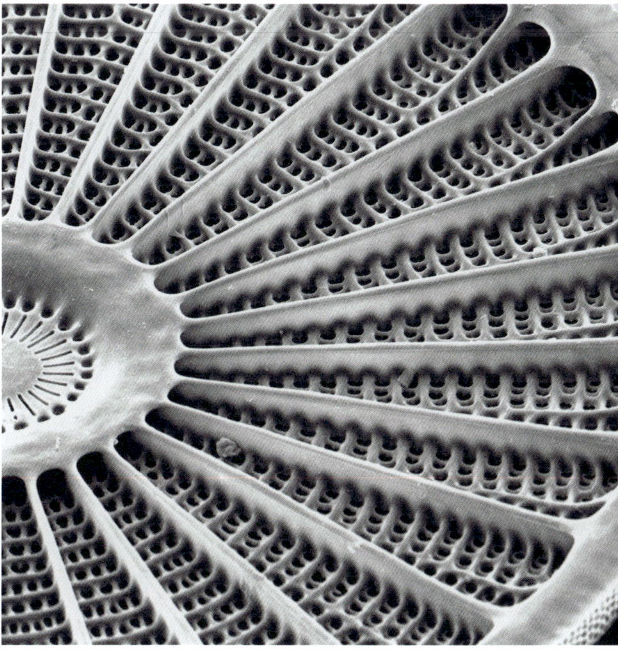

This article presents the philosophical discussion that is part of the art installation **INSTRUCTIONS TO BUILD A SPECIES** produced by the Mexican group Bios ex Machina. The artwork is the result of a transdisciplinary theoretical seminar where problems of ontology, biopolitics, taxonomy, epigenetics, and evolution were discussed. The artwork deconstructs the biological category of species through different classifications of fern: evolutive, medical, ornamental, transgenic, philosophical (from antiquity), and magical.

Meta Instructions to Instructions to Build a Species: Performing Philosophy through Arts

María Antonia González Valerio and Rosaura Martínez Ruiz

The artwork INSTRUCTIONS TO BUILD A SPECIES by Bios ex Machina aims to deconstruct the violence inherent in the epistemological practice of categorization. To our way of thinking, violently enforcing singularity limits our understanding of any given entity, which could otherwise be interpreted in accordance with different regimes and disciplines. Violence cancels out the discursive plurality whereby an entity emerges and is determined.

INSTRUCTIONS TO BUILD A SPECIES was exhibited from 2018 to 2019 at the Centro de Cultura Digital in Mexico City as part of N Festival's exhibition SPACES OF SPECIES. The installation encouraged spectators to walk through a montage of instruction manuals, all of which were replicated in small printed cards that could be taken home (FIGURES 1—4). Drawings were displayed alongside the sets of instructions (FIGURES 5—8), as well as some images from archives and laboratories of the National Autonomous University of Mexico, some green pigments used in ancient, modern and classical artworks, and plastic ferns in laboratory flasks.

INSTRUCTIONS TO BUILD A SPECIES deconstructs the biological category of species by displaying how different systems of classification account for ferns—organisms that have been widely employed in human practices and understood through multiple paradigms of thought.

Because biology's episteme configures a certain world order, its deconstruction unveils different interpretations of an organism as well as the knowledge and practices in which each is embedded. The category of species is of special interest because it gives way to the classification and differentiation of living entities. But this very category is neither fixed nor absolute; rather, it is constantly negotiated within biologies.

These sets of instructions question the order of things, of a world embroiled in taxonomies and of the codes that lay at the basis of certain epistemes and their hierarchies. Entities have been classified since the beginning of philosophy (as sensible/intelligible, natural/artificial, etc.) but these classifications have shifted through time and space.

Thus, these instructions raise a number of questions: What is the meaning of a classification of sensible entities according to different epistemes? What does it mean to order the world in accordance with

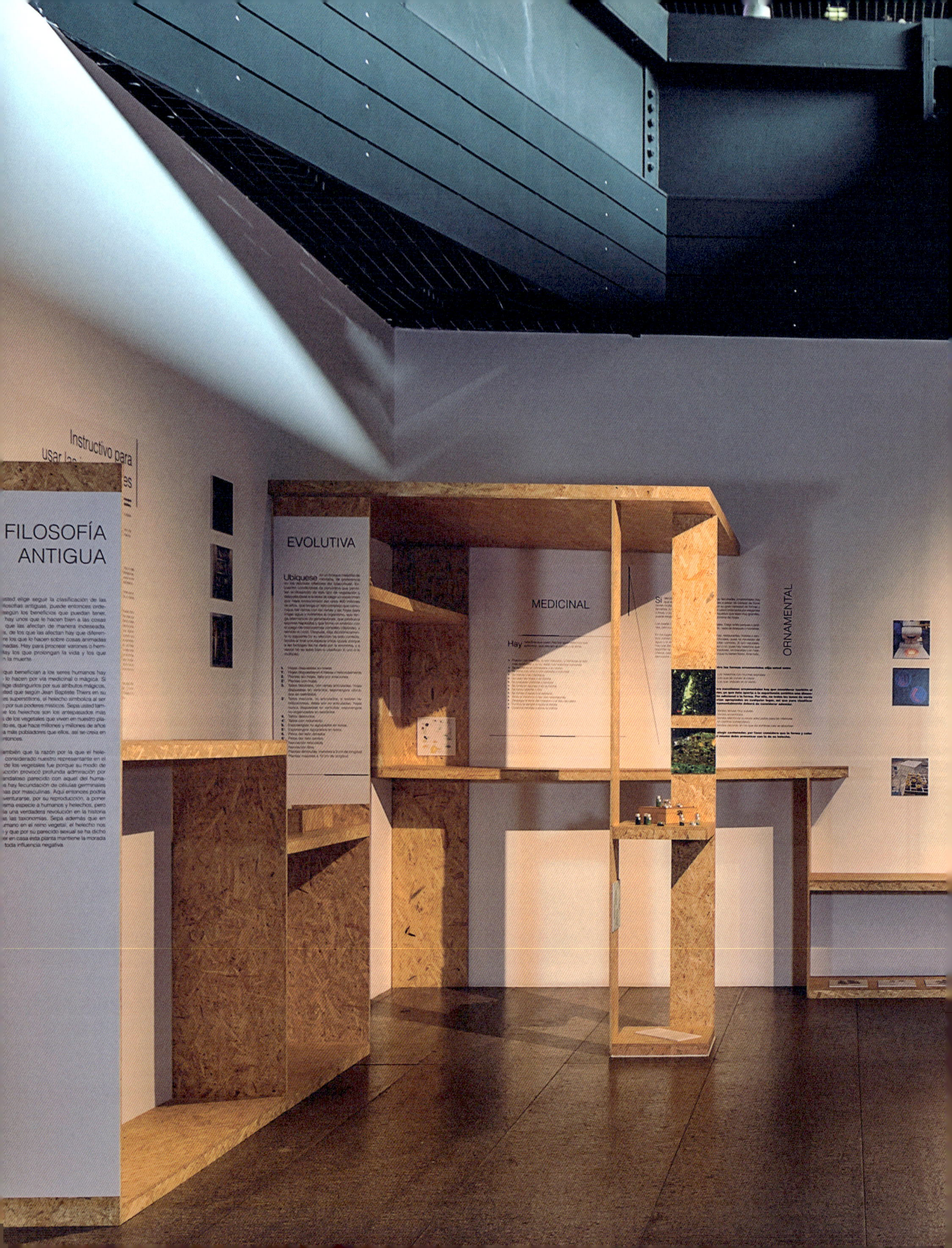

Instructivo para
usar las
FILOSOFÍA
ANTIGUA
EVOLUTIVA
Ubíquese
MEDICINAL
Hay
Si
ORNAMENTAL

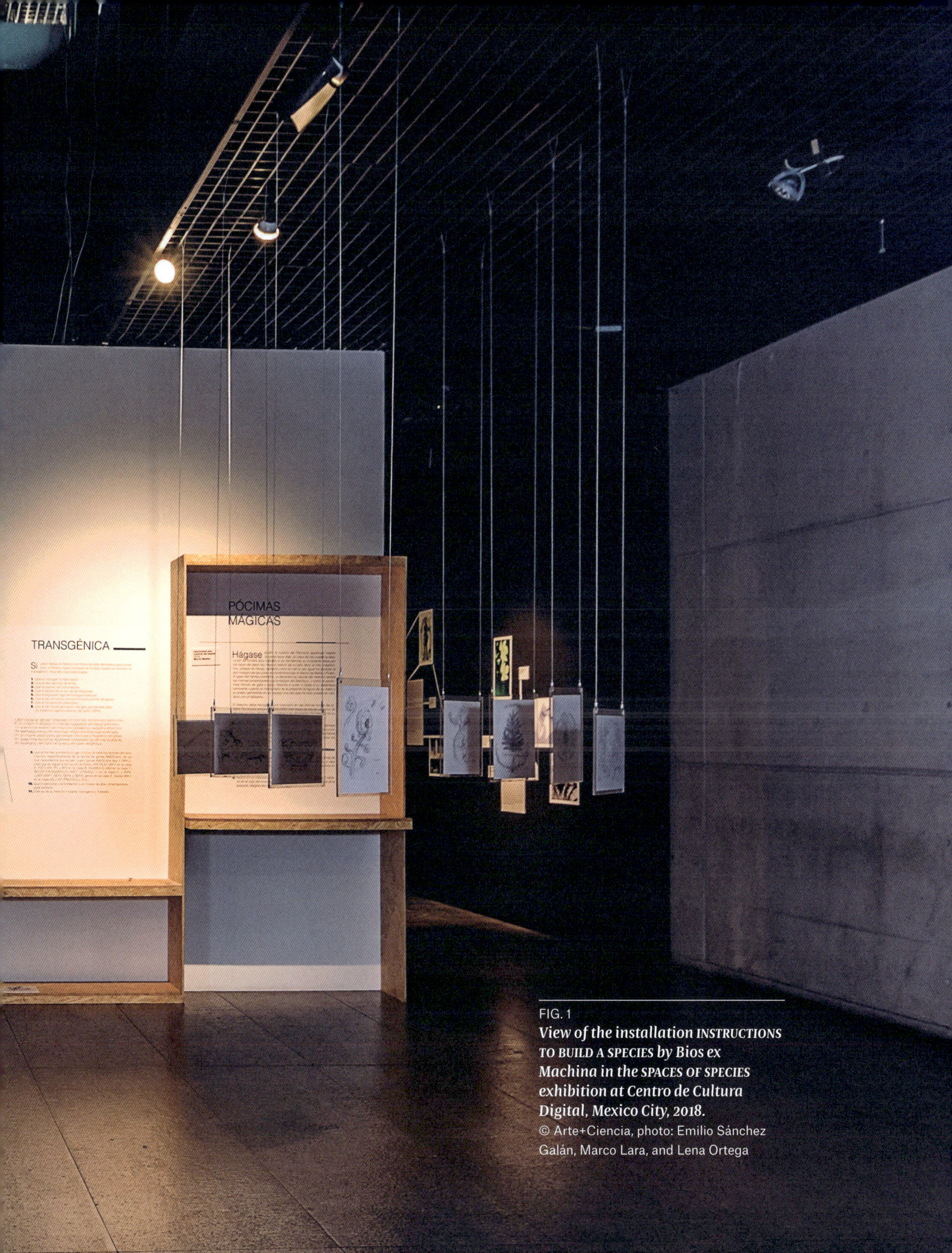

FIG. 1
View of the installation INSTRUCTIONS TO BUILD A SPECIES by Bios ex Machina in the SPACES OF SPECIES exhibition at Centro de Cultura Digital, Mexico City, 2018.
© Arte+Ciencia, photo: Emilio Sánchez Galán, Marco Lara, and Lena Ortega

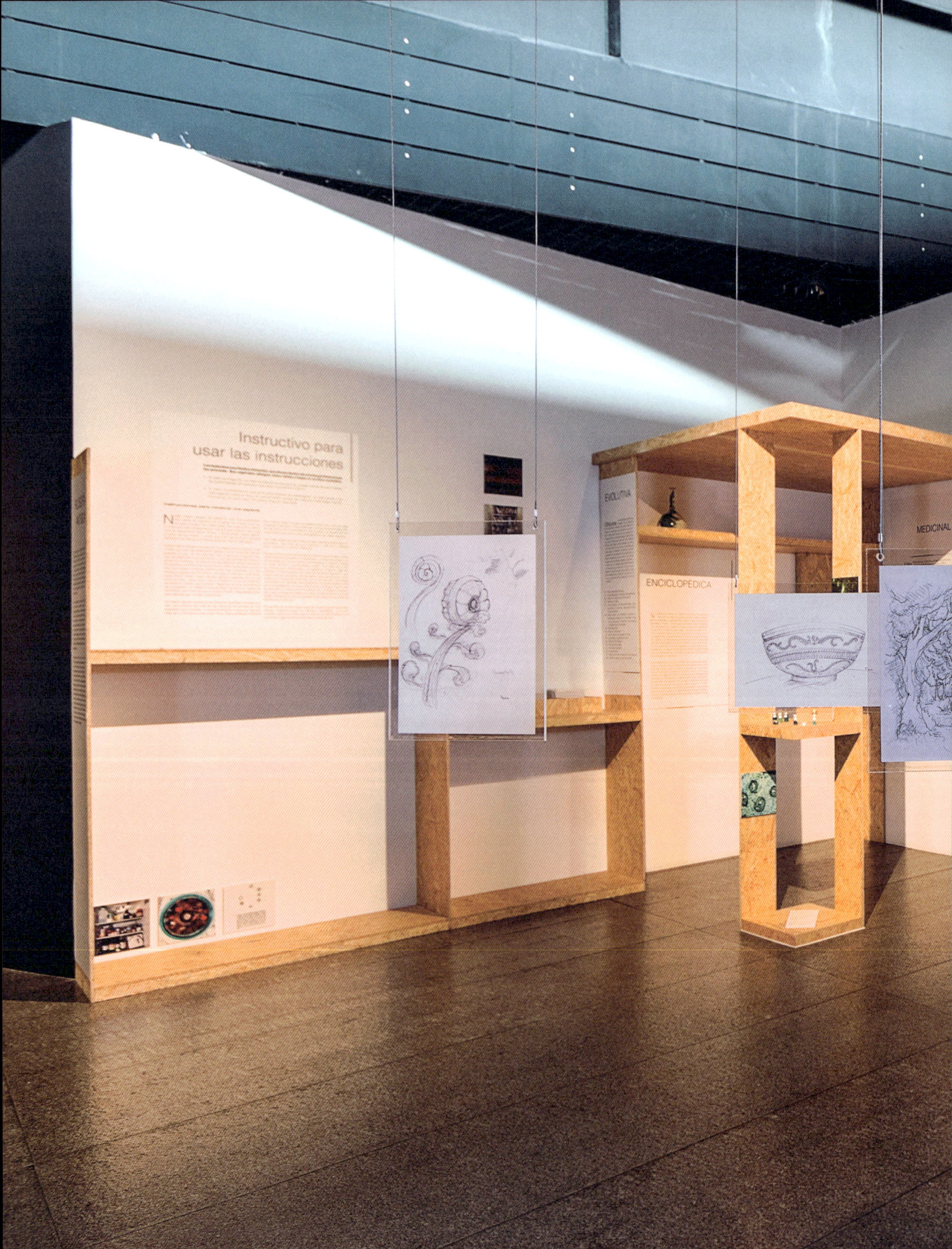

Instructivo para
usar las instrucciones
EVOLUTIVA
ENCICLOPÉDICA
MEDICINAL

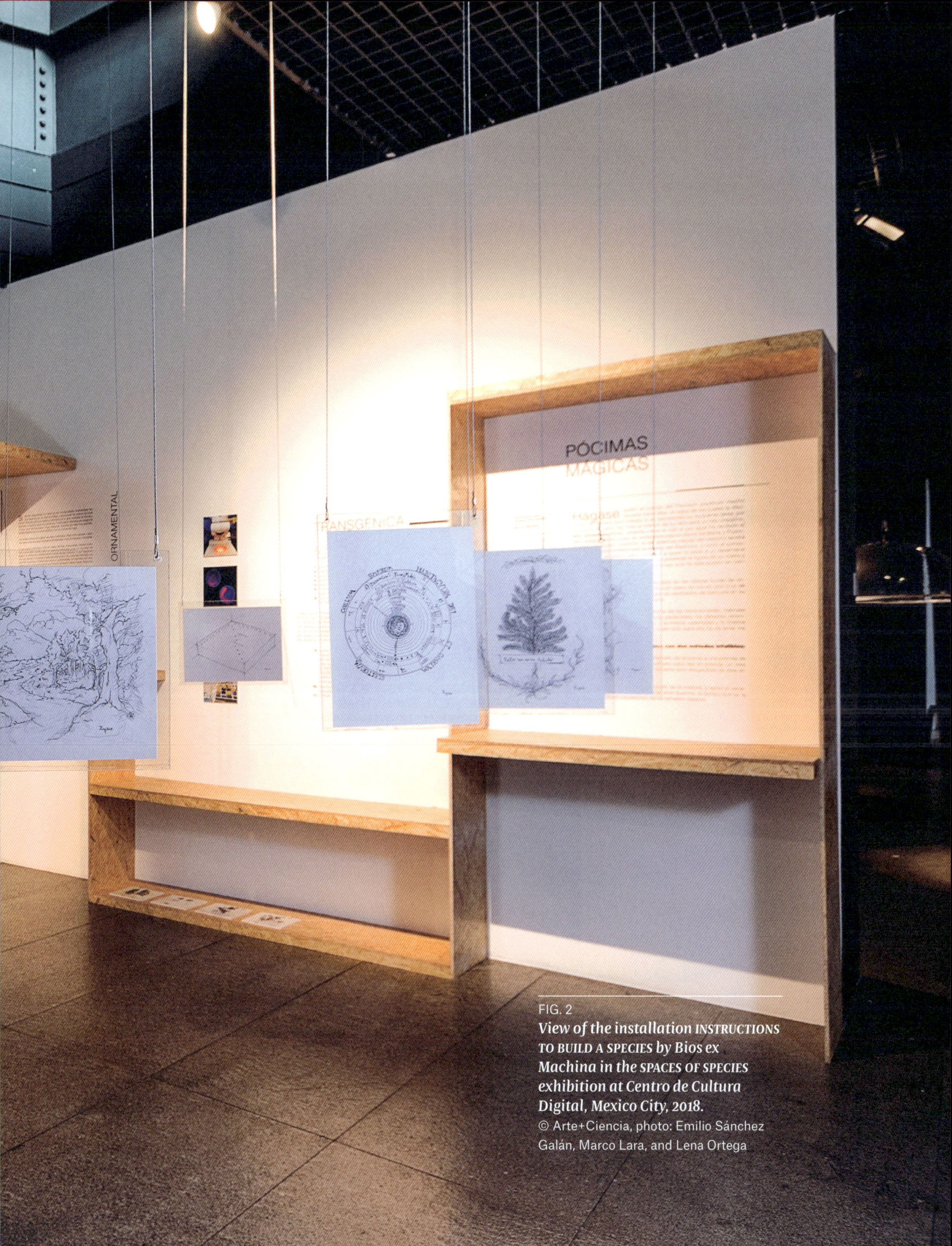

FIG. 2
View of the installation INSTRUCTIONS
TO BUILD A SPECIES *by Bios ex*
Machina in the SPACES OF SPECIES
exhibition at Centro de Cultura
Digital, Mexico City, 2018.
© Arte+Ciencia, photo: Emilio Sánchez
Galán, Marco Lara, and Lena Ortega

234

FIG. 3
View of the installation INSTRUCTIONS TO BUILD A SPECIES by Bios ex Machina in the SPACES OF SPECIES exhibition at Centro de Cultura Digital, Mexico City, 2018.
© Arte+Ciencia, photo: Emilio Sánchez Galán, Marco Lara, and Lena Ortega

cultural codes, and which ones ought to be followed? We wanted to stress that an organism is an archive in itself as, for example, when considered from an evolutionary perspective where phylogeny is embedded in each organism. The history of what was there before it and of what has happened to account for it is therefore entrenched within the organism itself. Can the organism be classified as a result of, and in accordance with, that history? Does this give the organism unity? Is the order of species the result of biological analysis? Or does the order of species—taxonomy— shed light on the tension that looms every time the order of things itself is at stake? What traditions should we look to when targeting an organism? How can theoretical frames and concepts provided by different epistemes be considered?

To tackle these matters, we draw inspiration from literature. Based on Julio Cortazar's tale »Instructions on How to Climb a Staircase,«[1] we have written six different »instructions« to build different species from ferns. The piece is also inspired by the short story »The Analytical Language of John Wilkins« by Jorge Luis Borges,[2] where the author develops a poetical classification of entities which is not based on any taxonomy or classical hierarchy, but rather on categories which, put together, are inconceivably linked. There is also an echo of the opening of Foucault's *The Order of Things*[3] where, quoting Borges, the philosopher addresses the problems of classification and order.

Furthermore, the notion of archive is central to the piece. We draw on what Derrida called »archive fever,« a notion that entails not just suffering from a malady but from a compulsion: to pursue the archive, to fervently seek out origins and ultimate commencements. Fever, explains Derrida, is understood as »burning with passion.«[4] In other words, archive fever is a drive to archive, a compulsion to produce categories, representations, names, to keep the living as living and the dead as dead—or perhaps even to produce such categories. Hence, the piece seeks to put forward the drive to archive as a compulsion to identify and classify organisms as simple, sound and indivisible unities, while also exposing the artificiality, arbitrariness and authoritarianism (or authoritativeness at best) of any given category.

Through the piece we address different archives of ferns. They include scientific ones, of course, but only alongside other archives that have not been as epistemologically privileged—in other words, that have been dismissed for not holding any truth whatever about ferns. To build a species of fern in accordance with botanical instructions, we worked with the idea of varieties of ferns that are »undecidables.« In Derridean terminology, undecidables are indeterminate concepts that point to the boundaries where a given classificatory order falls down.[5] That is, they mark the limit of ordering and disturb the logic of binary oppositions. Depending on the archon,[6] some varieties of ferns have either been classified in more

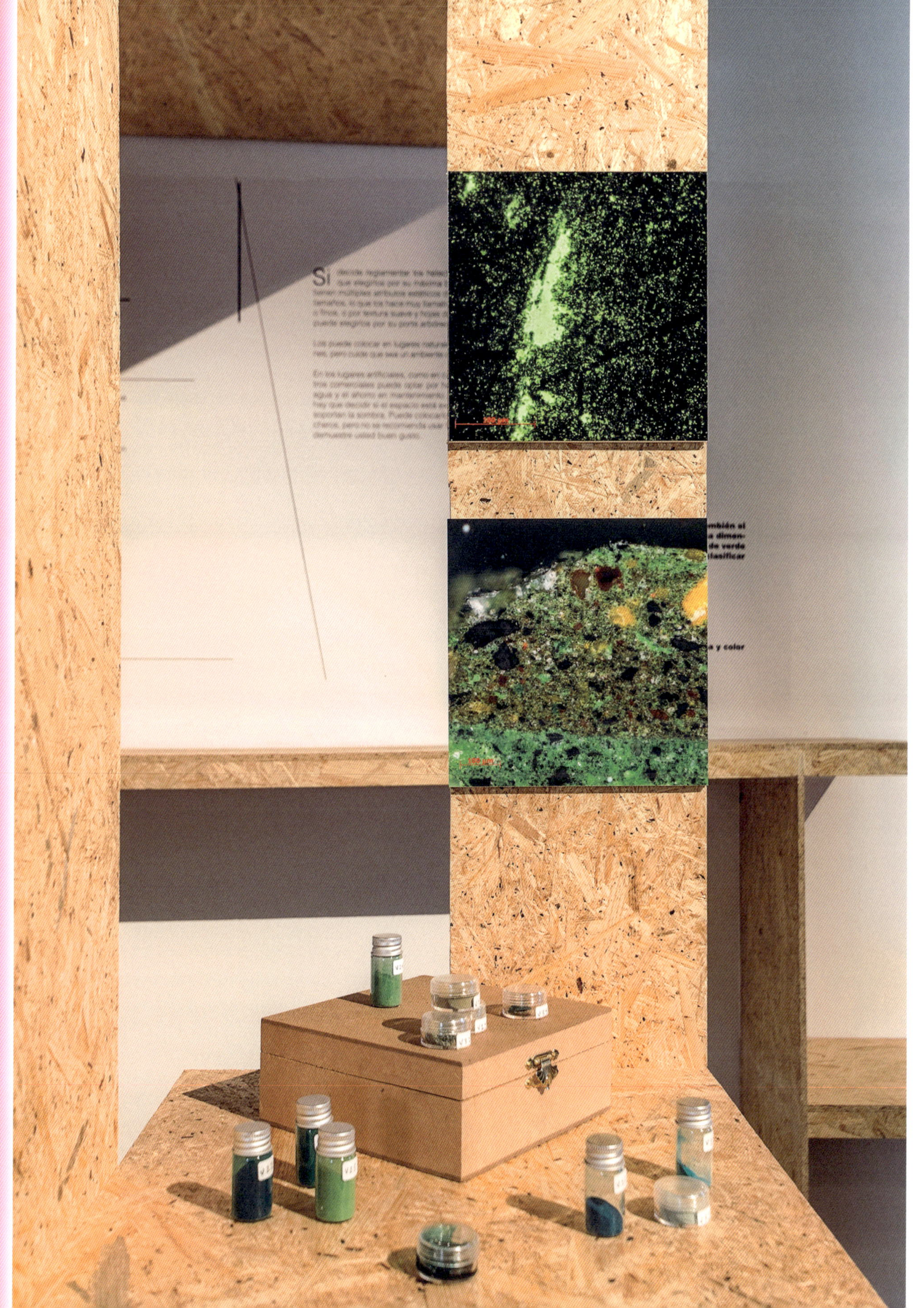

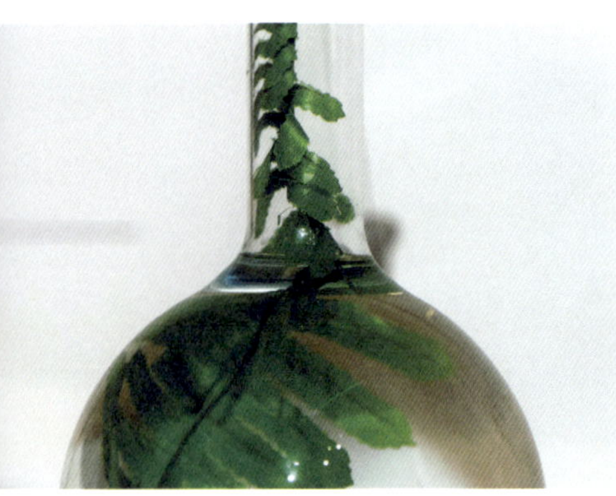

FIG. 4
View of the installation INSTRUCTIONS TO BUILD A SPECIES *by Bios ex Machina in the* SPACES OF SPECIES *exhibition at Centro de Cultura Digital, Mexico City, 2018.*
© Arte+Ciencia, photo: Emilio Sánchez Galán, Marco Lara, and Lena Ortega

than one subcategory or have been named more than once.[7] Thus, the taxonomy displays synonymy—which, by the way, is how botany refers to different categories that name one and the same organism. Thus, in a deconstructive gesture, these instructions have not been designed as an effort to go beyond undecidability, but to render it visible.

Through INSTRUCTIONS TO BUILD A SPECIES we question some ideas about unity and multiplicity, since some plant entities are not clearly identifiable as unitary—such is the case with grasses, for instance. Any notion of identity is the result of historical processes, of a momentary synthesis produced not by a single definition and not by one archive with one archon, but rather by something that is being constantly decided—and, one should add, not by one isolated and sovereign subject. Any category or any form of the singular has *become* so, and this process of becoming has been enforced through some form of violence and some force of authoritarianism. There is always an archon, even where a group of archons authorizes the category or, in other words, answers for the archive. As Borges observes, »it is clear that there is no classification of the Universe that is not arbitrary and full of conjectures.«[8]

Are biological entities and taxonomies—with their imprecisions, synonymies, and undecidables—the token of an episteme that operates through an allegedly fixed identification, which works not even on organisms but on fractions of organisms that want to be taken as the explanation for life? That is, can we deconstruct the hegemonic biological episteme as knowledge that—following an analysis of features initially understood through phenotypes, which are now being explored through the sequencing of DNA—orders and classifies what is? Molecular biology is, in a way, an episteme that does not look at the entity in its appearance but rather measures its molecules.[9] Is molecular biology called into question by taxonomy given that the latter always needs a reference to appearance through phenotypes, sizes, forms, colors, and so on?

The empirical does not appear by itself, but emerges within a given order. The ancient idea of the natural order has been obliterated by the explanation of functions in organisms and by the classification of molecules. Although contemporary biology works pretty much within the genetic paradigm, it cannot exclude the problem of classification (to discover order in nature). Because even if it analyzes small pieces of reality—a certain path of a gene and its sequence, for instance—it has to classify organisms, and the limit of classification is *species*. This category has been questioned many times in contemporary debates because of how difficult it proves to insert any organism squarely within it.[10]

Broader categories—such as kingdom, phylum or family—pose no problem in this respect. Why is the singular, that which can identify something as different from another, precisely what resists fixed order and necessary classification? Why is *this* the realm that opens up possibilities, negotiations and undecidability? The notion of species along with its critiques has shown us that the space for classification, hierarchization and identification cannot be subsumed within one code or one tradition. The organism resists.[11] And it invokes environment, circumstance, histories, and pluralities.

We attempted to rebuild and deconstruct the story of a single organism, the fern, and thus open a horizon for some ethical and political precautions that could intervene in the construction of more inclusive and democratic archives, be they botanical, decorative or philosophical, as the instructions themselves show. While there is no single archive that holds an all-encompassing, ultimate truth, we archive and must archive. Stating the impossibility of the archive is not meant as a precaution against archiving in the sense of holding back from classification and naming, but rather a radicalization of archiving. In other words, archiving is a radical political and ethical task. Therefore, the question of species opens up as a political matter—a matter of responsibility. This means that we must archive collectively, as in (Gr.) *polemos*, and with precautions and care for that which has been left outside the category, outside the archive, even if this be a wholly different archive in itself. The excluded must haunt us.

Where the archive is impossible, we must archive; but we must do so while displaying the task as inherently unfinished, open to changes to come.

A User's Guide to the Instructions by Bios ex Machina   If you want to build a species of fern, you need only to follow these instructions closely.

Ferns are living fossils that embody the very history that precedes them; their spirals are an evolutionary archive that wraps in on itself.

If you are wondering *why* a fern is what it is, perhaps you might consider framing the question within the possibilities of variation offered by these organisms themselves.

To follow the instructions accurately you do not really need an in-depth understanding of your own drive to classify, to speciate or to relentlessly question life and the knowledge that pivots on it— but bear in mind that no answers on these matters are provided here.

Instructions to Build a Species ＿＿ It cannot have gone unnoticed that we are often overcome with an urge to organize the things in this world. Out of all of them, some die while some merely vanish. Out of the ones that do die, there are those that move up and down, into and out of the seas, and those that reach upward towards the sun (a few of which even do so underwater).

You may also have noted that, as a consequence of this impulse to order and arrange, the things that reach towards the sun are often called *plants*, and that out of all the plants, there are several ancient ones which

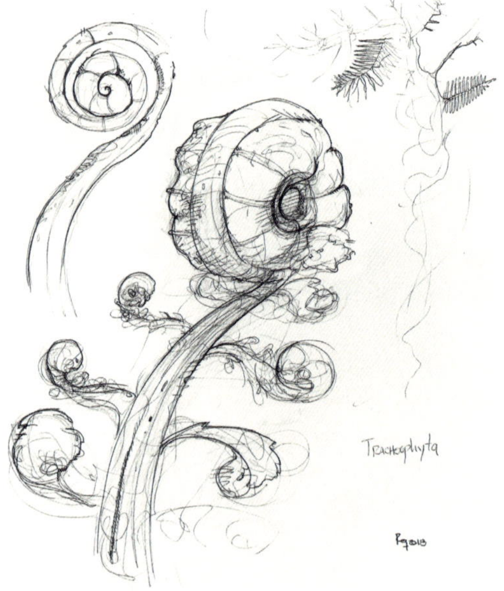

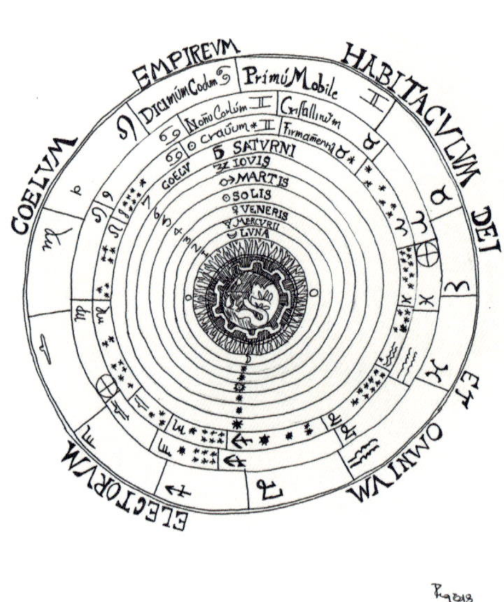

FIG. 5—8
Rodrigo Ramírez working with Bios ex Machina. Drawings for the installation INSTRUCTIONS TO BUILD A SPECIES *in the* SPACES OF SPECIES *exhibition at Centro de Cultura Digital, Mexico City, 2018.*
© Rodrigo Ramírez

History has chosen to call *ferns*, since they can be distinguished from all others by a significant number of visible and invisible characteristic traits. Their scientific name is usually more complicated because it relies on innumerable distinctions organized in a specific set of categories that begins with kingdoms and usually, but not always, follows this path: phylum, class, order, family, genus, species. At a certain stage in this itinerary, a fern becomes a fern. (Albeit one must not forget that even having reached the taxonomic rank of species, a fern may cease to be so at any moment—remember what happened to Pluto, recently demoted to a celestial orb of the superlunary world.)

You have probably also realized that this human drive to order and arrange has more often than not been paired with an indistinguishable impulse to *name* and even, surprisingly, to give more than one name to each specific thing. Multiple words that refer to singular entities are called *synonyms*. Please take heed of the fact that not even scientists—with their admirable vested interest in making knowledge a matter of universal value—have managed to resolve on a singular word for each singular fern. There are more words than things and more classification ranks than there are organisms.

There are many different traditions, and there are several different intentions behind the naming and sorting of ferns.

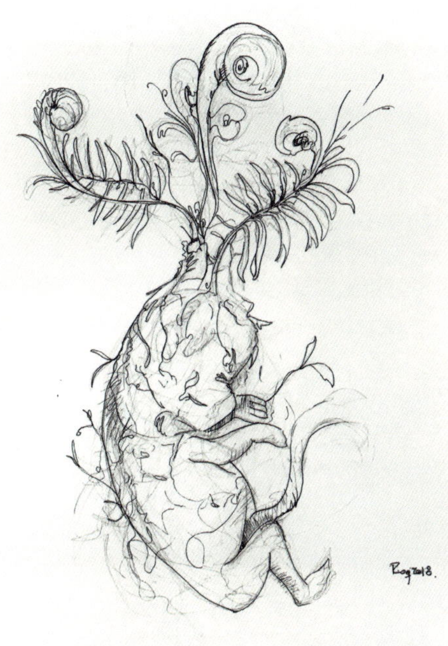

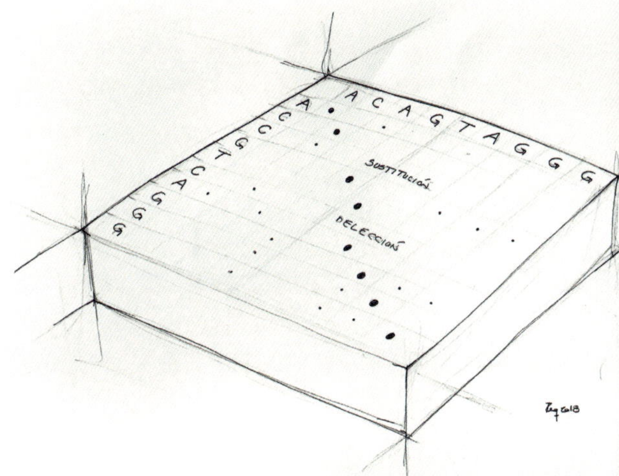

However, one could also leave words aside and focus instead on *visual forms* since, as is fairly clear, sometimes things which are similar in morphological structure, design and even color tend to be grouped together, in which case their names cease to be as binding as, perhaps, their figures or shapes. In the past, some artists and art historians have suggested there are some elemental natural forms to which human plastic creations all adjust: undulating, zig-zagged, parallel and spiraling lines are amongst the most elegant items of this repertoire. Telling forms apart is key to the human endeavor to sort things because forms—along with the names of forms—compose some of the many structures that attempt to explain how the world is laid out.

What follows are some sets of instructions to differentiate and sort ferns out, each befitting specific interests and necessities. But, you may wonder, what if there is something missing from the lists? Well, then you may either resort to a different archive, or you may very well make up a new one.

Evolutionary

Find a mountain mesophyll forest, preferably one on the lower slopes of the Iztaccíhuatl mountain. Then, find the shady habitat that most favors the development of this kind of vegetation, and now commit every effort to finding a specific specimen of fern: one that has evolved for 400 million years, with a stem that connects the roots to all branches and leaves to ensure the irrigation of water and nutrients throughout the organism; one displaying alternation of generations, that is, the production of haploid spores that grow into a gametophyte, and then of gametes that form an embryo with a full set of chromosomes (diploid), restarting the cycle. Later, pick one element from each of the following dichotomies (given the correct choices, you might perhaps be left with a unique species—but there is no guarantee, since biologists have a penchant for synonymity and are sometimes unsure of whether they are classifying singular entities or multiplicities):

1. Leaves arranged in a rosette
1'. Leaves arranged in four rows or in a helicoid
2. Plants with no leaves; stem with enations
2'. Plants with leaves
3. Tubular stems with articulated branches; leaves arranged in a whorl; sporangia-bearing strobili
3'. Woody nonarticulated stems; branches, if any, are not articulated; leaves never arranged in a whorl; sporangia not arranged in strobili
4. Naked stems
4'. Stems with indument
5. Sporangia not clustered in sori
5'. Sporangia clustered in sori
6. Golden trichome
6'. Brown trichome
7. Reticulate venation
7'. Open-system venation
8. Minute plant, smaller that 3 cm in length
8'. Larger plants, 10 cm in length

Encyclopedic

To make a fern (that is, a vascular plant with alternation of generations—diploid sporophytic and haploid gametophytic alternating phases—which produces neither seeds, flowers nor fruits; with megaphyllic leaves and circinate vernation; with sporangia grouped in sori on the underside of leaves), we must take a plant that shares its ancestry, such as Selaginella, the fern's closest living relative. We must then incorporate some of the characteristics of ferns, which Selaginella currently lacks, such as rhizomes, megaphylls, independent gametophytes (because Selaginella gametophytes depend on fungi for nutrition), a ramified vascular system, and nonlateral sporangia.

Then, we must produce hundreds of genetic mutations and gradually select the specimens that show, however incipiently, the characteristics that make up a fern. Put in another way, by using physical or chemical agents to incite mutations on hundreds of Selaginella spores and letting them reproduce, one may select the plants that show a modified vascular system, modified roots (to form rhizomes), and a gametophyte able to survive on its own without the intervention of fungi, etc. And, thus, after many generations, we would get a fern.

Medicinal

___ Some ferns heal, but some bring about disease of the body or the soul.
___ A fern may be male if it is sturdy, or female if more delicate, but a fern could also merely be a common fern.
___ It may be collected in the spring or in the fall.
___ It may be drunk with wine, barley or honey.
___ It may be boiled or mashed.
___ It may be picked for its leaves or for its rhizome.
___ It can be taken in pill or tonic form.
___ It can be taken on an empty stomach or right before bedtime.
___ It can be taken hot or cold.
___ One can take some fern extract or drink three cups.
___ It cures insomnia or conditions of the respiratory tract.
___ It removes a tapeworm from the intestines or a fetus from the uterus.
___ It purifies blood or lowers stress.
___ It helps with wounds or relieves sciatica.

Ornamental

If you choose to sort out ferns on the basis of their ornamental faculties then you must select them carefully, first and foremost, for their beauty. This is not a difficult task because in spite of their lack of flowers, the astounding variety of shapes available amongst ferns makes them particularly eye-catching and endows them with a wide range of aesthetic attributes.

Thus, you may choose to sort them out on the basis of their elegance or delicacy, or perhaps according to the suppleness of their texture or the slenderness of their leaves. The larger varieties can be selected for their tree-like poise or for the lusciousness of their crown of fronds.

You can place them in either natural or artificial settings. You may choose a garden or even a park, but make sure you pick a domesticated environment—forests or mountains will not do.

With regard to artificial spaces, such as houses, office buildings, restaurants, hotels, or shopping malls, you may opt for artificial ferns, which do not need any water and are comfortably low-maintenance. If you picked a natural fern, you must pay heed to whether the space you put it in has any sunlight, as some ferns cannot withstand shade.

You may grow ferns in containers, pots and planters, but flowerpots are best avoided namely because ferns do not have flowers. Above all, you must display good taste. From the ornamental fern shapes available, you may choose amongst the following:

a Those with many spirals.

b Those that curl upwards as they grow.

c Those that sway with the wind.

In ornamental matters it is always advisable to consider *color*, which is a crucial dimension of the aesthetic experience and is altogether different from the business of shape. This means that not all shades of green will suit any fern anywhere, so *do* make sure you consider your options:

a Light, very mellow greens.

b Bright greens.

c Neon greens (often appropriate for shopping mall interiors).

d Deep, dark, shade-absorbing greens.
 When you choose a container, make sure that its shape and color thoroughly suit your fern.

Transgenic

If you want a fern with flowers (perhaps you wish to place it in a flowerpot, or showcase it in a museum as part of an exhibition on scientific knowledge and species mutation), then you could make a transgenic one. To do so, follow these steps:

1. Get a lab.
2. Get permission to use the lab.
3. Get access to the equipment.
4. Get someone to help you use the machines and instruments.
5. Get someone to explain some basic molecular biology.
6. Get taught at least one gene-editing technique.
7. Get hold of some plasmids.
8. Get a gene sequence, for instance, this one that belongs to *Adiantum capillus-veneris*, taxon identifier 13818:

   1 atgccgacaa accaacaact tattaggaag gcgaggcaac ggctggagag tgggacgaaa

   61 tcccccgctc ttcgtgggtg cccccaacgg cggggagtgt gtactagggt gtatctcaca

   121 actacaccga aaaaacccaa ctctgcccta cgtaaagtcg ccagggttcg attaacctcc

   181 aaatttgagg taacggctta cataccaggt attggccaca atttgcagga acattcggtg

   241 gtcctcgtaa gaggcggaag ggtcaaagac ctacccggcg ttaggtatca catcgttaga

   301 ggagccttag atgctgtagg agtgaaagat cgtaaaaaag ggcgttccag tgcgttgcag

   361 agcattgtcg caactcgtat catcgcaacg attccgtatc aattgttctg a

9. Get your gene synthesized by introducing transcription factors into it, specifically those of the gene family MADS-box, since we will need the incidence of four MADS-box type II genes (MIKC)

in order to regulate the formation of flowers: PETALA1 (AP1) for class A, PISTILATA (PI) and AP3 for class B, AGAMOUS (AG) for class C, SEEDSTICK/AGAMOUS-LIKE11 (STK/AGL11) for class D, and SEPALLATA (SEP1, SEP2, SEP3, and SEP4) genes for class E. MIKC genes for class AG and AP1/FRUITFULL (FUL).

10. Enjoy your mutant, transgenic, flowering fern.

243

Ancient philosophy

If you chose to abide by the classifying systems of ancient philosophers, you can sort ferns according to the beneficial properties they may or may not have, because some do good whereas some bring about unwanted effects. Furthermore, from those pernicious ferns, we must sort the ones that affect animate beings from those that act upon inanimate beings.

Some can engender males or females.

Some prolong life and some cause death.

Out of the ones that bring benefits to humans, some do so through medicine and some through magic. If you choose to sort ferns according to their magical properties, be aware that Jean-Baptiste Thiers (1636–1703) states, in his *Traité des superstitions*, that ferns symbolize humans because of their mystical powers. Bear in mind that ferns are the most remote relatives of all vegetable forms now inhabiting our planet, which means that millions and millions of years ago, there were no other Earth dwellers but them. Take heed, too, of the reason the fern became our representative in the vegetable kingdom: their reproductive process aroused the deepest awe due to its scandalous similarities with human reproduction, since female germ cells are fertilized by males. Here you might even venture to classify humans and ferns as one and the same species, but that would be nothing short of a cataclysm in the history of taxonomy. Note that, as befits its role as the human of the vegetable kingdom, the fern protects us, and because of the sexual resemblance mentioned above, this plant guards homes from any and all negative influences.

Magic Potions

Leave the body of a male *Pteridum aquilinum* to rot for thirteen days; after this time, the dilution will turn cobalt blue if blessed in its incipient task by the Eye of Heaven. But pay heed, for the inner part of the stem will have turned into a clear, fine, pleated thread of ruby-like radiance if, in addition, it has received malignant gifts from the November Moon, under the influence of Mars and Pluto. In the course of this time period, the fern's natural decay will imbue it with much the same properties of Basilisk blood and, as is the case with the venomous hybrid of rooster and toad, its juices will bring instant death to any unfortunate soul who, catching a last glimpse of his face rising on the mirrored surface of the balm, drinks it.

Ferns must be picked during the last October rains. It is crucial to impale
the shrub with a gold nail on a cross at a forest clearing and imperative to
soak it with mercury distillments when mounting it, and with fresh honey
when removing it.

During the course of the first and last leg of the thirteen-day
period, it is necessary to weep on the mixture while mourning any
sorrows suffered, any slights borne, any wretched loves and any disillusion,
or else one could mourn the very calamity and infamy wished upon
the person who is bound to meet their fateful sleep that knows no end.

To apply it, there are two infallible methods.

1. Thread the plant's delicate, serpentine, ruby stem on a token:
   a pair of gloves, perhaps, or a handkerchief that will soon serve
   as a shroud, or maybe a ribbon to be bestowed upon someone
   at a bittersweet farewell of affected embraces.
2. Gather together the liquid elixir of the mixture and pour the
   venom into an enemy's ear followed by a stream of swift,
   gleeful whispers that impart the reasons for his impending fate.

1   Julio Cortázar, »Instructions on How to Climb a Staircase« in *Cronopios and Famas,* trans. Paul Blackburn, New York: New Directions (1999): 21–22.

2   Jorge Luis Borges, »The Analytical Language of John Wilkins« in *Other Inquisitions, 1937–1952*, trans. Ruth L.C. Simms, Austin: University of Texas Press (1964).

3   Michel Foucault, *The Order of Things: An Archaeology of the Human Sciences*, New York: Pantheon Books (1970).

4   Jacques Derrida, *Archive Fever: A Freudian Impression*, trans. Eric Prenowitz, Chicago: The University of Chicago Press (1998), 91.

5   See Jacques Derrida, *Limited Inc*, trans. Samuel Weber, Evanston: Northwestern University Press (1988), 148.

6   For a more profound understanding of how Derrida interprets the figure of the archon, see Jacques Derrida, *Archive Fever* (1998).

7   The history of fern classification is long and old, starting with Linnaeus (1753) who recognized a mere 15 fern genera in contrast to the 18 lycophyte and 319 fern genera in 51 families, 14 orders and two classes according to current taxonomy; see »PPG I 2016. A community-derived classification for extant lycophytes and ferns,« *Journal of Systematics and Evolution* 54 (2016): 563–603. Online: https://doi.org/10.1111/ jse.12229. However, this is a matter of considerable debate: Maarten J.M. Christenhusz and Mark W. Chase, »PPG Recognises Too Many Fern Genera,« *Taxon* 67, 3 (2018): 481–87. Online: https://doi.org/10.12705/673.2.

8   Jorge Luis Borges, *Other Inquisitions* (1964). Online: www.entish.org/ essays/Wilkins.html (last accessed September 14, 2020).

9   To a certain extent, molecular biology is the hegemonic episteme for the life sciences, even though in the past years, evolutionary developmental biology has gained more and more importance, along with the discussions about epigenetics; see Maurizio Meloni, *Political Biology: Science and Social Values in Human Heredity from Eugenics to Epigenetics*, London: Palgrave Macmillan (2016).

10   Darwin said, »I look at the term ›species‹ as one arbitrarily given, for the sake of convenience, to a set of individuals closely resembling each other,« Charles Darwin, *The Origin of Species* (1859). Online: http://darwin-online.org.uk/ Variorum/1866/1866-61-c-1872.html (last accessed September 14, 2020). From then onward, there are at least three different important definitions that keep the debate inconclusive and with biodiversity at the center of the problem: the Biological Species Concept, the Phylogenetic Species Concept and the Evolutionary Species Concept. See Quentin D. Wheeler and Rudolf Meier, *Species Concepts and Phylogenetic Theory: A Debate*, New York: Columbia University Press (2000). There, the authors state that of the 1.75 million species discovered so far, »the vast majority of the species come from a single point in space and time« (p. 14). There is a gap between the »real world« and the species concepts.

11   The tradition of an organicist understanding compared to a logical-mathematical one to explain life is explained well in Oliver Rieppel, *Phylogenetic Systematics: Haeckel to Hennig*, Boca Raton: CRC Press, Taylor & Francis Group (2016).

*My intention with this essay is to lay out the
potential of industrial microbiology to revolutionize the production of plastics, particularly having
in mind the detrimental effects these materials
have on marine environments. While the potential
of biotechnology to innovate our industry is
tremendous, it turns out that it is not so much the
development of new materials which is the hurdle
to create environmentally benign plastics.
It is much more that we do not know how to define
and test if a material is environmentally benign.
In this essay, I will try to follow the path of what we
need to know and define before we can develop
and introduce materials brought to us by synthetic
biology and industrial microbiology.*

About »Bio-« to Alleviate the Detrimental Impacts of Plastics on the Seas

Michael Sauer

Ramifications of a Material Revolution ___ With the invention of Bakelite in the early twentieth century, a revolution in industrial production and design began, based on artificial materials with fantastic new properties (see FIGURES 1—3 for examples). Low price, durability, favorable strength-to-weight ratios, and innovative methods for forming objects led to a huge success of plastics—and to all their consequences for our societal and industrial development. Since the real onset of plastics mass production in the 1950s, about 8.3 billion metric tons of plastic materials have been produced.[1] Such an enormous success story does not come without a downside, however.

These materials based on petroleum contribute to greenhouse gas emissions and therefore to climate change. And hugely problematic is the prolonged persistence of plastics in the environment—be it as managed waste in landfills or as pollution. Both problems grow with the scale of production. Bioplastics promise to alleviate these issues, but a critical perspective on what »bio-« means and which problems bioplastics can or cannot solve seems advisable.

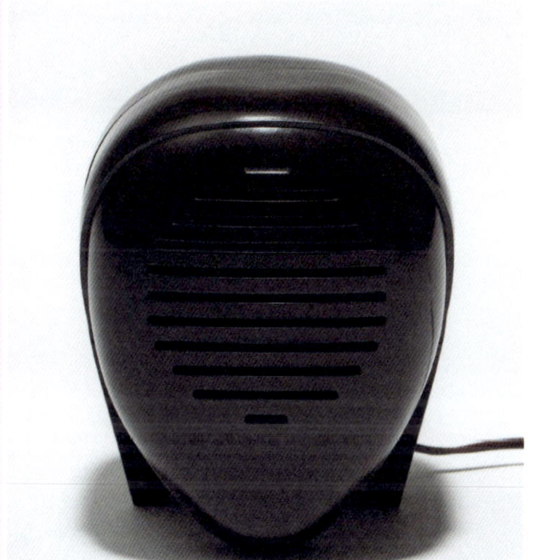

FIG. 1
Baby monitor made of Bakelite, the Zenith Radio Nurse, 1938, from the Georg Kargl Collection. Designed by Isamu Noguchi, manufactured by the Zenith Radio Corporation, USA.
© The Isamu Noguchi Foundation and Garden Museum, Bildrecht, Wien 2021, photo: MAK Museum Vienna/ Georg Mayer

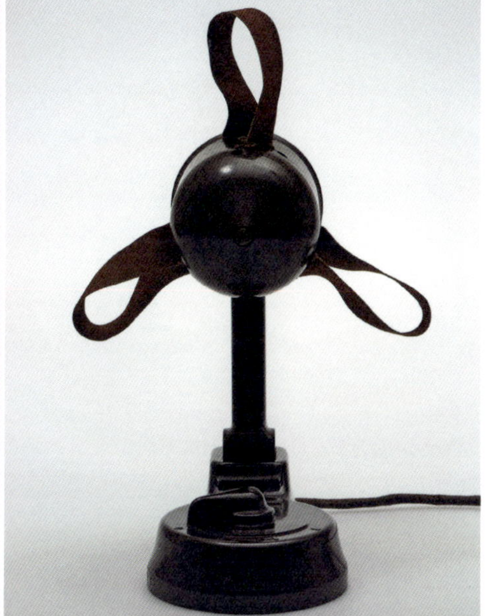

FIG. 2
Table fan made of Bakelite and leather, 1930s, courtesy Georg Kargl Collection. Manufactured by AEG, Germany or Austria.
© Photo: MAK Museum Vienna/Georg Mayer

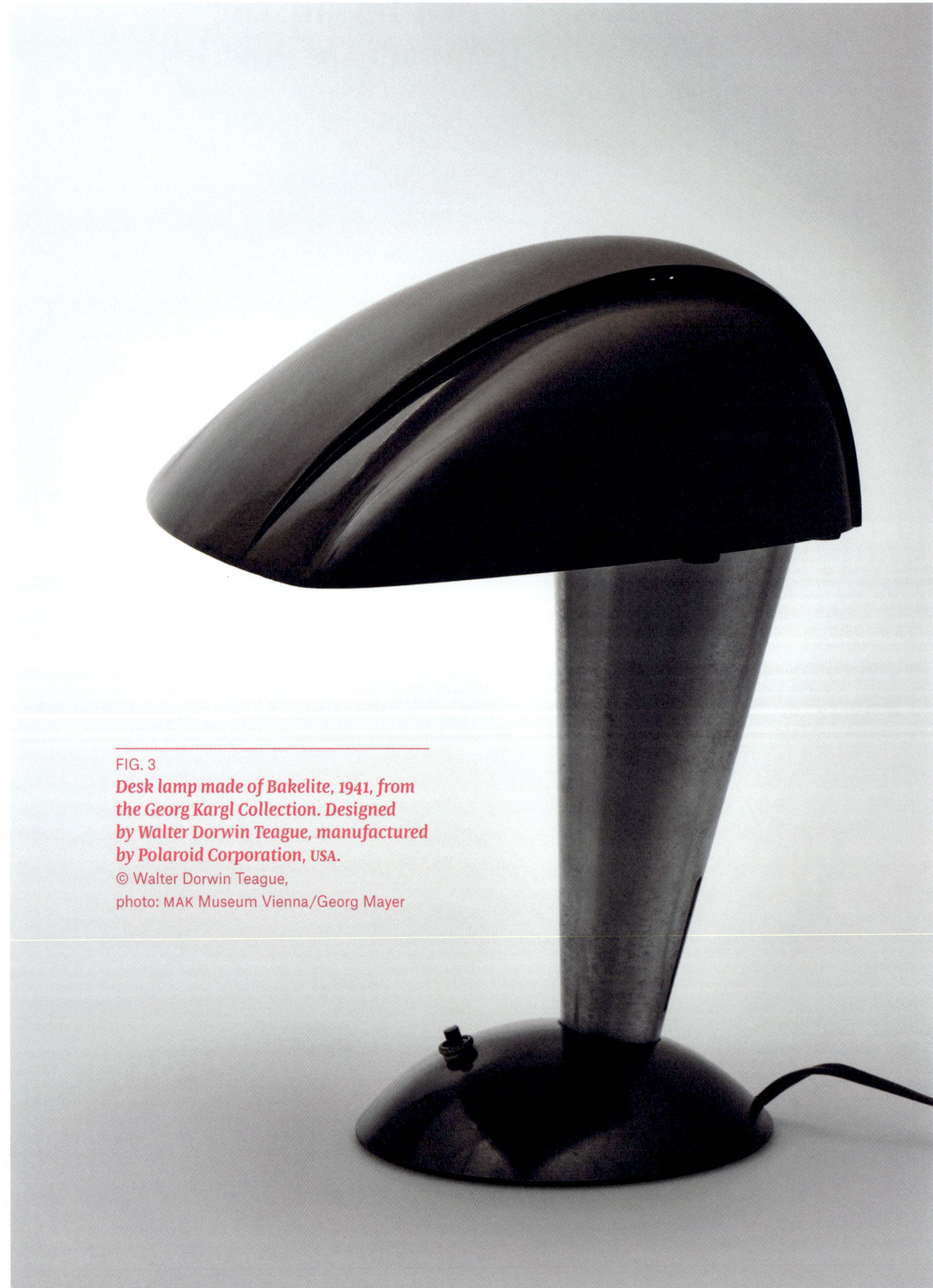

FIG. 3
Desk lamp made of Bakelite, 1941, from the Georg Kargl Collection. Designed by Walter Dorwin Teague, manufactured by Polaroid Corporation, USA.
© Walter Dorwin Teague,
photo: MAK Museum Vienna/Georg Mayer

Bioplastics ___ A wide variety of materials are named »bioplastics,« depending on who is using the term and what their intention may be. »Bio-« can mean that the material is derived from renewable resources instead of petroleum. Modern biotechnology is able to provide a wide array of such materials, which could readily substitute petroleum-derived plastics. To describe these materials more precisely, a better name would be »bio-based« plastics. Bio-based plastics have great potential to lower the greenhouse gas emissions connected with plastics. Bio-based does not imply, however, that such materials are readily degraded in various natural environments. Polyethylene (PE), for example, can nowadays be produced entirely bio-based from bioethanol, but the resulting material is chemically identical to the widely used petroleum-based PE connected with concerns of its persistence in nature.

A proper end-of-life management of these materials is required to keep them from ending up in the environment. Burning and recycling regimes address these issues; however, one has to consider that consumers' compliance and worldwide reliability of waste management systems are required if the appearance of these materials in the natural environment is to be avoided. Even in the best-case scenario, fractions of all materials will, to a certain extent, end up in unintended places, be it due to technical reasons, such as fiber particles deriving from machine washing of textiles, due to the negligence of consumers who discard objects improperly or through natural disasters. So, as long as such materials are produced and used, they will inevitably end up in nature, and we have to be aware that by now, we are talking about millions of tons spread across our planet. The marine environment is particularly vulnerable to plastic littering because most plastic materials are chemically quite stable in seawater. This means that once plastic has arrived in the oceans, it is doomed to remain there. Physical forces will break the pieces into ever smaller ones — until they are invisible to the naked eye (microplastics). However, these very small pieces will not disappear completely. Even worse, once small enough, they end up everywhere, even inside of sea creatures, thereby inevitably entering the worldwide food web. As of now, there is not one square kilometer of ocean anywhere on Earth that is free from plastic pollution. Trillions of pieces of plastic are floating in the world's oceans. A highly terrifying thought is that at current rates, plastics are expected to outweigh all the fish in the sea by the year 2050.[2]

What to Do? ___ First and foremost, we need to aim for materials and objects which do not *bioaccumulate* — in the sense that they do not accumulate in natural environments because they are entirely degraded (mineralized) by naturally occurring processes. Furthermore, the materials and objects should not be toxic. Toxic substances may enter the environment through leaching of constituents, such as plasticizers, or through the process of decomposition, which may bring forth toxic intermediates or products.

Biodegradability versus Ecocyclability __ At first glance, biodegradability – another aspect the term »bioplastics« is connected with – seems promising to avoid bioaccumulation. FIGURE 4 summarizes the four steps of biodegradation, where microorganisms are involved in the mineralization of plastics. A closer analysis of what »biodegradability« means, however, tells us that it is not necessarily helping to avoid bioaccumulation. Biodegradability as a term does not give any information about the timescale or the extent to which an object is decomposed, nor does it define where and under which conditions this can happen or not happen.[3] Polylactic acid (PLA), for example, is a bio-based plastic, which is compostable.

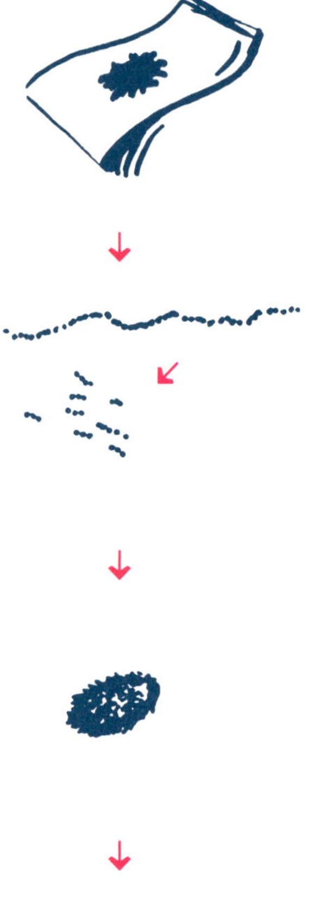

FIG. 4
*The four steps of
biodegradation of plastics.*
Illustration: Giuseppe Rizzo

This means that PLA is totally biodegradable under industrial composting conditions, which allows the material to be labeled »biodegradable.« However, under nonideal conditions, such as in a typical private household compost, PLA turns out to be quite resistant to degradation. Further, in cold seawater, it is entirely durable. Thus, the problem of bioaccumulation in the seas is not addressed at all by the biodegradability of PLA.

Whether a certain kind of plastic is biodegradable in a certain environment depends on its characteristics, such as chemical composition and crystallinity, but also on the presence of additives, such as plasticizers, dyes or the like. Of course it also depends on the environmental conditions in question: the presence of adequate microorganisms, temperature or pH, just to name a few. Plastics that are biodegradable in one environment might be essentially nondegradable in another environment (illustrated in FIGURE 5). Seawater, which is generally cool and has a low abundance of microorganisms, is an environment which is particularly ineffective for such degradation tasks.

253

FIG. 5
Plastics can be degradable in one environment but very stable in another – the physical and biological characteristics of an environment are decisive.
Illustration: Giuseppe Rizzo

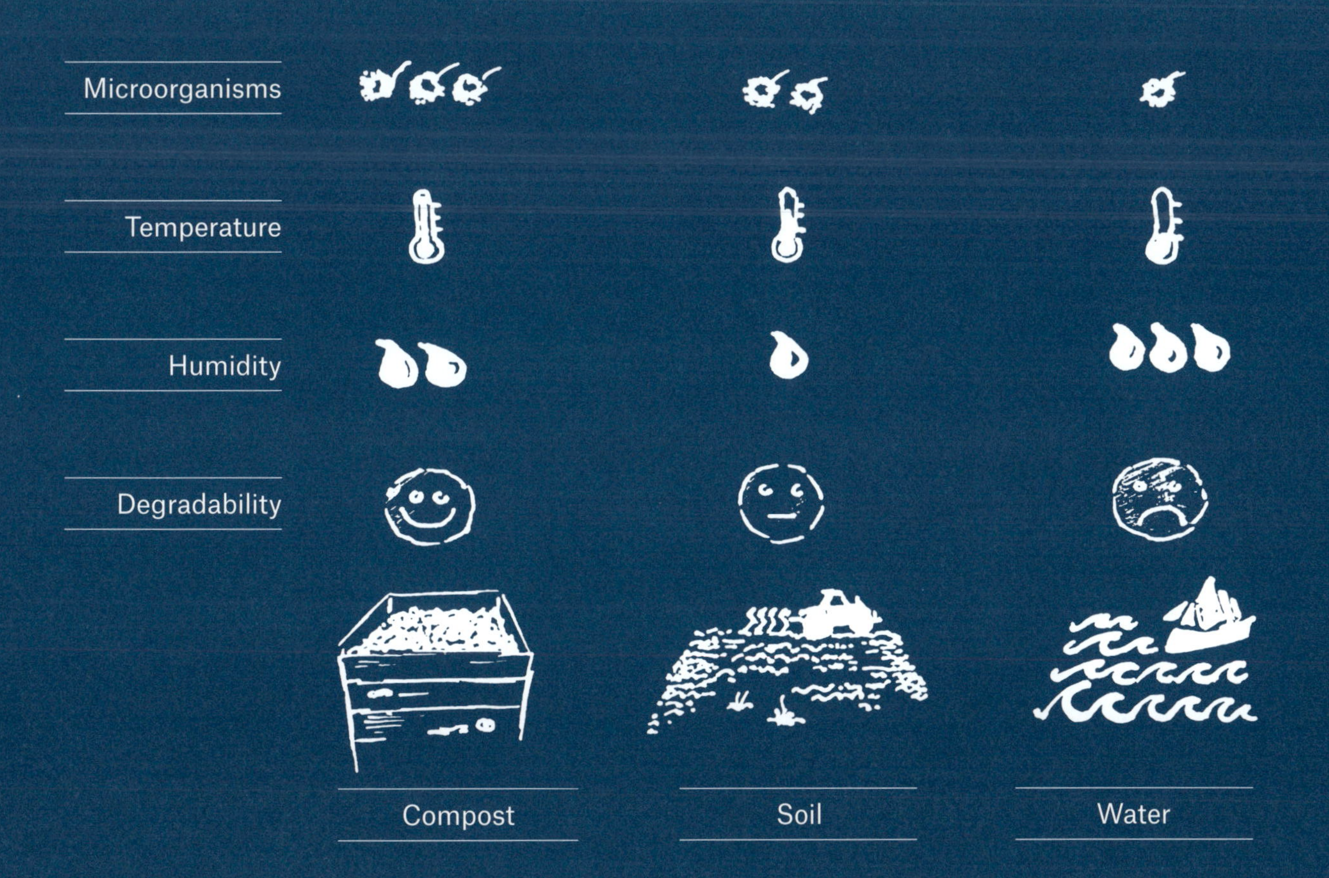

A further challenge in this context is the limited availability of test
methods to evaluate if a material is (bio)degradable under a certain condition within a defined period of time or not. Reliable methods are essential
to understand if a plastic material is biodegradable, for example, in the
sea. Without such understanding, no useful evaluation if a material should
be used on a large scale or not, is feasible. While testing of compostability
is standardized, fairly easy and widespread, methods for testing biodegradability in seawater environments are not.[4]

Due to the fact that »seawater« is not an environment which is very
defined and due to many technical issues, there are hardly any accepted
definitions for biodegradability in seawater, let alone methods for measuring this. One challenge relates to the microorganisms involved in the
degradation process: We do not know which microorganisms are present
to which extent in which part of the seas. Fresh seawater samples are
therefore required for the tests. However, it has been found that the microbial community present in a seawater sample changes profoundly as
soon as the water has been sampled by filling it into a bottle, for example.
A few hours after sampling, species of microbes predominate which were
scarce at the moment of sampling and many species have already been
lost. Due to such difficulties, so far no standardized test is widely accepted
and therefore the marine environment is, more often than not, not even
considered when a product is characterized as »biodegradable« for market-

---

FIG. 6
*Ecocyclability has been suggested
as the new standard to be developed
in order to define environmental
friendliness of materials.*
Plastic bottle image: freepik.com

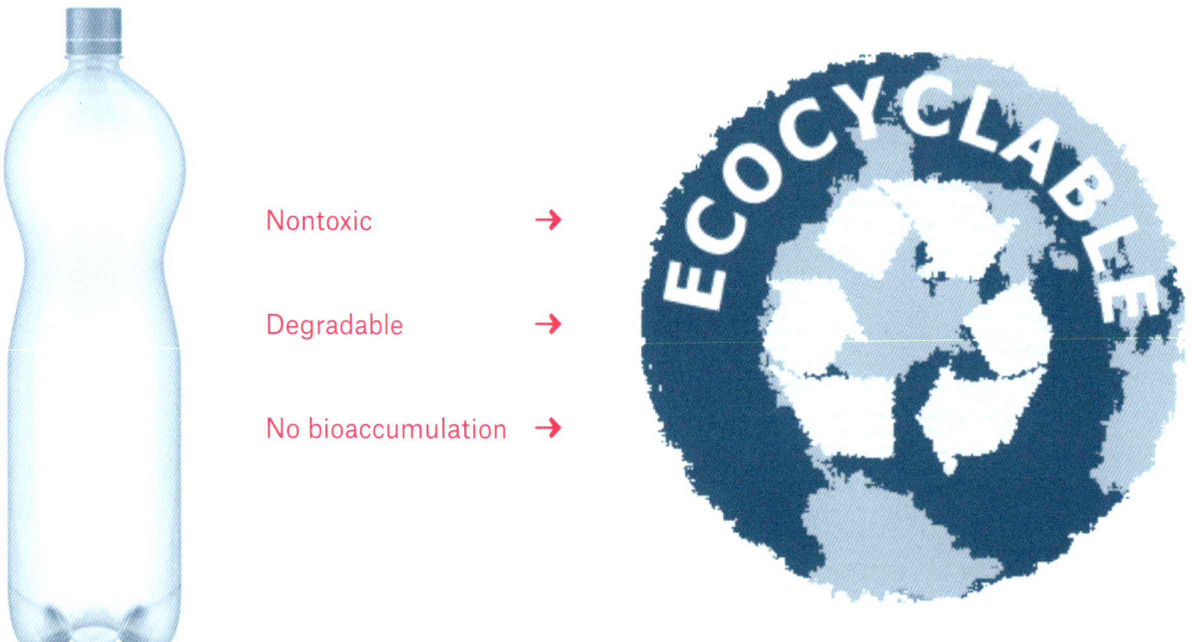

ing. I appeal, therefore, to the relevant authorities and branches of industry to include information about which tests have been used to arrive at the conclusion that a material is biodegradable. Even more, I appeal to the consumers to demand information about whether the marine environment has been appropriately considered.

All in all, the concept of biodegradability does not seem really helpful in trying to understand whether a material will become problematic for the environment or not. In a 2017 paper, Jason P. McDevitt and his coauthors proposed a new standard which helps to address this issue: They define »ecocyclable« as an appropriate standard for environmentally safe materials (FIGURE 6). To this end, ecocyclable materials must be nontoxic, degradable and not be prone to bioaccumulate.[5] Jason P. McDevitt and his coauthors also came up with an interesting testing method: the definition of degradability and nontoxicity is not given in absolute terms, but in relative terms, comparing the behavior of the material to two benign reference materials, one of them being cotton. How does this help? The absolute absence of a characteristic feature—such as the absence of toxicity—is basically impossible to prove, unless all possible conditions have been tested. Furthermore, Paracelsus already knew that »sola dosis facit venenum«—it is only the dose that makes the poison. This means that depending on how toxicity is tested (concentration of substance, length of contact, etc.), everything can be toxic. So, the simple—or rather simplified—definition of environmental toxicity, is not helpful. However, if we can agree that cotton is a material which is not problematic in the environment according to our current knowledge, we can define cotton as a standard and compare any new material to it. Does the material degrade like cotton? Is it as little toxic as cotton? Finally, a definition of certain types of environment to be tested would be a big step forward and of great help in the development of new materials. Once more, I appeal to customers to demand such information just as much as I appeal to policymakers to require such information from producers.

Biotechnology on the Front Line: New Plastics ___ Once the aim has been defined and test methods established, new materials can be developed. Reaching the goal is—like so often—a community effort. Polymer scientists need to collaborate with scientists from many other fields; microbiology and biotechnology are key technologies for the provision of new sustainable polymers. A more profound knowledge about the microorganisms present in different environments and their enzymatic possibilities for biodegradation will be required for tailoring new macromolecules accessible for microbial biodegradation. This underlines the fact that microbial ecology or geomicrobiology are of utmost importance: not only for our general understanding of nature, but also for very applied questions, such as the development of plastic materials.

Biotechnology offers microbial processes to produce plastic precursors or readymade polymers from plant-derived renewable resources or even from carbon dioxide directly. In any case, a »one-size-fits-all« solution in the form of a single polymer, which (bio)degrades rapidly in any kind of ecosystem and is useful for any kind of application, will not be found. Different applications have different requirements – in terms of material characteristics, but also in terms of the compromise regarding the degree of biodegradability. For example, the requirements for beverage containers stand in contrast to a fast biodegradation in water environments. To address such issues, proper life cycle analyses and risk assessments are inevitable when developing new ecocyclable polymers, which calls for the involvement also of specialists in these fields.

Many approaches are currently being pursued to provide new environmentally benign plastics. One approach is the direct provision of polymers which meet the requirements. Among them are polyhydroxy-alkanoates (PHAs).[6] These polymers were discovered a long time ago and are produced naturally by many bacteria as carbon and energy storage (similar to starch in plants, see FIGURE 7). The polymerization reactions are natural processes, and also the degradation has evolved to be quick and efficient. PHAs are interesting because structurally, many different versions exist with many different properties. The basic common feature of plastics is that they consist of long chains of polymers. The actual characteristics then depend on the precise structure of the chains, on the exact chain length, but also on the chain length distribution of the material, meaning: Are all molecules of the same length or is the material made of a mixture of molecules with varying chain lengths? Depending on the combination of all of these traits the ratio of the crystalline fraction

FIG. 7
Polyhydroxyalkanoates (PHA) are naturally formed plastics. The material is accumulated as carbon and energy storage in the form of granules inside bacterial cells.
Illustration: Giuseppe Rizzo

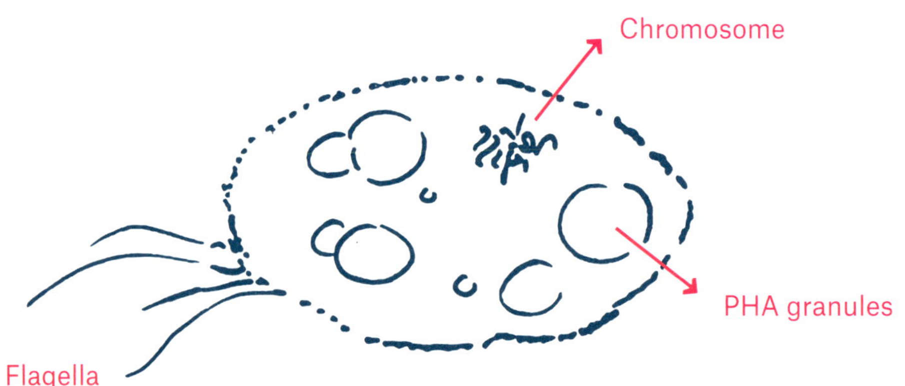

to the amorphous fraction changes, again modulating the characteristics — as outlined in FIGURE 8. All of these features can be biologically controlled when bacteria synthesize PHA. The formation of PHAs is quite well understood and thus synthetic biology approaches allow us to alter the polymer structure rationally by manipulating the metabolism of the bacteria. FIGURE 9 shows the basic structure of the PHA molecule and three examples of side chains. The precise structure of those side chains determines the physical characteristics of the material. Which side chains are formed depends on the metabolism of the bacteria. This opens a playground for the industrial microbiologist, because very different materials can be produced microbiologically and tested afterwards. This field has developed for a long time and some approaches are quite successful. However, at the time being, the plastics market is simply price-driven and PHAs are still more expensive than petroleum-derived materials, which limits their use. If the polymer market was more rationally regulated, they would be in a good position to take a significant share of the market.

FIG. 8

Characteristics defining the properties of plastics. Plastics are organic polymers composed of long, chain-like molecules with a high average molecular weight.

Illustration: Giuseppe Rizzo

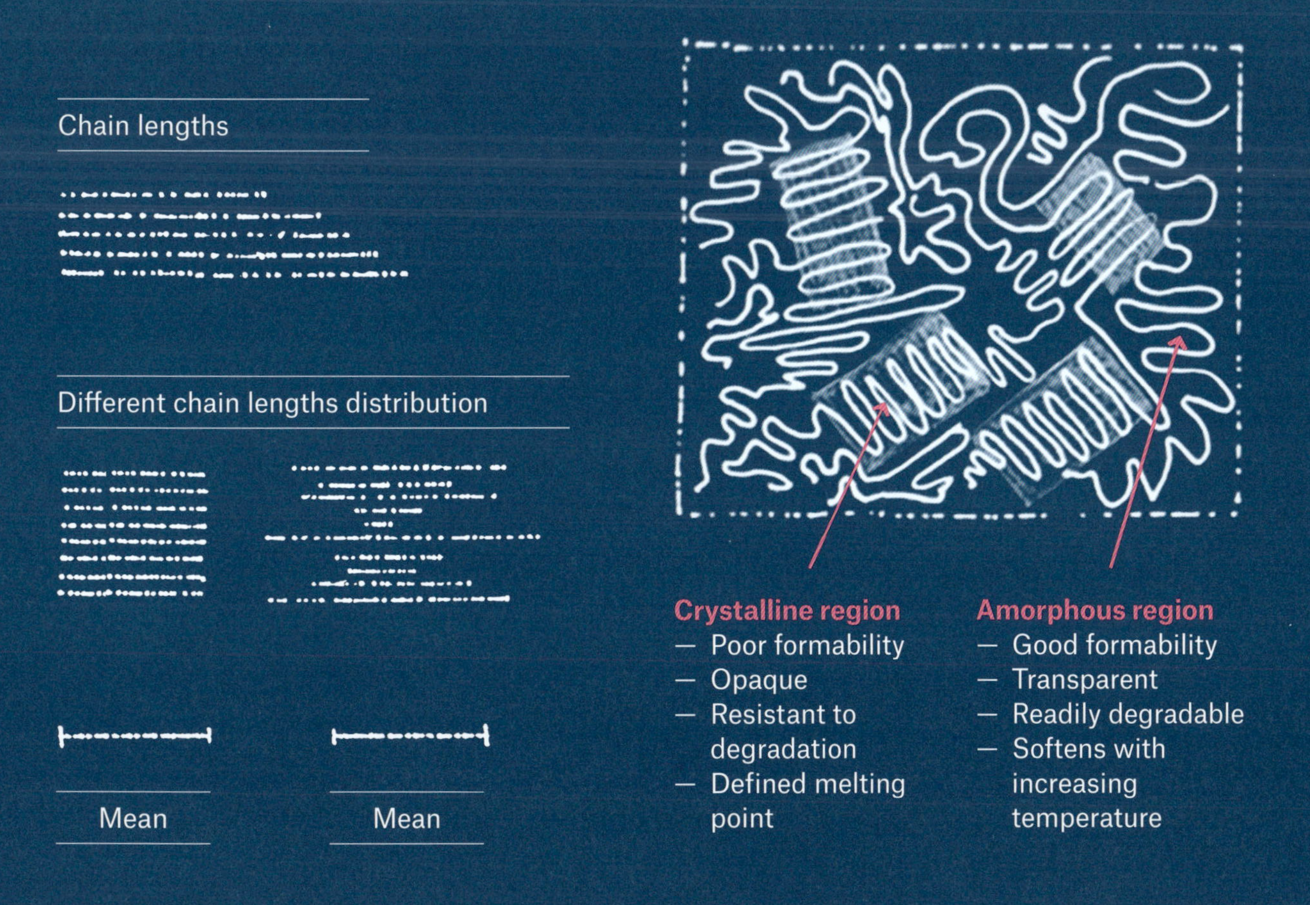

258

Polycaprolactam (PCL) is a plastic which was developed back in the 1930s. It is readily biodegradable under various conditions.[7] However, currently, PCL is produced exclusively from petroleum. PCL is used as a synthetic fiber (known as Perlon) and, due to its toughness and wear resistance, for gears and screws in mechanical engineering. Synthetic biology allows us to construct microorganisms which produce caprolactam, a precursor molecule of this plastic, from renewable resources, which will add the feature »bio-based« to the feature »biodegradable« in the future.

Very important for obtaining useful and biodegradable materials is the proper blending of different compounds. Blending of different polymers alters the characteristics of a material decisively. This is useful with respect to physical characteristics, which are important for the use envisaged. This is also useful, however, because it modulates biodegradability significantly. Polylactic acid (PLA) is a material, which has the potential to revolutionize the plastic market. It can be produced bio-based, because the starting material is lactic acid, which can be produced by microbial fermentation (very similar to the fermentative production of yoghurt). Thus, it can be produced independently from fossil resources. At the same time, PLA is biodegradable in the sense that enzymes and microbes can attack the material. However, as outlined above, the physical characteristics of pure PLA are such that it cannot be degraded at ambient temperatures. This means that in industrial composting plants, the material is very quickly mineralized (which allows labeling the material correctly as »biodegradable«). However, at ambient temperature or in seawater, the material is essentially nondegradable and stable like traditional PET. Proper blending of PLA with other polymers can solve this issue. The resulting plastics can be ecocyclable and very useful to substitute plastics which are currently produced from petroleum.

259

Another quite interesting approach, which shall be mentioned here, is the design of biodegradability into the material by embedding degrading enzymes into the plastic at the time of production. Enzymes are proteins catalyzing chemical reactions. They guide every cell's metabolism, but also play a crucial role in all natural degradation processes. With their help, macromolecules can be chopped into small pieces which can then be used by microorganisms. These characteristics are, for example, useful in detergents, where enzymes are added to degrade stains of protein or starch rapidly. Enzymes are also the basis for microbial degradation of plastics. Recently, a kind of enzyme has been discovered that is able to degrade PET—a plastic which had been considered nonbiodegradable before. In fact, processes have been developed which allow the complete mineralization of PET by microbial or enzymatic means. However, the conditions must be very specific. So, the discussion is the same as for PLA: under specific conditions, biodegradability is given, yet not under most natural conditions in the environment. Nevertheless, these enzymes can be embedded into the material, allowing the attack of the material from inside—and thus degradation—in places where such enzymes would not naturally occur. A plastic from PLA, for instance, with incorporated enzymes, which is fully biodegradable in many environments, has already been produced. Similar approaches are conceivable for many other plastic versions in the future.

A final interesting finding is that it has been discovered that certain mealworms (FIGURE 10) have the ability to live on plastic. In fact, they are able to degrade mixtures of plastics which have been deemed absolutely nonbiodegradable.[8] A very active and so far not fully understood microbial community in their intestines allows the worms to degrade complex plastic mixtures. At first sight, this does not help in the context of marine plastic pollution. Yet, it points once more to the power of nature and to the fact that nature itself may bring about solutions to problems we create.

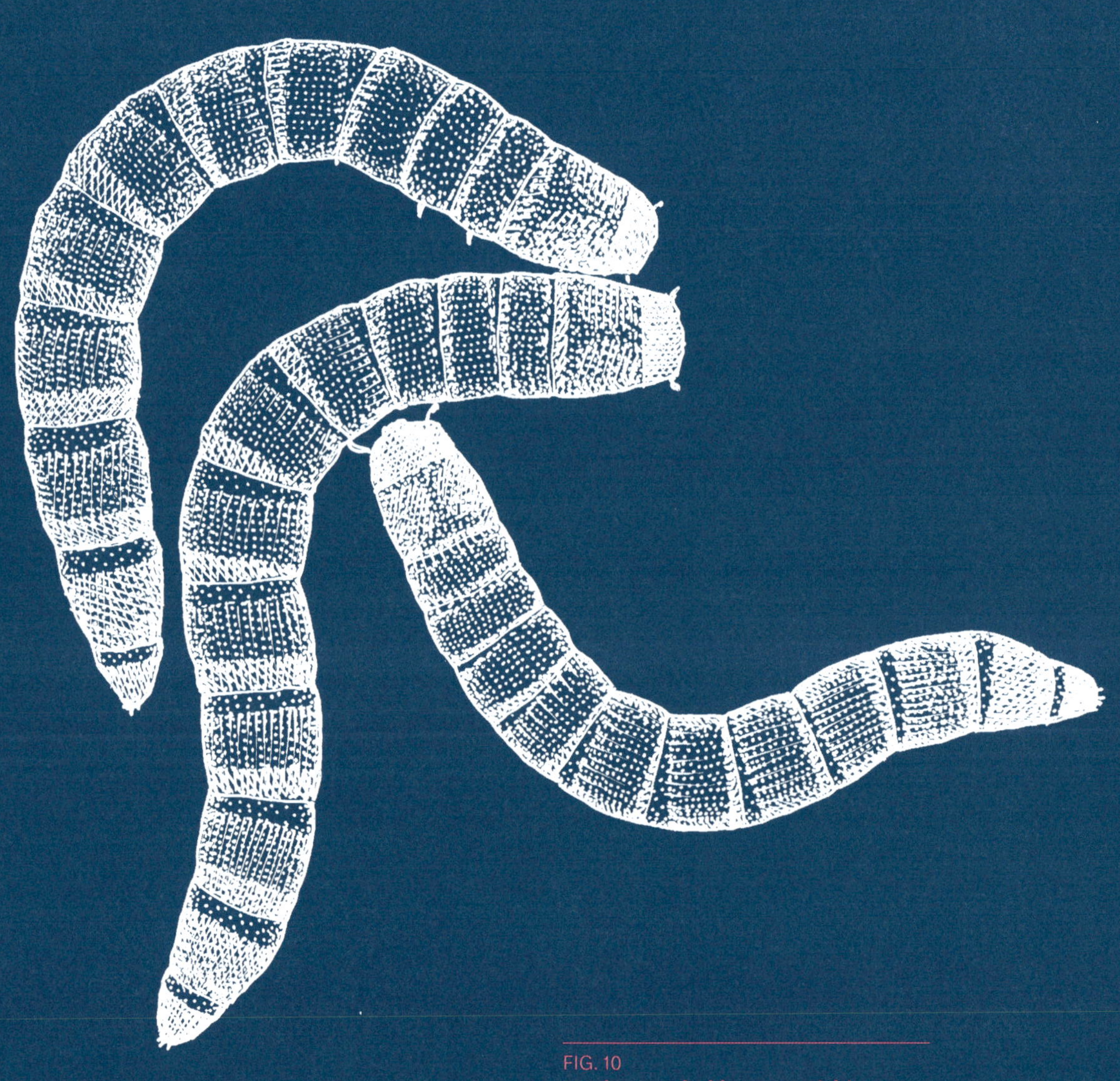

FIG. 10
*Mealworms hold an astonishing
microbial community in their
intestines that can degrade plastics.*
Illustration: larvae of *Tenebrio molitor*
by Des Helmore, Des Helmore/
Manaaki Whenua—Landcare Research,
Wikipedia, CC BY 4.0

1 Roland Geyer, Jenna R. Jambeck, and Kara Lavender Law, »Production, Use, and Fate of All Plastics Ever Made,« *Science Advances* 3, no. 7 (2017), e1700782, doi: 10.1126/sciadv.1700782.

2 www.biologicaldiversity.org/campaigns/ocean_plastics (last accessed November 4, 2020).

3 Jesse Patrick Harrison, Carl P. Boardman, Kenneth O'Callaghan, Anne-Marie Delort, and H. J. Song, »Biodegradability Standards for Carrier Bags and Plastic Films in Aquatic Environments: A Critical Review,« *Royal Society Open Science* 5, no. 5 (2018), 171792, doi: 10.1098/rsos.171792.

4 Tobias P. Haider, Carolin Völker, Johanna Kramm, Katharina Landfester, and Frederik R. Wurm, »Plastics of the Future? The Impact of Biodegradable Polymers on the Environment and on Society,« *Angewandte Chemie International Edition* 58, no. 1 (2019): 50–62, doi: 10.1002/anie.201805766.

5 Jason P. McDevitt, Craig S. Criddle, Molly Morse, Robert C. Hale, Charles B. Bott, and Chelsea M. Rochman, »Addressing the Issue of Microplastics in the Wake of the Microbead-Free Waters Act: A New Standard Can Facilitate Improved Policy,« *Environmental Science & Technology* 51, no. 12 (2017): 6611–6617, doi: 10.1021/acs.est.6b05812.

6 João Medeiros Garcia Alcântara, Francesco Distante, Giuseppe Storti, Davide Moscatelli, Massimo Morbidelli, and Mattia Sponchioni, »Current Trends in the Production of Biodegradable Bioplastics: The Case of Polyhydroxyalkanoates,« *Biotechnology Advances* 42 (2020), 107582, doi: 10.1016/j.biotechadv.2020.107582.

7 Tanja Narancic and Kevin E. O'Connor, »Plastic Waste as a Global Challenge: Are Biodegradable Plastics the Answer to the Plastic Waste Problem?« *Microbiology* 165, no. 2 (2019): 129–137, doi: 10.1099/mic.0.000749.

8 Anja Malawi Brandon and Craig S. Criddle, »Can Biotechnology Turn the Tide on Plastics?« *Current Opinion in Biotechnology* 57 (2019): 160–166, doi: 10.1016/j.copbio.2019.03.020.

Notes on Contributors

Brandon Ballengée is an artist and biologist and creates multimedia artworks inspired by his ecological fieldwork and laboratory research. Ballengée's art has been exhibited internationally in over 20 countries. He has received numerous awards and fellowships, including a Creative Capital Award (2019). Currently, he is a Research Associate at Louisiana State University studying the impact of the 2010 Gulf of Mexico oil spill on fish. In 2019, he delivered a TEDxLSU talk on his trans-species public art series »Love Motel for Insects.« In 2016, Ballengée cofounded the Atelier de la Nature, an eco-education campus and nature reserve in Arnaudville, Louisiana.

Dianna Cohen is a Los Angeles-based visual artist, as well as CEO (since 2014) and co-founder of Plastic Pollution Coalition (PPC). For twelve years, PPC has grown into a global alliance of over 1,200 organizations, businesses and thought leaders working toward a world free of plastic pollution and its toxic impacts on humans, animals and the environment. Cohen continues to be exhibited internationally and explore the resources of art-as-activism and art as a powerful tool of science communication through TED and INK talks and appearances worldwide. www.plasticpollutioncoalition.org

Martina Rosina Fröschl is a digital artist and scientist with a focus on computer-generated imagery that incorporates the immediate inputs of nonvirtual reality. She holds a PhD from the University of Applied Arts Vienna where she currently works as a senior scientist at the Science Visualization Lab. She studied media technique and design and gained her PhD with research on computer-animated scientific visualizations of computed tomography (CT) scans of microscopic organisms. Since that time, the depiction of realities and biological phenomena has driven her creations. She has worked on various documentary and fiction productions for TV and cinema as a visual effects and computer graphics artist. Martina is on the core team of PIXELvienna Society, the *NOISE AQUARIUM* collective and was guest lecturer and artist in residency at the Art|Sci Center UCLA.

María Antonia González Valerio is a philosopher and professor in the Faculty of Philosophy and Literature at the National Autonomous University of Mexico (UNAM). She works within the research fields of ontology and aesthetics and interdisciplinary research in arts, sciences and humanities, specifically in the field of art that uses biomedia. She is the head of the research group Arte+Ciencia (Art+Science) and author of the books: *Cabe los límites. Escritos sobre filosofía natural desde la ontología estética* (Mexico: UNAM/Herder, 2016), *Un tratado de ficción* (Mexico: Herder, 2010) and *El arte develado* (Mexico: Herder, 2005).

Stephan Handschuh is a biologist and obtained his PhD from the University of Vienna. Since 2012, he works as a staff scientist at VetCore Facility for Research, University of Veterinary Medicine Vienna, where he has a strong technical focus on quantitative microscopic x-ray imaging and 3D data visualization and analysis. Additionally, he has a broad biological background in diverse fields such as evolutionary biology, comparative and functional morphology, microscopic anatomy, developmental biology, and theoretical biology. Since 2010 he is also a member of the Science Visualization Lab at the University of Applied Arts Vienna, where he works on creating scientifically meaningful 3D models of microscopic animal samples.

Christian de Lutz is a curator, cofounder and codirector of Art Laboratory Berlin, where he has curated over 40 exhibitions, including the series »Time & Technology,« »Synaesthesia,« »[macro]biologies & [micro]-biologies,« and »Nonhuman Subjectivities.« His curatorial work focuses on the interface of art, science and technology in the 21st century, with special attention given to BioArt, DIY science initiatives, and facilitating collaborations between artists and scientists. His interest is in building multidisciplinary networks and unleashing their creative potential. He is currently involved in collaborative cultural projects connecting Berlin with other cities in Europe and Asia, building international networks for art-science and DIWO (do-it-with-others) communities.

Mary Maggic is a nonbinary artist researching hormone biopolitics and environmental toxicity, and how the ethos and methodologies of biohacking can serve to demystify invisible lines of molecular (bio)power. After completing a master's degree at the MIT Media Lab (Design Fiction research group), Maggic is currently a member of the online network Hackteria: Open Source Biological Art and the laboratory theater collective Aliens in Green. In 2017, their project *Open Source Estrogen* received an Honorable Mention at Prix Ars Electronica Hybrid Arts. In 2019 Maggic completed a 10-month Fulbright residency in Yogyakarta, Indonesia, investigating the role of Javanese mysticism in the plastic pollution crisis.

Rosaura Martínez Ruiz is a philosopher and professor in the Faculty of Philosophy and Literature at the National Autonomous University of Mexico (UNAM). She served as coordinator of the research projects »Philosophers after Freud« and »Philosophy and Psychoanalysis as Critical Borders of the Political Sphere.« She is the author of the books: *Freud y Derrida: escritura y psique* (2013) and *Eros: Más allá de la pulsión de muerte* (2018). She serves on the advisory board of the International Consortium of Critical Theory Programs.

Reiner Maria Matysik is an artist and professor of three-dimensional design at the Burg Giebichenstein University of Art and Design Halle. He studied fine arts at the Braunschweig University of Art and at the Ateliers Arnhem. He works in manifold ways with concepts for future landscapes and organisms such as postevolutionary life forms. Working specifically in object art, installation, and video he has developed a dynamic scenario of future landscapes and organisms. In this way he creates an area of conflict between promise and failure in a potential future. Both visual implementation and linguistic form can be recognized here as the essential artistic strategies that he uses as his own interface between the worlds of scientific research and pseudoscientific fiction.

Regine Rapp is an art historian, curator, and director of Art Laboratory Berlin. Her current research focuses on installation art, artist books, hybrid art, and art-science collaborations. She has taught art history at the Burg Giebichenstein University of Art and Design Halle. As cofounder and director of Art Laboratory Berlin, she researches, curates, and publishes on 21st century art at the interface of (natural) science and technology. In this context she conceived the international conferences »Synaesthesia. Discussing a Phenomenon in the Arts, Humanities and (Neuro)Science« (2013), »Nonhuman Agents« (2017), and THE CAMILLE DIARIES (2020). In the project *Mind the Fungi*, in collaboration with the Technische Universität Berlin, she worked at the interface between artistic and scientific research.

Ingeborg Reichle is an art historian, media theorist, and professor in the Department of Media Theory at the University of Applied Arts Vienna. She served as founding chair of the Department of Cross-disciplinary Strategies (CDS) from 2017 until 2018, where she designed an integrated BA study program on applied studies in art, science, philosophy, and global challenges. Her current area of research and teaching is the encounter of the arts with cutting-edge technologies such as biotechnology and synthetic biology, taking also into account artistic responses to systemic risks and global challenges such as climate change and ecological collapse in order to develop a critical understanding of the role of twenty-first century arts. She is the author of more than 50 scientific articles and a number of books including *Kunst aus dem Labor: Zum Verhältnis von Kunst und Wissenschaft im Zeitalter der Technoscience* (2005) and *Art in the Age of Technoscience: Genetic Engineering, Robotics, and Artificial Life in Contemporary Art* (2009) both published by Springer.

Michael Sauer is Associate Professor at the Institute of Microbiology and Microbial Biotechnology at the University of Natural Resources and Life Sciences (BOKU), Vienna. He is an industrial microbiologist who focuses on sustainable microbial chemical production. After receiving a diploma in biotechnology at the Swiss Federal Institute of Technology (ETH) in Zurich, he pursued his doctoral studies in biochemistry at the University of Vienna. Postdoc studies at the University of Milano-Bicocca brought him in close contact with industry—a contact that he has maintained ever since. He obtained his habilitation and *venia docendi* in industrial microbiology at BOKU. His research is based on exploring biodiversity to find nature's answers to industrial problems, combined with synthetic biology, when required, to augment the capabilities of microorganisms.

Thomas Schwaha is a zoologist with a particular focus on animal morphology and evolution. Since 2011 he holds a postdoc position at the Department of Evolutionary Biology, University of Vienna, where he is responsible for all the microscopes and imaging facilities of the department. His active research fields include various aspects of bryozoan biology, including soft body morphology evolution, systematics and biodiversity of ctenostomes and, more recently, boring bryozoans. The research involves multiple morphological techniques such as serial sectioning, confocal laser-scanning microscopy or microCT that each yield 3D data sets useful for 3D imaging techniques. His particular methodological background has led to multiple collaborative projects involving a plethora of various organisms, as well as projects involving science education and visualization.

Robertina Šebjanič is an artist and independent researcher based in Ljubljana, Slovenia. Her work deals with the cultural, (bio)political, and ecological realities of aquatic environments. Her projects engage with philosophical questions situated at the intersection of art, technology and science, and are often realized in collaboration with others, through interdisciplinary and informal integration in her work. She received an Honorable Mention at the Prix Ars Electronica 2016, and was nominated for the STARTS 2016 and STARTS 2020 Prize. In 2017 she was SHAPE artist and, in 2018, took part (with Gjino Šutič) in a residency program at Ars Electronica (Emap/Emare). Her artwork AURELIA 1+HZ / PROTO VIVA GENERATOR became part of the BEEP Electronic Art Collection, Spain, in 2019.

Alfred Vendl is a film director/producer, writer, and a scientist with a background in chemistry. He studied materials science at the Technical University of Vienna, where he gained his PhD. He was a research scientist at Imperial College, University of London, U.K.; the University of Freiburg im Breisgau, Germany; the University of California San Diego, La Jolla, USA; and the Max Planck Institute for Metals Research, Stuttgart, Germany. From 1981 until 2014 he was a professor at the University of Applied Arts Vienna and Chair of the Institute of Art and Technology, as well as Associate Professor at the Technical University Vienna. Since 2016 he is Director of the Science Visualization Lab at the Department of Digital Art, University of Applied Arts Vienna. He has authored around 80 scientific publications in materials science, archaeometry and art technology.

269

Victoria Vesna is an artist and professor in the UCLA Department of Design Media
Arts and Director of the Art|Sci Center at the UCLA School of the Arts
& Architecture and California NanoSystems Institute (CNSI). She gained
a PhD from CAiiA_STAR, University of Wales (2000), and a diploma
from the Academy of Fine Arts, University of Belgrade (1985). Her work
involves long-term collaborations with composers, nanoscientists,
neuroscientists, and evolutionary biologists. She is the North American
editor of *AI & Society* journal (Springer Publishing, UK) and in 2007
published an edited volume, *Database Aesthetics: Art in the Age of Informa-*
tion Overflow (University of Minnesota Press), and another in 2011,
Context Providers: Conditions of Meaning in Media Arts, co-edited with
Christiane Paul and Margot Lovejoy (Intellect Ltd UK, 2011).

Jennifer Wagner-Lawlor is Professor of Women's, Gender, and Sexuality Studies
(WGSS) at Pennsylvania State University, an ambassador of Plastic Pollution
Coalition (PPC), and former President of the Society for Utopian Studies.
She co-curated the exhibition »Plastic Entanglements: Ecology, Aesthetics,
Materials« (2018–2020), a major traveling exhibition of nearly thirty
international artists featuring plastic as material artifact and cultural
metaphor. Her publications focus on utopian fiction and theory, as
well as plastic pollution, the concept of plasticity, plasticity, and the
concept of »plaesthetics,« all comprising a larger project, *Becoming Plastic*.

Pinar Yoldas is an artist, infradisciplinary architect, neuroenthusiast, and an
assistant professor at University of California San Diego in the Design
and Visual Arts Department. Her work develops within biological
sciences through architectural installations, kinetic sculpture, and multimedia interventions. She works within the themes of techno-feminism,
ecological activism, and critical futurism. Her most known works
are *Kitty AI: Artificial Intelligence for Governance* and AN ECOSYSTEM OF EXCESS.
While the former examines the future of AI and urbanism, the latter
imagines a whole new ecosystem that emerges out of pelagic plastics.
She is a Guggenheim fellow, a MacDowell fellow, and a recipient of
the FEAT art and technology award. She holds a bronze medal in organic
chemistry from the National Science Olympiad.

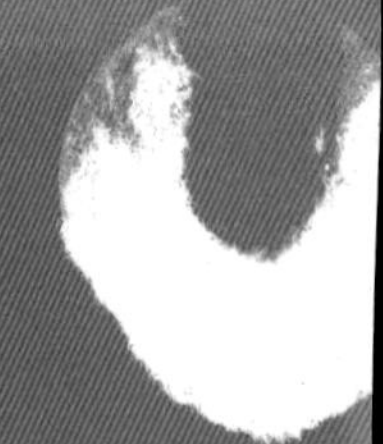

Ingeborg Reichle (Ed.)
Univ.-Prof. Dr. Ingeborg Reichle
University of Applied Arts Vienna | Department for Media Theory
Oskar-Kokoschka-Platz 2 | 1010 Vienna, Austria
www.medientheorie.ac.at

With contributions by Brandon Ballengée, Dianna Cohen and Jennifer Wagner-Lawlor,
Martina R. Fröschl and Alfred Vendl, María Antonia González Valerio and Rosaura Martínez Ruiz,
Mary Maggic, Reiner Maria Matysik, Regine Rapp and Christian de Lutz, Ingeborg Reichle,
Michael Sauer, Thomas Schwaha and Stephan Handschuh, Robertina Šebjanič, Victoria Vesna,
and Pinar Yoldas.

Project Management »Edition Angewandte« on behalf of the University of Applied Arts Vienna:
Stefanie Schabhüttl and **Barbara Wimmer**, Vienna, Austria
Content and Production Editor on behalf of the Publisher: **Katharina Holas**, Vienna, Austria

Proofreading/Copyediting: **Gloria Custance**, Berlin, Germany | **Scott Clifford Evans**, Vienna, Austria |
Stefanie Schabhüttl and **Barbara Wimmer**, Vienna, Austria
Cover Image: **Martina R. Fröschl** and **Alfred Vendl**, *CGI of PARAMECIUM and inner simulations*
© Science Visualization Lab Angewandte, Vienna, Austria
Graphic Design, Cover Design, and Typography: **Andrea Neuwirth**, Vienna, Austria
Layout Assistant: **Gabriel Fischer**, Vienna, Austria
Type Fonts: **Bionik** by Arne Freytag | **Atlas Grotesk** by Kai Bernau and Susana Carvalho
with Christian Schwartz
Paper: Colibri lagoon | Pergraphica Classic Rough 120 g/m²
Copyright Licensing: **Zahra Mirza**, Vienna, Austria
Image Editing: **pixelstorm**, Vienna, Austria
Printing: **Holzhausen**, die Buchmarke der Gerin Druck GmbH, Wolkersdorf, Austria

Library of Congress Control Number: 2021933996

Bibliographic information published by the German National Library
The German National Library lists this publication in the Deutsche Nationalbibliografie;
detailed bibliographic data are available on the Internet at http://dnb.dnb.de.

ISSN 1866-248X
ISBN 978-3-11-074472-9
e-ISBN (PDF) 978-3-11-074477-4

www.degruyter.com